AF412969

ELECTRODYNAMICS OF DENSITY DUCTS IN MAGNETIZED PLASMAS

ELECTRODYNAMICS OF DENSITY DUCTS IN MAGNETIZED PLASMAS

Igor G. Kondrat'ev
University of Nizhny Novgorod, Russia

Alexander V. Kudrin
University of Nizhny Novgorod, Russia

and

Tatyana M. Zaboronkova
Technical University of Nizhny Novgorod, Russia

GORDON AND BREACH SCIENCE PUBLISHERS
Australia • Canada • China • France • Germany
India • Japan • Luxembourg • Malaysia • The Netherlands
Russia • Singapore • Switzerland

Amsteldijk 166
1st Floor
1079 LH Amsterdam
The Netherlands

British Library Cataloguing in Publication Data

Kondrat'ev, Igor G.
    Electrodynamics of density ducts in magnetized plasma
    1.Electromagnetic waves - Mathematics 2.Plasma (Ionized gases) 3.Density matrices
    I.Title II.Kudrin, Alexander V. III.Zaboronkova, Tatyana M.
    539.2´0151

ISBN: 90-5699-200-7

Contents

Chapter 4. Modes in Axially Uniform Ducts

Chapter 5. Integral Representation of Source-excited Fields on a Duct

Chapter 6. Modal Representation of Source-excited Fields on a Duct

Chapter 7. Wave Propagation Along Axially Nonuniform Ducts

Chapter 8. Wave Re-emission from a Density Duct

Preface

This book is intended to provide a systematic, self-contained treatment of excitation, propagation and re-emission of electromagnetic waves guided by density ducts in magnetized plasmas. Its object is to set out the theoretical basis of electrodynamics of ducts as fully as possible. In this book, the classical dielectric-waveguide theory is generalized to the case of open guiding systems in a magnetoplasma and the conceptual physical and mathematical aspects of the theory, together with its applications to problems encountered in actual practice, are emphasized.

The book is partly based on lectures delivered by the authors at the Radiophysics Department of the Nizhny Novgorod University and is mainly supported by original works of the authors in the subject matter. It is therefore hoped that this book will serve both as a textbook for a novice theorist, and as a reference source for more experienced research workers. The reader is assumed to be familiar with calculus, vectors analysis, the theory of complex variables, and some background in electromagnetic theory.

To assist in understanding, especially for those comparatively new to the subject, we begin with a discussion of the underlying physical phenomena and then provide relevant theoretical fundamentals. This material is presented in Chapters 1, 2 and 3 which have three principal objectives: (i) to offer an outline of some of the results of observations and formation mechanisms of natural and artificial density ducts; (ii) to give some of the basic theory for electromagnetic waves in plasmas needed throughout the remainder of the book; and (iii) to discuss a few specialized topics, primarily concerning radiation from given sources in a magnetoplasma, which may not be familiar to all readers, but are crucial to an understanding of the further material. Chapter 4 deals with the theory of guided wave propagation along axially uniform ducts. It also serves to furnish the reader with sufficient references in relevant areas of study. The purpose of Chapters 5 and 6 is to show how to find the field excited by a given source in the presence of a duct. The theory of these two chapters is set out very fully because the authors know of no other complete treatment on that topic. The emphasis is placed on excitation of guided modes discussed earlier. Chapter 7 is concerned with the asymptotic theory of wave propagation along axially nonuniform ducts. In Chapter 8 mode re-emission from a duct is considered.

Throughout the book we are dealing with the full wave theory to make our exposition self-consistent and rigorous. We attempt to start with simpler cases, gradually increasing the complexity of our treatment. Some of the more advanced material as well as specific details are printed in smaller type and can be skipped on the first reading. It is inevitable that many important topics have been omitted. Among them is a detailed comparison of the theoretical results with existing experimental data. This really needs separate consideration in itself.

We are indebted to many colleagues at the Nizhny Novgorod University; the Radiophysical Research Institute, Nizhny Novgorod; the Institute of Applied Physics, Nizhny Novgorod; and elsewhere, who gave us valuable advice and help. It is impossible to list them all, but we wish especially to mention Professor G. A. Markov of the Nizhny Novgorod University, and Dr A. V. Kostrov of the Institute of Applied Physics, Nizhny Novgorod, whose experimental works have inspired some of our own research. We are also grateful to Mrs L. R. Semenova of the Radiophysical Research Institute, Nizhny Novgorod, and Dr L. L. Popova, Mrs L. E. Kurina, and Mrs E. V. Yurasova of the Radiophysics Department of Nizhny Novgorod University for the assistance that they provided in the preparation of the manuscript.

Chapter 1

Introduction

1.1. Density ducts in the earth's magnetosphere

In recent decades, a great deal of information has been accumulated on the properties of the very low-frequency (VLF) electromagnetic signals known as whistlers, or whistling atmospherics (see Storey, 1953; Gershman and Ugarov, 1961; Helliwell, 1965; Walker, 1976), which travel in the near-Earth environment roughly in the direction of the geomagnetic field. These signals, at frequencies below the local electron gyrofrequency (which is usually less than the electron plasma frequency in the ionosphere and magnetosphere), propagating over long distances in the earth's magnetosphere, play an important role in space plasma physics and may serve as a diagnostic tool for investigating the distribution and dynamics of the magnetospheric plasma (Brice and Smith, 1971; Sazhin *et al.*, 1992). It is now believed that whistlers can be guided through the earth's magnetosphere by ducts, tubes of enhanced or reduced ionization which are aligned with the geomagnetic field (Helliwell, 1965; Walker, 1976). The detailed discussion of characteristics of ducts and their formation mechanisms is beyond the scope of this book. Only a brief outline is given here of the major features of ducts and ducted whistlers.

It is well established that whistlers are waves from lightning flashes near the earth's surface. These are impulse signals which can penetrate the ionosphere and enter the magnetosphere. It has been shown by Storey in his pioneer work (Storey, 1953) that the ray direction of a whistler must be within about 20° of the direction of the earth's magnetic field.* Any wave propagating in the whistler mode will therefore tend to travel almost along the geomagnetic field line. It was suggested that the whistler originating in a lightning flash in one hemisphere travels roughly along the geomagnetic field line, over the equator, and reaches earth in the other hemisphere somewhere near the opposite end of the line (a 'short whistler'). It can be reflected by the earth and return along nearly the same path to a point near its origin (a 'long whistler'). Sometimes multiple reflections of whistlers back and forth along the same line may occur. This is the mechanism proposed by Storey (1953) for

---

* This will be explain in detail in §§ 2.6 and 3.7.

guidance of the low-frequency electromagnetic waves through the magnetosphere. The mechanism is illustrated schematically in Figure 1.1. The curved line shows the geomagnetic field line and the forked arrow indicates a lightning flash, representing the source of the whistlers.

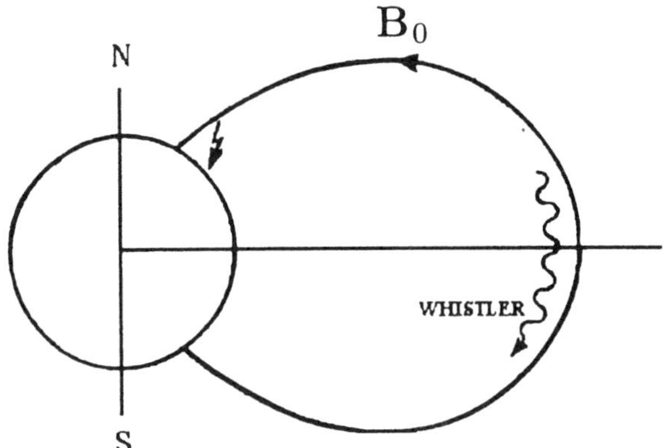

Figure 1.1. Sketch (not to scale) to illustrate whistler wave propagation in the magnetosphere.

Soon after publication of Storey's work, theoretical studies of whistler ray paths in the magnetosphere (Yabroff, 1961) showed that this guiding mechanism is not sufficient to explain the arrival of a whistler on the ground. Only the whistlers whose wavenormals made a small angle with the earth's magnetic field were observed from the ground. On the other hand, the whistler wavenormal may generally make a large angle with the magnetic field even if the ray direction is almost along the magnetic field line (see, for details, Booker and Dyse (1965) and §2.6 of this book). To explain the ground whistler observations, it has been suggested that the whistler must be guided by a field aligned duct of enhanced ionization over at least part of its path (Smith *et al.*, 1960; Smith, 1961; Helliwell, 1965). Such a duct, due to the specific dependence of the whistler refractive index on plasma density, can keep the ray direction as well as wavenormal angle confined within a small cone along the magnetic field line. There is now strong experimental and theoretical evidence that the great majority of ground whistlers are guided through the magnetosphere along field aligned enhancements of ionization, that is in whistler ducts.

Indirect evidence for the existence of ducts capable of guiding whistler waves between conjugate hemispheres, together with full details of work done prior to 1965 is given in the book by Helliwell (1965). Direct *in situ* observations showing that such ducts actually exist have been made by satellites on relatively few occasions (Smith and Angerami, 1968; Angerami, 1970; Ondoh, 1976). Angerami (1970) has studied the properties of whistler ducts by means of whistlers received on an inward pass of the OGO 3 satellite while it was between L shells with $L = 4.7$ and 4.1.* From the satellite whistler data presented by Angerami (1970), the following conclusions may be summarized for the ducts which he observed near $L = 4$. The L shell thicknesses of the ducts at the equator ranged from 223 to 430 km, that is between 0.035 and 0.07 earth radii; at ionospheric heights the range was estimated from 15 to 27 km. Effective separation of the ducts in L value was comparable with their thickness and ranged from 110 to 1140 km at the equator, that is between 0.017 and 0.18 earth radii; at ionospheric heights the estimated range was 6–73 km. The longitudinal width of the ducts was about $4°$, corresponding to about 1900 km, or 0.3 earth radii, at the equator. This means that duct extents in the equatorial plane are four to eight times greater in longitude than L space.† The ducts are thus inferred to be generally of elliptical cross-section in the equatorial plane. At ionospheric heights the ducts are probably of circular cross-section. Angerami (1970) also noted that the whistler ducts are much more likely to be ionization enhancements than troughs. The enhancement factor was unlikely to be above 22% and a probable value was about 10%. Ondoh (1976) deduced the presence of ducts from whistler observations made on board the ISIS satellite; he found the average radial width of the ducts in the equatorial plane to vary from 10 km at $L = 1.33$ to 80 km at $L = 2.00$. From whistler data recorded by the FR-1 satellite, it was found that duct foot-points were usually located above the altitude 750 km (Cerisier, 1974). Sometimes duct ends extend to lower altitudes (Cerisier, 1974) and the foot-points can stretch below the altitude 300 km at night.

Other properties of ducts can be inferred from ground whistler observations. In particular, based on these observations, the lifetime of the whistler ducts can be deduced. The typical duct lifetimes are estimated to be of the order of one day (Park and Carpenter, 1970; Sagredo and Bullough, 1973). On some occasions, the observations indicate smaller lifetimes, with the values of a few minutes to hours (Hansen *et al.*, 1983).

The fine structure of ducts is not yet well known in much detail. Some ground observations of whistlers reveal the presence of multiple and complex duct structures in which fine structure can be superimposed on a broader main duct (e. g., Lester and Smith, 1980). Evidence for a conglomeration of ducts is confirmed by

---

* The parameter L usually used to specify a particular line in the geomagnetic field (McIlwain, 1961) is defined as the distance measured in earth radii from the earth center and the point where the particular field line meets the magnetic equator. Rotation of the L line about the earth magnetic-dipole axis produces the so-called L shell.

† Some ground observations which are in agreement with relative equatorial-duct dimensions deduced from OGO 3 data (Angerami, 1970), have recently been reviewed by Strangeways (1996).

in situ observations using rockets (Scarf and Chappel, 1973). For the multiple duct structure, it is shown that the number of simultaneously existing ducts can vary appreciably with time, for example, from 4 to 31 in about four hours (Lester and Smith, 1980). Ducting of whistlers within such multiple structures can, in general, proceed in a rather complicated manner. Thus, the guided waves can become re-trapped in the duct from which they had previously leaked; the waves which are first trapped in the main enhancement as a whole can be further trapped and ducted inside fine structure enhancements (Strangeways, 1982).

An important feature of ducts near the equator is that their electron density profiles are smooth enough so that the density does not vary appreciably within a local wavelength. This allows employing comparatively simple approaches for describing the electromagnetic field in the ducts, for example, ray tracing, the WKB approximation, the parabolic equation method, etc.

There are some interesting and important topics relating to guided whistler propagation which have been omitted in this brief outline. Among them are the effect of curvature of the geomagnetic field lines on ducting of whistlers, propagation of low-latitude whistlers in the Earth-ionosphere waveguide and their trapping in filamentary field-aligned ionospheric irregularities, and others. These topics need a special discussion which is beyond the scope of this book.

For completness, let us now delineate the mechanisms involved in the formation of ducts. It is now believed that quasi-stationary electric fields existing in the magnetosphere and perturbating the magnetospheric plasma play a dominant role in duct formation. Various mechanisms for the formation of ducts have been proposed by Cole (1971), Park and Helliwell (1971), Walker (1978) and other authors, while mathematical models for the evolution of ducts have been discussed by Bernhardt and Park (1977), Thomson (1978), Richards and Cole (1979), and Wang *et al.* (1984). The theories of Cole (1971), Park and Helliwell (1971) and Walker (1978) all suppose that field aligned irregularities are formed by motion of tubes with different plasma content under the action of an irregularity in the magnetospheric electric-field component perpendicular to the magnetic field direction. Park and Helliwell (1971) used a model in which the localized electric field in the magnetosphere, together with the geomagnetic field, produced field-aligned density irregularities by mixing tubes of ionization through the plasma drift occurring in the crossed electric and magnetic fields. Within this model, a $0.1\,mV/m$ electric field in the equatorial plane was found to be able to produce density enhancements and depressions of the order of 5% at $L = 4$ in about half an hour. They also showed that a very complex structure can evolve from a very simple initial condition and a very simple electric field distribution.

The theories mentioned above differ in the way in which the electric field irregularity creating the ducts is produced. Cole (1971) suggests a tube interchange process driven by electrostatic polarization fields which originate in conductivity irregularities in the E-region of the ionosphere and which map up to the equatorial plane. Park and Helliwell (1971) suggest that thundercloud electricity could be the source suitable to provide the formation of ducts. In this mechanism, the electric field in the magnetosphere arises from a mapping upwards (with substantial attenuation) of the strong electric field associated with thunderclouds. Park and Dejnakarintra (1973) studied the penetration of thundercloud electric fields into the ionosphere and magnetosphere at middle and subauroral latitudes and concluded that thunderclouds may provide adequate electric fields to form whistler ducts. It is of interest to note that recent studies (Hegai *et al.*, 1990) have indicated that the thundercloud electric field penetrates into the ionosphere even more effectively than found by Park and Dejnakarintra (1973). Some calculations (Velinov and Tonev, 1994, 1995) show that giant thundercloud electric fields can have a significant effect on the density profiles in the E- and F-regions of the ionosphere, strongly influencing even the main ionospheric maximum.

Discussion of other duct formation theories close to those mentioned above may be found, for example, in works of Thomson (1978), Walker (1978), and Lester and Smith (1980).

1.2. Artificial density ducts created with strong electromagnetic fields in magnetized plasmas

If the electromagnetic field is strong enough, it can produce artificial irregularities in space or laboratory plasmas due to various nonlinear effects (see, for example, Tsytovich, 1970; Gurevich, 1978; Golant and Fedorov, 1989). The study of nonlinearity in the formation of such structures is now being carried forward energetically. Even a brief review of the works relevant to this realm would take up much space, and is beyond the scope of this book. Only illustrative considerations are given here of the most important nonlinear mechanisms of creating artificial density ducts capable of guiding antenna-launched whistler-mode waves. Such ducts arising in the vicinity of antennas in a magnetoplasma are now believed to be of considerable importance for many promising applications, and therefore they will receive much attention in our future treatment. Some related topics on the artificial modification of the ionosphere by large-amplitude waves from ground-based transmitters will be omitted in this brief outline.

In a collisionless magnetoplasma, the formation of ducts is due to the ponderomotive force. Its action gives rise to a redistribution of plasma, thereby causing a variation of the plasma density. As a result, a filament forms, which confines and guides the wave-beam whose field creates the ponderomotive force (see Karpman and Kaufman, 1982 a; Karpman *et al.*, 1984; and references therein). In the whistler band under the action of the ponderomotive-force nonlinearity, both field-aligned density crests and troughs may be realized, with the enhancement or depression factor of about a few percent.

More appreciable variations of plasma parameters are caused by Joule heating of the electrons, which usually dominates when the plasma is collisional enough. This occurs in the majority of the laboratory experiments and in some active ionospheric experiments on the formation of artificial ducts. As a rule, electron heating is mainly caused by the antenna near-zone field. If the local heating at the antenna is not strong enough to produce an additional ionization, it gives rise to a thermal-diffusion-driven redistribution of plasma. Because of a rapid plasma transport along an ambient magnetic field, this process creates a duct within which the density is usually diminished, with the depression factor of about 10–20% (Stenzel, 1976, 1977; Sugai *et al.*, 1978; Egorov *et al.*, 1988; Zaboronkova *et al.*, 1992 a). Typically, such thermal ducts are transient and may have a fine structure superimposed on a broader main density trough (Zaboronkova *et al.*, 1992 a).

Beginning with a certain value of an amplitude of the field, a breakdown and additional ionization of the background plasma can occur. The ionization may be caused by the antenna-launched wave field as well as the field in the near-zone of the radiator (Markov *et al.*, 1979, 1981; Vdovichenko *et al.*, 1986; Markov, 1988 a; Agafonov *et al.*, 1990; Golubyatnikov *et al.*, 1995). Note that ionization effects may create strong, elongated density perturbations in which the enhancement factor ranges from several times up to a few orders of magnitude.

Since thermal and ionization nonlinearity mechanisms seem to be the most suitable to produce noticeable variations in plasma density, it is desirable to elucidate, using at least a simplified treatment, when each of the mechanisms will be dominant in the formation of artificial ducts. This may be clarified as follows.

Let the following conditions be satisfied for the plasma parameters and the amplitude of an alternating (antenna-launched) electromagnetic field: (i) their spatial variations along and across an ambient static magnetic field are much greater than the electron mean free path and the ion gyroradius, respectively; (ii) their temporal variations are slow on the scale of the mean time between two collisions of one particle; (iii) the pressure of the plasma is much less than that of the ambient magnetic field. Besides, we assume that the quasi-neutrality is fulfilled to a good accuracy, so that one can write for the plasma density, N,

$$N = N_e \approx N_i, \tag{1.1}$$

where the quantities N_e and N_i denote the number of electrons and ions per unit volume, respectively.

Under the above-mentioned assumptions, the equation of continuity for the plasma (Gurevich, 1978) is

$$\frac{\partial N}{\partial t} + \nabla \cdot \mathbf{\Gamma} = q_{\text{ext}} + \nu_{\text{ion}} N - \nu_a N - \alpha_r N^2, \tag{1.2}$$

where

$$\nabla \cdot \boldsymbol{\Gamma} = \nabla \cdot \boldsymbol{\Gamma}_e = \nabla \cdot \boldsymbol{\Gamma}_i,$$

$$\boldsymbol{\Gamma}_e = -\frac{1}{e}\,\hat{\sigma}_e \cdot \mathbf{E}_s - \hat{D}_e \cdot \nabla N - \frac{N}{T_e}\,\hat{D}_{Te} \cdot \nabla T_e, \tag{1.3}$$

$$\boldsymbol{\Gamma}_i = \frac{1}{e}\,\hat{\sigma}_i \cdot \mathbf{E}_s - \hat{D}_i \cdot \nabla N.$$

Here q_{ext} is the rate of production of electrons and ions by external sources of ionization, ν_{ion} the frequency of ionization by energetic electrons, ν_a the attachment frequency, α_r the electron-ion recombination coefficient, e the magnitude of the electron charge, $\mathbf{E}_s$ the quasi-stationary electric field arising due to the plasma nonuniformity, $\hat{\sigma}_e$ ($\hat{\sigma}_i$) the conductivity tensor for electrons (ions), $\hat{D}_e$ ($\hat{D}_i$) the diffusion tensor for electrons (ions), $\hat{D}_{Te}$ the thermal-diffusion tensor for electrons, T_e the electron temperature. The transport of particles is represented by the diffusion flux vectors $\boldsymbol{\Gamma}_e$ and $\boldsymbol{\Gamma}_i$. The explicit expressions for the quantities occurring in (1.2) and for the tensors $\hat{\sigma}_{e,i}$, $\hat{D}_{e,i}$ and $\hat{D}_{Te}$ are given elsewhere (see, for example, Braginskii, 1965; Krall and Trivelpiece, 1973; Gurevich, 1978). In the following, we shall employ the longitudinal (transverse) unipolar transport coefficients $\sigma_{e,i\,\parallel}$, $D_{e,i\,\parallel}$ and $D_{Te\,\parallel}$ ($\sigma_{e,i\,\perp}$, $D_{e,i\,\perp}$ and $D_{Te\,\perp}$) which are the diagonal elements of the tensors $\hat{\sigma}_{e,i}$, $\hat{D}_{e,i}$, and $\hat{D}_{Te}$, respectively. The symbols $\parallel$ and $\perp$ denote the directions parallel or perpendicular to the ambient magnetic field.

It is worthwhile to stress that the term q_{ext}, representing the contribution to ionization from external (non-electromagnetic) sources, is responsible for maintaining the background plasma, whereas the term $\nu_{\mathrm{ion}} N$ in (1.2) gives the rate of production of charged particles due to the electron impact ionization of neutral molecules in the strong electromagnetic field. The remaining terms in the right-hand side of (1.2) give the rate of removal of electrons. The term $\nu_a N$ shows how the attachment of the electrons to neutral molecules contributes to the removal, while the term $\alpha_r N^2$ represents the contribution from recombination of the electrons with positive ions.

If the spatial scales of the plasma perturbation is small enough, $\nabla \cdot \boldsymbol{\Gamma}$ predominates over the right side of Equation (1.2), and we then have

$$\frac{\partial N}{\partial t} + \nabla \cdot \boldsymbol{\Gamma} = 0. \tag{1.4}$$

This is the equation governing the dynamics of the thermal-diffusion redistribution of the plasma. To simplify matters, we shall assume that all the perturbed plasma parameters are independent of the azimuth about the ambient magnetic field. If the longitudinal size, $L_\parallel$, and the transverse size, $L_\perp$, of the heated region satisfy the inequality

$$L_\parallel^2/L_\perp^2 \gg D_{e\parallel}/D_{A\perp}, \tag{1.5}$$

where $D_{A\perp}$ is the coefficient of the ambipolar diffusion across the ambient magnetic field, the first equation in (1.3) reduces to

$$\boldsymbol{\Gamma}_{e\perp} = \boldsymbol{\Gamma}_{i\perp}$$

(see, for example, Gurevich, 1978). This, in turn, gives

$$\nabla \cdot \mathbf{\Gamma} = -\nabla_\perp \cdot \left(D_{A\perp} \nabla_\perp N + D_{TA\perp} \frac{N}{T_e} \nabla_\perp T_e \right). \tag{1.6}$$

Here the ambipolar coefficients $D_{A\perp}$ and $D_{TA\perp}$ are expressed by

$$D_{A\perp} = \frac{\sigma_{i\perp} D_{e\perp} + \sigma_{e\perp} D_{i\perp}}{\sigma_{e\perp} + \sigma_{i\perp}}, \qquad D_{TA\perp} = \frac{\sigma_{i\perp} D_{Te\perp}}{\sigma_{e\perp} + \sigma_{i\perp}}. \tag{1.7}$$

In the simplest, steady-state case ($\partial N/\partial t = 0$), the solution to Equation (1.4) can be written

$$N\, T_e^{k_\perp} = \mathrm{const}, \tag{1.8}$$

where $k_\perp = D_{TA\perp}/D_{A\perp}$.

The relation (1.8) is the usual diffusion equilibrium condition. As seen from (1.8), the density variation is determined by the sign of $k_\perp$. Under comparatively strong plasma magnetization, as is typical of the ionosphere, the quantity $k_\perp$ is usually positive for a weakly ionized plasma, in which the electron-neutral collisions predominate over the electron-ion ones. In this case the electron heating causes a decrease in plasma density within the perturbed region. On the contrary, when the plasma is more heavily ionized and the electron-ion (Coulomb) collisions predominate, $k_\perp$ is usually negative, so that the heating leads to an increase in density (see, for details, Kostrov and Kim, 1988).

In the other limiting case, when

$$L_\parallel^2/L_\perp^2 \ll D_{A\parallel}/D_{i\perp} \tag{1.9}$$

($D_{A\parallel}$ is the coefficient of the ambipolar diffusion along the ambient static magnetic field), it is possible to deduce that

$$\mathbf{\Gamma}_{e\parallel} = \mathbf{\Gamma}_{i\parallel}$$

and hence

$$\nabla \cdot \mathbf{\Gamma} = -\nabla_\parallel \cdot \left(D_{A\parallel} \nabla_\parallel N + D_{TA\parallel} \frac{N}{T_e} \nabla_\parallel T_e \right). \tag{1.10}$$

The expressions for $D_{A\parallel}$ and $D_{TA\parallel}$ are obtainable from (1.7) by making the subscript replacement '$\perp$' $\rightarrow$ '$\parallel$'. Now the steady-state solution to Equation (1.4) is

$$N\, T_e^{k_\parallel} = \mathrm{const}, \tag{1.11}$$

with $k_\parallel = D_{TA\parallel}/D_{A\parallel}$. Since the quantity $k_\parallel$ is almost always positive here, the heating of the plasma electrons causes the formation of a density trough.

In the case when neither (1.5) nor (1.9) are satisfied, the thermal diffusion proceeds in general in a rather complicated manner. This topic was discussed by Voskoboinikov *et al.* (1989), who theoretically investigated the thermal-diffusion redistribution caused by electron heating in the field of a source of small size. As they noted, for such a source, the heat conduction must essentially be taken into account in order to accurately determine the shape and size of the heated region. These authors found that the one-dimensional ambipolar-diffusion approximation is not sufficient in that case. In a weakly ionized plasma, the electrons diffuse in a unipolar manner away from the perturbed region along the ambient magnetic field, while the ions diffuse unipolarly across it. To maintain quasi-neutrality, the currents then appear in the surrounding plasma (see also Zhilinsky and Tsendin (1980) and references therein). The currents are supported in the background medium by the longitudinal electron flux and the transverse ion flux, which close the current circuit. The presence of these eddy 'short-circuit' currents leads to the appearance of certain regions with reduced density in the background plasma. In a core of the density irregularity, this process results in the formation of a field-aligned channel within which the plasma density also falls below the background value. As the channel is created, it is accompanied by a surrounding layer with a plasma density somewhat higher than the background. Similar behavior has also been shown to occur in the partially ionized plasma.

When the sizes $L_{\perp,\parallel}$ of a heated region are sufficiently greater than the characteristic diffusion scales, $D^{1/2}_{\perp,\parallel}/(\nu_a+\alpha N)^{1/2}$, it is permissible to neglect, to a first approximation, the diffusion effects. Then, the steady-state density profile can be obtained from the equation

$$(\nu_{\text{ion}} - \nu_a)\, N - \alpha_r N^2 + q_{\text{ext}} = 0, \tag{1.12}$$

which is a simplified version of Equation (1.2). If the ionization frequency ν_{ion} grows rather rapidly with the electron temperature, the ionization mechanism may become apparent and consequently the formation of a density enhancement will occur. A more realistic situation, however, is that the additional ionization is accompanied by some diffusion transport. Very often, because of either a rapid transport along the ambient magnetic field or the trapping of ionizing radiation in the creating irregularity, the enhancement takes the form of a field-aligned density perturbation adjacent to the electromagnetic source. The simplest model of the density distribution which is possible in such a perturbation is shown in Figure 1.2, a. In some cases, more complicated structures can appear. One such structure, occurring when the diffusion near the ends of a density perturbation proceeds by the electron unipolar mechanism along the ambient magnetic-field direction and by the ion unipolar mechanism across it, is presented in Figure 1.2, b (see, for details, Zhilinsky and Tsendin, 1980). As pointed out earlier, this type of diffusion is accompanied by eddy currents that leads to the formation of density troughs in the background plasma. Since the troughs in the outer region cannot be very deep, the background currents are no longer able to support the unipolar diffusion process described above when the enhancement in a perturbed region becomes very strong. Therefore, a strong enhancement of density will likely be similar to that shown in Figure 1.2, a.

It is to be noted that although some progress toward the basic understanding of the thermal-diffusion formation and the ionization formation of field-aligned near-antenna ducts has been made in recent years, a satisfactory solution to this problem is yet to be found. Therefore, the outline presented above should be considered as a very much oversimplified picture of the actual mechanisms of duct formations, which only gives a general guide as to how artificial density structures may be created by strong electromagnetic fields.

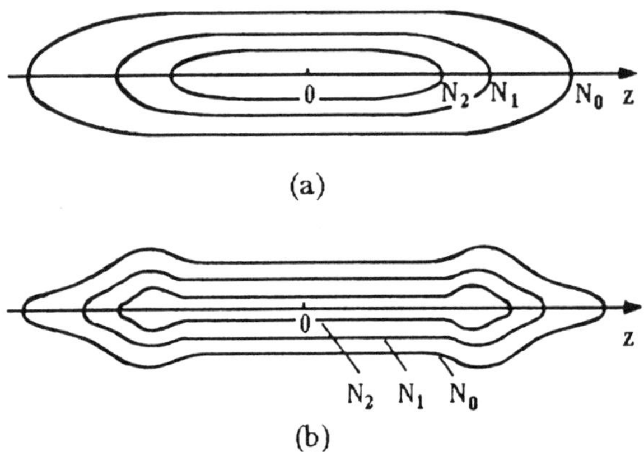

Figure 1.2. Sketches (not to scale) showing typical structures of equal
density lines in an artificial plasma perturbation. $N_2 > N_1 > N_0 > N_a$
(N_a is an ambient plasma density, not shown); see text for discussion.

1.3. The results of model laboratory and active ionospheric experiments on observation of artificial ducts

In the present section, an outline of recent experimental advances concerning the
formation of artificial ducts in the vicinity of electromagnetic sources will be given.
The principal purpose of this section is to discuss the results of the relevant experiments which played an important role in the theoretical developments.

We begin with a thermal-diffusion mechanism of the formation of ducts that was
realized in a number of the laboratory experiments (Stenzel, 1976, 1977; Sugai *et
al.*, 1978; Egorov *et al.*, 1988; Zaboronkova *et al.*, 1992 a). For example, Zaboronkova
et al. (1992 a) observed the near-antenna ducts in the experiments which were conducted in a vacuum chamber 150 cm in length and 80 cm in diameter. A background
argon plasma was created at a pressure 5×10^{-3} Torr by a radio frequency (rf)
pulsed discharge in a uniform static magnetic field 240 G; this plasma was shaped
like a quasi-uniform column of length 100 cm and diameter 40 cm. Under the conditions of the experiment, the electron temperature and the ion temperature in this
plasma column coincided and came to $\simeq 0.4$ eV. The experiments were carried out
in a decaying plasma whose characteristic decay time was about 2 ms. Whistler
waves were launched by the antenna, which was a loop of radius 2.5 cm. The antenna was placed on the axis of the plasma column, with the plane of the loop

perpendicular to the ambient magnetic field. After switching off the pulse source creating the plasma column, during the decay stage of a plasma with its initial density $N = 4 \times 10^{12}\,\mathrm{cm}^{-3}$, a rf voltage was fed to the antenna with fixed angular frequency $3.9 \times 10^8\,\mathrm{s}^{-1}$, length up to 2 ms, and amplitude 2 to 150 V. The experiments were carried out with both small and large power levels P fed to the antenna. In the case of comparatively small power levels ($P \leq 2\,\mathrm{W}$), there was no density perturbation, and electromagnetic waves excited in the uniform plasma propagated from the source roughly along the external magnetic field. Results of phase measurements, which were carried out along the loop axis at the distances 6 to 32 cm from the antenna, showed that the wavelength of the wave propagating in the plasma was $\lambda \simeq 6\,\mathrm{cm}$, while its amplitude fell off as z^{-1} with distance z from the source. This wave is easily identified as a non-ducted quasi-longitudinal whistler propagating almost along the magnetic field line. For sufficiently high power levels fed to the antenna ($P \geq 80\,\mathrm{W}$), it was observed that a marked increase of the electron temperature occurred near the radiator due to Joule heating of the plasma in the quasi-static field of the loop. At time $500\,\mu\mathrm{s}$ after the start of the rf voltage pulse, this temperature reached a value of 1.5 eV. As a result of electron heating and the thermal-diffusion redistribution of the plasma, a perturbed density profile formed near the axis of the system. The profile is shown in Figure 1.3. It is clear from Figure 1.3 that the perturbed density distribution was characterized by the formation of a channel with a decreased density on the axis of the system and an annular layer with an enhanced density surrounding this channel. We notice that it is the formation of such density perturbations that was discussed qualitatively in § 1.2. One should only take into account that in the experiments described above Coulomb (electron-ion) collisions predominated over electron-neutral ones, the collision frequency being much less than the angular frequency of the whistler-mode waves excited by the loop.

Measurements of the spatial distribution of the fields showed that the picture of electromagnetic field propagation is much more complicated in the nonlinear case than it is at linear power levels. Within the channel with reduced plasma density (see Figure 1.3, $\rho \leq 5\,\mathrm{cm}$) Zaboronkova et $al.$ (1992 a) recorded ducted propagation of electromagnetic radiation with a longitudinal wavelength of the order of $\lambda_1 \simeq 9.5\,\mathrm{cm}$. Ducted propagation of electromagnetic radiation was also observed in the region of enhanced plasma density (see Figure 1.3, $\rho \simeq 7\,\mathrm{cm}$); the longitudinal wavelength in this case came to $\lambda_2 \lesssim 5.5\,\mathrm{cm}$. It will be explained theoretically later, in § 4.9, that in the central part of the low-density channel the ducted propagation of oblique whistler-mode waves does occur, while the high-density annular layer that surrounds this central part supports ducted propagation of quasi-longitudinal whistlers.

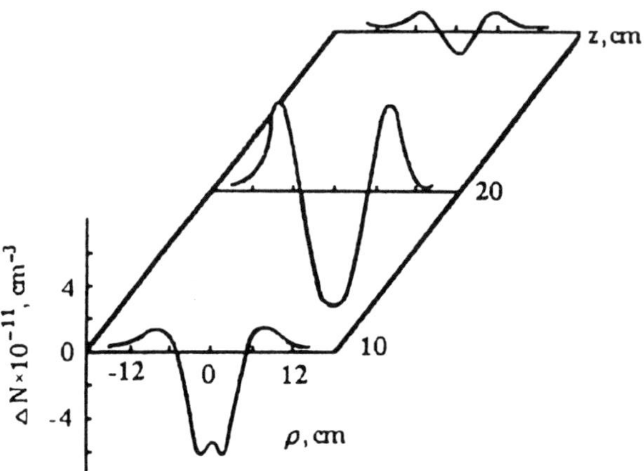

Figure 1.3. Typical profiles of the density variation ΔN across the magnetic field sampled at different distances from the source, under the conditions of experiments described by Zaboronkova *et al.* (1992 a); the background density $N \simeq 4 \times 10^{12}\,\text{cm}^{-3}$.

An experimental observation of a whistler duct caused by additional ionization of the background plasma in the antenna-launched electromagnetic field was reported by Vdovichenko *et al.* (1986) and Markov (1988 a). Their experiments were carried out in a chamber 150 cm in length and 20 cm in diameter, in which the pressure of the filling gas (air) was $6 \times 10^{-3}\,\text{Torr}$, and the static magnetic field was 160 G. A quasi-uniform background plasma with density $N \gtrsim 2 \times 10^{10}\,\text{cm}^{-3}$ was created by an oscillator with power 360 W and angular frequency $10^8\,\text{s}^{-1}$. The waves in this plasma were excited by the antenna which was a helical coil of radius 3 cm coaxial with the chamber axis. The coil was fed by another oscillator with power $P \leq 25\,\text{W}$ and angular frequency $10^9\,\text{s}^{-1}$. It is evident from the indicated parameters that this frequency lies in the whistler band.

At lower levels of the power supplied to the antenna ($P \leq 0.1\,\text{W}$), Vdovichenko *et al.* (1986) observed the field pattern typical of the helical-coil antenna in the uniform (background) magnetoplasma. As the power was raised ($P > 5\,\text{W}$), the spatial distribution of the plasma density and of the antenna-launched field changed substantially. A comparatively narrow plasma filament formed in the chamber, spanning the entire apparatus in the longitudinal direction. Figure 1.4 shows radial density profile for two distances z from the source at $P = 10\,\text{W}$. This elongated plasma structure was caused by additional ionization of the background plasma in the quasi-static and wave fields of the antenna. Direct phase measurements along the plasma filament revealed that a guided whistler mode with a wavelength $\lambda \simeq 20$ cm was propagating along the direction of the magnetic field. The observed value of the wavelength satisfied the conditions of trapping the whistler in the filament. This wavelength increased slightly with distance from the antenna because of a decrease in the plasma density. Based on the field measurements, Vdovichenko *et al.* (1986) concluded that the field was confined in the filament which thus guided the whistler-mode waves emitted from the antenna.

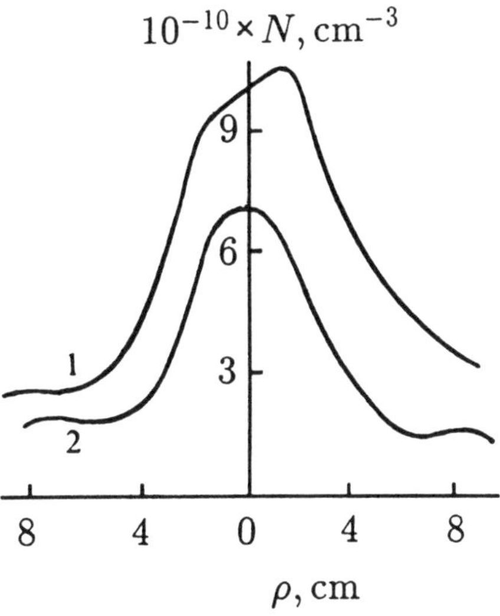

Figure 1.4. Density profiles across the magnetic field at different distances z from the source, measured in the experiments of Vdovichenko *et al.* (1986) and Markov (1988 a): 1 — $z = 30$ cm, 2 — $z = 85$ cm (reproduced from Markov, 1988 a).

The ionization formation of a field aligned structure due to breakdown and ionization of a neutral gas released in the antenna near-zone has recently been observed by Golubyatnikov *et al.* (1995) in a large plasma device 10 m in length and 3 m in diameter. Golubyatnikov *et al.* (1995) carried out their experiments in an argon plasma which was created at a pressure 5×10^{-5} Torr by an induction discharge in an external magnetic field $\simeq$440 G and had the form of a quasi-uniform column of radius 60 cm and length about 3 m. A large volume of the device made the effects of chamber walls on the results of measurements negligible. The background plasma density was 2×10^{11} cm^{-3}; the electron and ion temperatures were nearly the same and equal to 0.5 eV. The experiments were carried out in a decaying-plasma regime, with decay time 1.5 ms. As the source of waves in the plasma, a loop antenna of radius 9 cm was used, with its axis parallel to the external magnetic field. In order to produce a rather strong density enhancement, some portion of a neutral argon was released into the background plasma in the antenna vicinity and simultaneously, when a pressure of the filling gas reached the value $\gtrsim 10^{-4}$ Torr, a pulse with time length 100 μs (or 600 μs), angular frequency $3.1 \times 10^7\,\mathrm{s}^{-1}$, and power 0.5 MW was fed to the antenna. This caused the breakdown and ionization of the released gas, so that the relative variations in the electron temperature and the plasma density in the neighbourhood of the antenna increased up to $\Delta T_e/T_e \simeq 50$ and $\Delta N/N \simeq 100$, respectively. The transverse dimension of the perturbed region was close to the antenna radius, while the longitudinal dimension extended up to 2 m. Golubyatnikov *et al.* (1995) indicated that under the conditions of their experiments ionization was caused mainly by the antenna near-zone field and, presumably, by oblique (quasi-electrostatic) whistler-mode waves emitted from the loop; the authors also pointed out the possible role of energetic particles accelerated by the antenna field for ionizing the surrounding medium.

It is clear that such strong perturbations of the plasma parameters near the antenna should influence appreciably its radiation characteristics. It was suggested by Markov (1988 b) to use the near-antenna ducts of enhanced ionization for increasing the power radiated from sources in the whistler band. The convincing corroboration of this possibility has been demonstrated by Markov (1988 b) and Kudrin and Markov (1991) in laboratory experiments conducted with the use of various types of antennas. These experiments showed that when a near-antenna duct with enhanced density formed, the matching of an antenna with a feeder improved, so that the power reflection coefficient from the radiator decreased by a few times. Simultaneously, the corresponding increase in the radiated power was registered. The increase of radiation corroborates the effective excitation of the whistler-mode waves, some portion of which was trapped in the duct. Field measurements along the duct axis revealed the adiabatic propagation of the guided whistler modes whose wavelengths increased gradually as the density enhancement reduced smoothly with distance from the source. These two works (Markov, 1988 b; Kudrin and Markov, 1991) give evidence that a considerable increase of radiation can be achieved in the whistler band by using an artificial near-antenna duct with enhanced density. This topic will be considered from the theoretical viewpoint in Chapter 6.

It is important that the majority of modeling laboratory experiments reviewed above were carried out for the plasma similarity parameters typical of the earth's ionosphere. They provide indirect evidence that the artificial field-aligned density irregularities can be formed in the vicinity of powerful electromagnetic sources arranged aboard rockets or satellites. Direct *is situ* observations of corresponding plasma structures in the lower ionosphere as a result of operating a powerful source based on a rocket have been reported by Markov (1988 a) and Agafonov *et al.* (1989, 1990). According to their results, when a sufficiently high power is fed to a rocket-based antenna, additional ionization of the background plasma takes place, and a nonuniform channel forms which extends along the geomagnetic field, the plasma density within a channel exceeding considerably (by 1–2 orders of magnitude) the background value. Besides, it has been demonstrated that a further increase of plasma density can be achieved by the use of releases of neutral gas or chemical mixtures into the perturbed region (Markov, 1988 a; Agafonov *et al.*, 1989).

A discussion of experiments in which field-aligned density enhancements were created in rf discharge by an on-board electromagnetic source in the polar nocturnal ionosphere at altitudes of 100–150 km, was presented by Agafonov *et al.* (1990). As the source, the authors used a dipole radiator made up of a rocket body and a wire ring of diameter 2 m, held by an axial telescoping rod at a height of 2 m above the head of the rocket (the rocket diameter was 0.4 m and its length was 10 m). The plane of the ring was perpendicular to the rocket axis. The rf potential of amplitude $\simeq 1\,\mathrm{kV}$ and frequency 480 kHz was supplied between the ring and the body of the rocket. The maximum output power of an on-boad rf oscillator was about 1 kW. The rf signal was modulated in the telegraph regime according to a special cyclogram. When the oscillator was switched on, the radiating system described above emitted electrostatic waves capable of exciting the rf discharge near the rocket.

The results of processing the telemetric information for start up at 18:31 UT on October 9, 1989 in the northern part of the Norwegian Sea are given in Figure 1.5, where the flight time t and the corresponding values of the altitude H of the rocket above the Earth's surface are shown on the horizontal axis. On the vertical axis are the values of the plasma density N near the head of the rocket. The intervals where a decrease in density is seen, correspond to a pause in the operation of the on-board oscillator.

These experimental data indicate that it is possible to form and support a discharge in the ionosphere in the field of an electromagnetic source. Such a discharge is a strong local perturbation of the ionospheric plasma parameters. Indeed, the plasma density near the head of the rocket was $N \gtrsim 10^7\,\mathrm{cm}^{-3}$, and the electron temperature in the perturbed region was $T_e \gtrsim 10\,\mathrm{eV}$, while the background (unperturbed) values of these parameters were $N \simeq 5 \times 10^5\,\mathrm{cm}^{-3}$ and $T_e \simeq 0.1\,\mathrm{eV}$. The sharp peaks seen in Figure 1.5 ($150\,\mathrm{s} < t < 250\,\mathrm{s}$) are attributed to firing four explosive charges at the rocket. The form of the plasma perturbation was determined by several factors. Its longitudinal dimension (along the direction of the geomagnetic field) was about the damping length of the electrostatic waves excited by the antenna in the discharge plasma and reached the value $\simeq 1\,\mathrm{km}$ (see Kudrin *et al.* (1991) for more details). The transverse dimension was determined by diffusion processes and depended on the horizontal component of the rocket velocity. Under the conditions of this experiment its probable value was about 10 m and it was unlikely to be above 100 m.

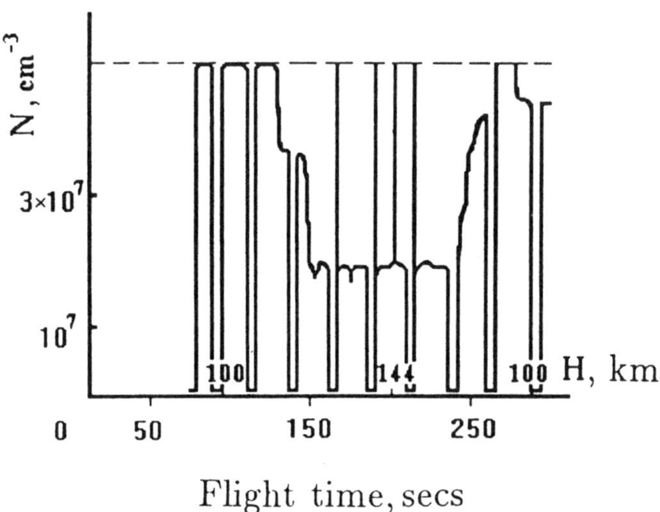

Figure 1.5. Variations in the plasma density N near the head of the rocket in the ionospheric experiment conducted by Agafonov *et al.* (1990). The dash line indicates a limit of telemetric signal. The diagram is courtesy of G. A. Markov, University of Nizhny Novgorod, Nizhny Novgorod, Russia.

Summarizing, there is strong evidence that artificial ducts with enhanced density can actually be created in the ionosphere. We do not dwell on the results of other active rocket and satellite experiments on changing the local parameters of the ionosphere, involving neutral material releases, critical ionization phenomena, etc. For review of these topics, and a bibliography, see Raitt (1996).

It is to be stressed that interpreting the results of the concrete experiments is beyond the scope of this book. However some theoretical models and simplifying assumptions employed in the book are based on this material. Therefore the experimental data given above will be referred to throughout the book for estimates and numerical calculations.

Chapter 2

The Basic Equations

2.1. Introduction

In this chapter some of the basic theory for electromagnetic waves in a magneto-plasma is given. The majority of this material should be familiar to most readers. It is included though to serve as the foundation for later works and also to define some of the notations to be used in what follows. The more experienced reader can proceed at once to §§ 2.5 and 2.6.

It is to be noted that SI units are used throughout, except in a few cases where different units are specified temporarily. For harmonically varying fields we use the time factor $\exp(i\omega t)$.

2.2. Maxwell's equations

We begin by writing the Maxwell equations governing the electromagnetic field in a medium. They are

$$\nabla \cdot \mathbf{D} = \rho^e, \tag{2.1}$$

$$\nabla \cdot \mathbf{B} = \rho^m, \tag{2.2}$$

$$\nabla \times \mathbf{E} = -\frac{\partial \mathbf{B}}{\partial t} - \mathbf{J}^m, \tag{2.3}$$

$$\nabla \times \mathbf{H} = \frac{\partial \mathbf{D}}{\partial t} + \mathbf{J}^e. \tag{2.4}$$

The field vectors $\mathbf{D}$, $\mathbf{B}$, $\mathbf{E}$, and $\mathbf{H}$ are the electric induction, the magnetic induction, the electric field intensity, and the magnetic field intensity, respectively. It is assumed that there are electric and magnetic sources, whose current densities, $\mathbf{J}^{e,m}$, and charge densities, $\rho^{e,m}$, are related by the following *equations of continuity*:

$$\nabla \cdot \mathbf{J}^e + \frac{\partial \rho^e}{\partial t} = 0, \tag{2.5}$$

$$\nabla \cdot \mathbf{J}^m + \frac{\partial \rho^m}{\partial t} = 0. \tag{2.6}$$

Although the magnetic charges and currents represented respectively by ρ^m and $\mathbf{J}^m$ do not exist in reality, it is convenient to introduce them to the Maxwell equations.

As is known from the electromagnetic theory, the use of magnetic charges and currents allows a simplification in the description of the electromagnetic field in many cases.

In all the problems discussed in this book, the sources of the electromagnetic field are assumed to be varying sinusoidally with time t, at a frequency f. The Maxwell equations are linear, and very often the relations between $\mathbf{D}$, $\mathbf{B}$ and $\mathbf{E}$, $\mathbf{H}$ in a medium are also linear. In such a *linear* medium it is convenient to describe the electromagnetic field in terms of *complex amplitudes*. At any instant the relations between the observed quantities, say $\mathbf{E}(t)$ and $\mathbf{H}(t)$, and their complex amplitudes, $\mathbf{E}(\omega)$ and $\mathbf{H}(\omega)$, are given by

$$\mathbf{E}(t) = \mathrm{Re}\left\{\mathbf{E}(\omega)\,e^{i\omega t}\right\}, \quad \mathbf{H}(t) = \mathrm{Re}\left\{\mathbf{H}(\omega)\,e^{i\omega t}\right\},$$

where Re denotes that the real part is to be taken, and $\omega = 2\pi f$ is called the *angular frequency*. We note that consideration of time-harmonic processes does not lead to loss of generality because in a linear medium an arbitrary signal can be resolved, by Fourier analysis, into spectral components harmonically varying in time. On finding the solution for each component, they all are resynthesized to give the resultant signal.

When all variables contain the time t only through the factor $e^{i\omega t}$, the operator $\partial/\partial t$ is equivalent to multiplication by $i\omega$. If this is used, the Maxwell equations are rewritten in terms of complex amplitudes thus:

$$\nabla \cdot \mathbf{D} = \rho^e, \tag{2.7}$$

$$\nabla \cdot \mathbf{B} = \rho^m, \tag{2.8}$$

$$\nabla \times \mathbf{E} = -i\omega\mathbf{B} - \mathbf{J}^m, \tag{2.9}$$

$$\nabla \times \mathbf{H} = i\omega\mathbf{D} + \mathbf{J}^e. \tag{2.10}$$

The continuity equations (2.5), (2.6) written in terms of complex amplitudes become

$$\nabla \cdot \mathbf{J}^e + i\omega\rho^e = 0, \tag{2.11}$$

$$\nabla \cdot \mathbf{J}^m + i\omega\rho^m = 0. \tag{2.12}$$

Throughout this book we employ the last two Maxwell equations (2.9), (2.10) written in either Cartesian or cylindrical polar coordinates. Therefore it is useful at once to present these equations in full, in both coordinate systems.

Let x, y, z be right-handed Cartesian coordinates, and let $\hat{\mathbf{x}}_0$, $\hat{\mathbf{y}}_0$, $\hat{\mathbf{z}}_0$ be unit vectors in the directions of the x-, y-, z-axes. Subscripts x, y, z will be used to denote the corresponding components of a vector. For example, the components of

$\mathbf{A}$ will be written A_x, A_y, A_z. In Cartesian coordinates, Equations (2.9), (2.10) become

$$\frac{\partial E_z}{\partial y} - \frac{\partial E_y}{\partial z} = -i\omega B_x - J_x^m, \qquad \frac{\partial H_z}{\partial y} - \frac{\partial H_y}{\partial z} = i\omega D_x + J_x^e,$$

$$\frac{\partial E_x}{\partial z} - \frac{\partial E_z}{\partial x} = -i\omega B_y - J_y^m, \qquad \frac{\partial H_x}{\partial z} - \frac{\partial H_z}{\partial x} = i\omega D_y + J_y^e, \qquad (2.13)$$

$$\frac{\partial E_y}{\partial x} - \frac{\partial E_x}{\partial y} = -i\omega B_z - J_z^m, \qquad \frac{\partial H_y}{\partial x} - \frac{\partial H_x}{\partial y} = i\omega D_z + J_z^e.$$

Now take cylindrical coordinates ρ, ϕ, z, with unit vectors $\hat{\boldsymbol{\rho}}_0$, $\hat{\boldsymbol{\phi}}_0$, $\hat{\mathbf{z}}_0$. These coordinates are connected to Cartesian coordinates via the following relations:

$$x = \rho \cos \phi \qquad \rho = (x^2 + y^2)^{1/2},$$

$$y = \rho \sin \phi \qquad \phi = \arctan(y/x), \qquad (2.14)$$

$$z = z \qquad z = z.$$

The components of any vector $\mathbf{A}$ in the cylindrical coordinate system are denoted as A_ρ, A_ϕ, A_z. In this system, Equations (2.9), (2.10) may be written

$$\frac{1}{\rho}\frac{\partial E_z}{\partial \phi} - \frac{\partial E_\phi}{\partial z} = -i\omega B_\rho - J_\rho^m, \qquad \frac{1}{\rho}\frac{\partial H_z}{\partial \phi} - \frac{\partial H_\phi}{\partial z} = i\omega D_\rho + J_\rho^e,$$

$$\frac{\partial E_\rho}{\partial z} - \frac{\partial E_z}{\partial \rho} = -i\omega B_\phi - J_\phi^m, \qquad \frac{\partial H_\rho}{\partial z} - \frac{\partial H_z}{\partial \rho} = i\omega D_\phi + J_\phi^e, \qquad (2.15)$$

$$\frac{1}{\rho}\frac{\partial}{\partial \rho}(\rho E_\phi) - \frac{1}{\rho}\frac{\partial E_\rho}{\partial \phi} = -i\omega B_z - J_z^m, \qquad \frac{1}{\rho}\frac{\partial}{\partial \rho}(\rho H_\phi) - \frac{1}{\rho}\frac{\partial H_\rho}{\partial \phi} = i\omega D_z + J_z^e.$$

The following are useful relations between transverse vector components in Cartesian and cylindrical coordinates. They are:

$$A_\rho = A_x \cos \phi + A_y \sin \phi, \qquad A_x = A_\rho \cos \phi - A_\phi \sin \phi,$$

$$A_\phi = -A_x \sin \phi + A_y \cos \phi, \qquad A_y = A_\rho \sin \phi + A_\phi \cos \phi. \qquad (2.16)$$

2.3. The constitutive relations

Before Equations (2.7) to (2.10) can be applied to the theory of wave-propagation in a magnetoplasma, it is necessary to express the electric induction $\mathbf{D}$ and the magnetic induction $\mathbf{B}$ in terms of the field intensities $\mathbf{E}$ and $\mathbf{H}$. The resulting expressions are called the constitutive relations of a magnetoplasma and derived in numerous books (see, for example, Ratcliffe, 1959; Budden, 1961 a; Stix, 1962; Ginzburg, 1970; Krall and Trivelpiece, 1973; Akhiezer *et al.*, 1975). These relations are written thus:

$$\mathbf{D} = \epsilon_0 \hat{\varepsilon} \cdot \mathbf{E}, \tag{2.17}$$

$$\mathbf{B} = \mu_0 \mathbf{H}, \tag{2.18}$$

where ϵ_0 and μ_0 are known, respectively, as the electric and magnetic constants, and $\hat{\varepsilon}$ is the dielectric tensor of a magnetoplasma. The relation (2.18) means that the magnetic permeability of the plasma is the same as that of free space. The justification for this is discussed, for example, by Ginzburg (1970). The expressions for the elements of the dielectric tensor $\hat{\varepsilon}$ and its properties are discussed thoroughly in the above-mentioned books on plasma physics and the theory of waves in plasmas. The treatment given here is necessarily much briefer.

In formulating the constitutive relation (2.17), it is convenient to choose the z-axis to be along the direction of the ambient static magnetic field $\mathbf{B}_0$. For the rest of this book, we will be using the two-component 'cold plasma' approximation in writing the dielectric tensor. In that approximation the tensor $\hat{\varepsilon}$, for a collisionless plasma, can be written in Cartesian coordinates as

$$\hat{\varepsilon} = \begin{pmatrix} \varepsilon & -ig & 0 \\ ig & \varepsilon & 0 \\ 0 & 0 & \eta \end{pmatrix}, \tag{2.19}$$

where

$$\varepsilon = 1 + \frac{\omega_{\mathrm{p}}^2}{\omega_{\mathrm{H}}^2 - \omega^2} + \frac{\Omega_{\mathrm{p}}^2}{\Omega_{\mathrm{H}}^2 - \omega^2},$$

$$g = -\frac{\omega_{\mathrm{p}}^2 \omega_{\mathrm{H}}}{(\omega_{\mathrm{H}}^2 - \omega^2)\omega} + \frac{\Omega_{\mathrm{p}}^2 \Omega_{\mathrm{H}}}{(\Omega_{\mathrm{H}}^2 - \omega^2)\omega}, \tag{2.20}$$

$$\eta = 1 - \frac{\omega_{\mathrm{p}}^2}{\omega^2} - \frac{\Omega_{\mathrm{p}}^2}{\omega^2},$$

and where ω_{H}, Ω_{H} and ω_{p}, Ω_{p} are the electron and ion gyrofrequencies and the electron and ion plasma frequencies, respectively. These quantities may be written

$$\omega_{\mathrm{p}} = \left(\frac{Ne^2}{\epsilon_0\,\mathrm{m}} \right)^{1/2}, \qquad \Omega_{\mathrm{p}} = \left(\frac{Ne^2}{\epsilon_0\,\mathrm{M}} \right)^{1/2},$$

$$\omega_{\mathrm{H}} = \frac{|e\mathbf{B}_0|}{\mathrm{m}}, \qquad \Omega_{\mathrm{H}} = \frac{|e\mathbf{B}_0|}{\mathrm{M}}, \tag{2.21}$$

where m and M are respectively the electron and ion masses. The terms in the first expression in (2.20) can be rearranged to give

$$\varepsilon = \frac{(\omega^2 - \omega_{\mathrm{UH}}^2)(\omega^2 - \omega_{\mathrm{LH}}^2)}{(\omega^2 - \omega_{\mathrm{H}}^2)(\omega^2 - \Omega_{\mathrm{H}}^2)} , \tag{2.22}$$

where ω_{UH} and ω_{LH} are the upper hybrid and lower hybrid frequencies, respectively. These frequencies can be written, to a good approximation, as

$$\omega_{\mathrm{UH}} = \left(\omega_{\mathrm{p}}^2 + \omega_{\mathrm{H}}^2\right)^{1/2}, \tag{2.23}$$

$$\omega_{\mathrm{LH}} = \omega_{\mathrm{H}} \left(\frac{\Omega_{\mathrm{p}}^2 + \Omega_{\mathrm{H}}^2}{\omega_{\mathrm{p}}^2 + \omega_{\mathrm{H}}^2}\right)^{1/2}. \tag{2.24}$$

When $\omega_{\mathrm{p}} \gg \omega_{\mathrm{H}}$, that is typical, for example, of the earth's ionosphere, expression (2.24) becomes

$$\omega_{\mathrm{LH}} = (\omega_{\mathrm{H}}\Omega_{\mathrm{H}})^{1/2}. \tag{2.25}$$

It is easy to verify that in the case $\omega \gg \omega_{\mathrm{LH}}$, which will be of special interest to us throughout this book, expressions (2.20) reduce to

$$\varepsilon = 1 + \frac{\omega_{\mathrm{p}}^2}{\omega_{\mathrm{H}}^2 - \omega^2}, \qquad g = -\frac{\omega_{\mathrm{p}}^2\omega_{\mathrm{H}}}{(\omega_{\mathrm{H}}^2 - \omega^2)\omega}, \qquad \eta = 1 - \frac{\omega_{\mathrm{p}}^2}{\omega^2}. \tag{2.26}$$

Sometimes the convenient quantities X and Y are introduced in the expressions (2.26). These quantities may be written

$$X = \frac{\omega_{\mathrm{p}}^2}{\omega^2}, \qquad Y = \frac{\omega_{\mathrm{H}}}{\omega}. \tag{2.27}$$

In the preceding formulas for the elements of the dielectric tensor, the effect of collisions of the plasma particles has been neglected. This is justified when the electron and ion collision frequencies, ν_{e} and ν_{i} respectively, are much less than the angular frequency ω. It is shown in books on plasma physics that if the electron-ion collisions may be neglected and only the electron-neutral and ion-neutral collisions are sufficient, the tensor elements may be written thus:

$$\varepsilon = 1 + \frac{\omega_{\mathrm{p}}^2(\omega - \mathrm{i}\nu_{\mathrm{e}})}{\left[\omega_{\mathrm{H}}^2 - (\omega - \mathrm{i}\nu_{\mathrm{e}})^2\right]\omega} + \frac{\Omega_{\mathrm{p}}^2(\omega - \mathrm{i}\nu_{\mathrm{i}})}{\left[\Omega_{\mathrm{H}}^2 - (\omega - \mathrm{i}\nu_{\mathrm{i}})^2\right]\omega},$$

$$g = -\frac{\omega_{\mathrm{p}}^2\omega_{\mathrm{H}}}{\left[\omega_{\mathrm{H}}^2 - (\omega - \mathrm{i}\nu_{\mathrm{e}})^2\right]\omega} + \frac{\Omega_{\mathrm{p}}^2\Omega_{\mathrm{H}}}{\left[\Omega_{\mathrm{H}}^2 - (\omega - \mathrm{i}\nu_{\mathrm{i}})^2\right]\omega}, \tag{2.28}$$

$$\eta = 1 - \frac{\omega_{\mathrm{p}}^2}{(\omega - \mathrm{i}\nu_{\mathrm{e}})\omega} - \frac{\Omega_{\mathrm{p}}^2}{(\omega - \mathrm{i}\nu_{\mathrm{i}})\omega}.$$

More general expressions for the tensor elements may be found elsewhere (Akhiezer *et al.*, 1975).

In the most of this book we consider only a collisionless (loss-free) magneto-plasma. For Chapter 4 onwards there is some case when taking account of collisions will be important.

2.4. The notations $\mathcal{H}$, $\mathcal{J}^e$ and $\mathbf{H}$, $\mathbf{J}^e$

Following Budden's argument (Budden, 1961 a) we find it convenient to adopt for the magnetic field $\mathbf{H}$ a different measure

$$\mathcal{H} = Z_0 \mathbf{H}, \tag{2.29}$$

where $Z_0 = (\mu_0/\epsilon_0)^{1/2}$ is the characteristic impedance of free space. Analogously, it is convenient to adopt for the electric current $\mathbf{J}^e$ a different measure

$$\mathcal{J}^e = Z_0 \mathbf{J}^e. \tag{2.30}$$

Then the last two Maxwell equations become

$$\nabla \times \mathbf{E} = -ik_0\mathcal{H} - \mathbf{J}^m, \qquad \nabla \times \mathcal{H} = ik_0\,\hat{\varepsilon} \cdot \mathbf{E} + \mathcal{J}^e, \tag{2.31}$$

where $k_0 = \omega/c$, the wavenumber in free space ($c = (\epsilon_0\mu_0)^{-1/2}$ is the velocity of electromagnetic waves in free space). The vectors $\mathcal{H}$ and $\mathcal{J}^e$ have the same physical dimensions as the electric-field intensity $\mathbf{E}$ and the magnetic-current density $\mathbf{J}^m$, respectively. The use of the notations $\mathcal{H}$ and $\mathcal{J}^e$ simplifies the equations and has the same effect as if $\mathbf{E}$, $\mathbf{H}$, $\mathbf{J}^e$, and $\mathbf{J}^m$ were measured in Gaussian units having many advantages in theoretical considerations. With these notations, it is convenient to come to Gaussian units in the field expressions. If one wishes, this may be done by making the replacements $\mathcal{H} \to \mathbf{H}$, $\mathcal{J}^e \to \dfrac{4\pi}{c}\mathbf{J}^e$, and $\mathbf{J}^m \to \dfrac{4\pi}{c}\mathbf{J}^m$.

The equations written in the form (2.31) will be used throughout this book. In Cartesian coordinates they become

$$\frac{\partial E_z}{\partial y} - \frac{\partial E_y}{\partial z} = -ik_0\mathcal{H}_x - J_x^m, \tag{2.32}$$

$$\frac{\partial E_x}{\partial z} - \frac{\partial E_z}{\partial x} = -ik_0\mathcal{H}_y - J_y^m, \tag{2.33}$$

$$\frac{\partial E_y}{\partial x} - \frac{\partial E_x}{\partial y} = -ik_0\mathcal{H}_z - J_z^m, \tag{2.34}$$

$$\frac{\partial \mathcal{H}_z}{\partial y} - \frac{\partial \mathcal{H}_y}{\partial z} = ik_0(\varepsilon E_x - ig E_y) + \mathcal{J}_x^e, \tag{2.35}$$

$$\frac{\partial \mathcal{H}_x}{\partial z} - \frac{\partial \mathcal{H}_z}{\partial x} = \mathrm{i}k_0(\mathrm{i}gE_x + \varepsilon E_y) + \mathcal{J}_y^e, \tag{2.36}$$

$$\frac{\partial \mathcal{H}_y}{\partial x} - \frac{\partial \mathcal{H}_x}{\partial y} = \mathrm{i}k_0\eta E_z + \mathcal{J}_z^e. \tag{2.37}$$

Here we made use of the constitutive relations (2.17), (2.18). In the cylindrical coordinate system Equations (2.31) may be written thus:

$$\frac{1}{\rho}\frac{\partial E_z}{\partial \phi} - \frac{\partial E_\phi}{\partial z} = -\mathrm{i}k_0\mathcal{H}_\rho - J_\rho^m, \tag{2.38}$$

$$\frac{\partial E_\rho}{\partial z} - \frac{\partial E_z}{\partial \rho} = -\mathrm{i}k_0\mathcal{H}_\phi - J_\phi^m, \tag{2.39}$$

$$\frac{1}{\rho}\frac{\partial}{\partial \rho}(\rho E_\phi) - \frac{1}{\rho}\frac{\partial E_\rho}{\partial \phi} = -\mathrm{i}k_0\mathcal{H}_z - J_z^m, \tag{2.40}$$

$$\frac{1}{\rho}\frac{\partial \mathcal{H}_z}{\partial \phi} - \frac{\partial \mathcal{H}_\phi}{\partial z} = \mathrm{i}k_0(\varepsilon E_\rho - \mathrm{i}gE_\phi) + \mathcal{J}_\rho^e, \tag{2.41}$$

$$\frac{\partial \mathcal{H}_\rho}{\partial z} - \frac{\partial \mathcal{H}_z}{\partial \rho} = \mathrm{i}k_0(\mathrm{i}gE_\rho + \varepsilon E_\phi) + \mathcal{J}_\phi^e, \tag{2.42}$$

$$\frac{1}{\rho}\frac{\partial}{\partial \rho}(\rho \mathcal{H}_\phi) - \frac{1}{\rho}\frac{\partial \mathcal{H}_\rho}{\partial \phi} = \mathrm{i}k_0\eta E_z + \mathcal{J}_z^e. \tag{2.43}$$

Note that in derivation of the components of $\hat{\varepsilon} \cdot \mathbf{E}$ in cylindrical coordinates the relations (2.16) were used.

2.5. Dispersion properties of characteristic modes in a magnetoplasma

A very thorough discussion of propagation of plane waves in a uniform, cold magnetoplasma may be found in monographs by Ratcliffe (1959), Ginzburg (1970) and other authors. Recall that, for a given wavenormal direction, there are in general two characteristic electromagnetic modes with two different refractive indices. Since some familiarity with their properties is essential to an understanding of the future treatment, we now briefly discuss the behavior of refractive indices for the conditions of interest to us.

2.5.1. Dispersion equation

To obtain the dispersion equation and find the refractive index for a wave of angular frequency ω and propagation vector $\mathbf{k}$, we seek a solution of source-free Maxwell's equations which represents a progressive plane wave, so that all field components vary in space only through the term $e^{-i\mathbf{kr}}$ where $\mathbf{r}$ denotes position vector. Then we have

$$\mathbf{n} \times \mathbf{E} = \mathcal{H}, \qquad \mathbf{n} \times \mathcal{H} = -\hat{\varepsilon} \cdot \mathbf{E}, \tag{2.44}$$

where $\mathbf{n} = \mathbf{k}/k_0$ is called the *wavenormal vector*. If $\mathcal{H}$ is eliminated

$$(\mathbf{n} \cdot \mathbf{n})\,\mathbf{E} - (\mathbf{n} \cdot \mathbf{E})\,\mathbf{n} - \hat{\varepsilon} \cdot \mathbf{E} = 0. \tag{2.45}$$

This equation may be written in matrix form, thus:

$$\left(n^2 \delta_{ij} - n_i n_j - \varepsilon_{ij}\right) E_j = 0, \tag{2.46}$$

where $i, j = x, y, z$, ε_{ij} are the Cartesian elements of the dielectric tensor (2.19), and δ_{ij} is the Kronecker symbol defined as

$$\delta_{ij} = \begin{cases} 1 & \text{if } i = j, \\ 0 & \text{if } i \neq j. \end{cases} \tag{2.47}$$

Equation (2.46) written in matrix notation forms a set of homogeneous equations in E_x, E_y, E_z which have a non-trivial solution only when

$$D \equiv \det \left| n^2 \delta_{ij} - n_i n_j - \varepsilon_{ij} \right| = 0. \tag{2.48}$$

The determinant D obtained after working out some algebra is expressed as

$$\begin{aligned} D = & -\eta n_z^4 - (\eta + \varepsilon)(n_x^2 + n_y^2)\, n_z^2 - \varepsilon(n_x^2 + n_y^2)^2 \\ & + 2\varepsilon\eta n_z^2 + (\varepsilon\eta + \varepsilon^2 - g^2)(n_x^2 + n_y^2) - \eta(\varepsilon^2 - g^2) = 0. \end{aligned} \tag{2.49}$$

This equation is called the *dispersion equation*. It implicitly defines the refractive indices of the characteristic modes of the plasma.

To discuss the properties of the refractive indices, we consider a three-dimensional space which will be called the *refractive index space* or, more concisely, the **n**-*space*. In it we use the quantities n_x, n_y, n_z as coordinates and choose the directions of the n_x-, n_y-, n_z-axes to be parallel to the x-, y-, z-axes, respectively, of ordinary observation space. Also, it is convenient to introduce the polar form through the definitions:

$$\begin{aligned} n_x &= n \sin\vartheta \cos\varphi, & n &= \left(n_x^2 + n_y^2 + n_z^2\right)^{1/2}, \\ n_y &= n \sin\vartheta \sin\varphi, & \vartheta &= \arctan \frac{\left(n_x^2 + n_y^2\right)^{1/2}}{n_z}, \\ n_z &= n \cos\vartheta, & \varphi &= \arctan\left(n_y/n_x\right). \end{aligned} \tag{2.50}$$

Here n is the magnitude of $\mathbf{n}$, and ϑ and φ are respectively the polar angle and the azimuthal angle of the vector $\mathbf{n}$. In terms of the spherical-polar coordinates n, ϑ, φ, Equation (2.49) becomes

$$D = -(\varepsilon \sin^2 \vartheta + \eta \cos^2 \vartheta)\, n^4 + \left[2\varepsilon\eta + (\varepsilon^2 - g^2 - \varepsilon\eta)\sin^2 \vartheta\right]n^2 - \eta(\varepsilon^2 - g^2) = 0, \quad (2.51)$$

and its solutions are

$$n_{1,2}^2(\vartheta) = \frac{2\varepsilon\eta + (\varepsilon^2 - g^2 - \varepsilon\eta)\sin^2 \vartheta \pm \left[(\varepsilon^2 - g^2 - \varepsilon\eta)^2 \sin^4 \vartheta + 4g^2\eta^2 \cos^2 \vartheta\right]^{1/2}}{2(\varepsilon \sin^2 \vartheta + \eta \cos^2 \vartheta)}.$$

$$(2.52)$$

It is now evident that D may be represented

$$D = -(\varepsilon \sin^2 \vartheta + \eta \cos^2 \vartheta)\left(n^2 - n_1^2(\vartheta)\right)\left(n^2 - n_2^2(\vartheta)\right). \quad (2.53)$$

This means that there are two different progressive plane waves which are the characteristic modes of a cold, uniform magnetoplasma. They have different refractive indices $n_{1,2}(\vartheta)$ independent of the azimuth about $\mathbf{B}_0$. If collisions are neglected, there is no absorption of energy in a cold plasma, and each quantity of $n_{1,2}$ is then either purely real or purely imaginary. If the refractive index is purely real, the wave is propagating. Conversely, if the refractive index is purely imaginary, the wave cannot be propagating, and it is evanescent. In general, the refractive indices $n_{1,2}(\vartheta)$ are complex. We note that when the plasma properties are unaffected by the ions, the expression (2.52) reduces to the well-known Appleton – Hartree formula for the refractive index (see Budden, 1961 a, Chapter 6).

It can be shown directly from (2.52) that if $\operatorname{sgn}\varepsilon \neq \operatorname{sgn}\eta$, so that $\varepsilon\eta < 0$, one refractive index of $n_{1,2}$ becomes infinite at the angle

$$\vartheta = \vartheta_r = \arctan\sqrt{-\eta/\varepsilon}. \quad (2.54)$$

The wavenormals which are at this angle to the static magnetic-field direction form in the $\mathbf{n}$-space the surface of a cone of half angle ϑ_r whose axis lies along the static magnetic-field direction. This cone is called the *resonance cone* in the $\mathbf{n}$-space and its half angle is called the *wavenormal resonance-cone half angle*. The plasma in which such an infinity for the refractive index of one of the characteristic modes occurs is called the *resonant plasma*. It is easy to find that the above stated condition $\varepsilon\eta < 0$, under which the plasma is resonant, is satisfied only in the following three frequency bands:

$$\omega < \Omega_{\mathrm{H}},$$

$$\omega_{\mathrm{LH}} < \omega < \min(\omega_{\mathrm{H}}, \omega_{\mathrm{p}}), \quad (2.55)$$

$$\max(\omega_{\mathrm{H}}, \omega_{\mathrm{p}}) < \omega < \omega_{\mathrm{UH}}.$$

These are sometimes called the *resonant bands* of a cold magnetoplasma.

In two important particular cases when the wavenormal is parallel ($\vartheta = 0$) or perpendicular ($\vartheta = \pi/2$) to the static magnetic field $\mathbf{B}_0$, the formula (2.52) gives

$$n_{1,2}^2(0) = \varepsilon \pm g, \qquad (2.56)$$

$$n_{1,2}^2\left(\frac{\pi}{2}\right) = \begin{cases} \eta, \\ (\varepsilon^2 - g^2)/\varepsilon. \end{cases} \qquad (2.57)$$

In many books on wave-propagation in the ionosphere, the wave which is unaffected by the ambient magnetic field for transverse propagation is called the *ordinary wave* and marked 'o'. The other wave, which is affected by the ambient magnetic field for all directions of the wavenormal, is called the *extraordinary wave* and marked 'x'.

In this book we find it more significant to effect a different definition for "ordinary" and "extraordinary" waves which is based on the analytic properties of the functions $n_{1,2}$ (see, for example, Felsen and Marcuvitz, 1973). Let the square root in (2.52) be chosen to have a positive real part. Then the wave for which the positive sign is taken before this root, will be called the "ordinary" wave, whereas the other wave will be called the "extraordinary" wave. Thus, we adopt that

$$n_\alpha^2(\vartheta) = \frac{2\varepsilon\eta + (\varepsilon^2 - g^2 - \varepsilon\eta)\sin^2\vartheta - \chi_\alpha\left[(\varepsilon^2 - g^2 - \varepsilon\eta)^2 \sin^4\vartheta + 4g^2\eta^2\cos^2\vartheta\right]^{1/2}}{2(\varepsilon\sin^2\vartheta + \eta\cos^2\vartheta)},$$
$$(2.58)$$

where $\alpha = $ 'o' or $\alpha = $ 'x', and $\chi_o = -1$, $\chi_x = 1$. It is worth noting that if $\varepsilon^2 - g^2 - \varepsilon\eta < 0$, the definition (2.58) does then coincide with that used in the theory of wave-propagation in the ionosphere.

The statement that the "ordinary" and "extraordinary" waves are the characteristic modes of a uniform magnetoplasma means that the electromagnetic field at each point wherein no sources exist may be sought in terms of the progressive waves which involve the dependence on spatial coordinates only through the factors

$$\exp\left[\mp ik_0\Big(xn_\alpha(\vartheta)\sin\vartheta\cos\varphi + yn_\alpha(\vartheta)\sin\vartheta\sin\varphi + zn_\alpha(\vartheta)\cos\vartheta\Big)\right].$$

The relations between the complex amplitudes of different field components in these waves are found directly from Maxwell's equations. Let $\mathcal{F}$ be some field component, and let r, θ, ϕ be spherical polar coordinates in the observation space; r is the magnitude of the vector $\mathbf{r}$, and θ and ϕ are its polar and azimuthal angles, respectively.

Then we may write

$$x = r \sin\theta \cos\phi, \qquad r = (x^2 + y^2 + z^2)^{1/2},$$

$$y = r \sin\theta \sin\phi, \qquad \theta = \arctan \frac{(x^2 + y^2)^{1/2}}{z}, \tag{2.59}$$

$$z = r \cos\theta, \qquad \phi = \arctan(y/x).$$

Using these relations, the field quantity $\mathcal{F}(\mathbf{r})$ within the source-free region may be represented formally by

$$
\mathcal{F}(r, \theta, \phi) = \sum_{\alpha=o}^{x} \int_0^{\pi} d\vartheta \int_0^{2\pi} \left\{ F_\alpha^{(+)}(\vartheta, \varphi) \right.
\tag{2.60}
$$
$$
\times \quad \exp\left[-\mathrm{i}k_0\, n_\alpha(\vartheta)\, r\left(\sin\theta \sin\vartheta \cos(\phi - \varphi) + \cos\theta \cos\vartheta\right)\right]
$$
$$
+ \quad \left. F_\alpha^{(-)}(\vartheta, \varphi) \exp\left[\mathrm{i}k_0\, n_\alpha(\vartheta)r\left(\sin\theta \sin\vartheta \cos(\phi - \varphi) + \cos\theta \cos\vartheta\right)\right]\right\} d\varphi,
$$

where the 'α' summation covers the "ordinary" and "extraordinary" waves. The problem of finding the field excited by an arbitrary (but prescribed) distribution of sources is thus reduced to determining first the quantities $F_\alpha^{(\pm)}(\vartheta, \varphi)$ and performing then the integrals such as in (2.60).

2.5.2. The functions $p_{\mathrm{o,x}}$ and $q_{1,2}$

Very often it is more convenient to represent the determinant (2.49), using the quantities q, φ, p which are related to the coordinates n_x, n_y, n_z in the **n**-space by

$$n_x = q \cos\varphi, \qquad q = (n_x^2 + n_y^2)^{1/2},$$

$$n_y = q \sin\varphi, \qquad \varphi = \arctan(n_y/n_x), \tag{2.61}$$

$$n_z = p, \qquad p = n_z.$$

Then (2.49) becomes

$$D = -\eta p^4 - \left[(\eta + \varepsilon)q^2 - 2\varepsilon\eta\right]p^2 - \varepsilon q^4 + (\varepsilon\eta + \varepsilon^2 - g^2)q^2 - \eta(\varepsilon^2 - g^2) = 0. \tag{2.62}$$

This quartic equation is a special case of the more general equation known as the *Booker quartic* (see Budden, 1961 a). Consider (2.62) as a quartic equation in p. The solutions of the quartic are

$$p_\alpha^2(q) = \varepsilon - \frac{1}{2}\left(1 + \frac{\varepsilon}{\eta}\right)q^2 + \chi_\alpha\left[\frac{1}{4}\left(1 - \frac{\varepsilon}{\eta}\right)^2 q^4 - \frac{g^2}{\eta}q^2 + g^2\right]^{1/2}, \tag{2.63}$$

where $\alpha = $ 'o' or 'x', and $\chi_o = -1$, $\chi_x = 1$. On the other hand, Equation (2.62) may be considered as a quartic equation in q, and its solutions are then

$$
\begin{aligned}
q_k^2(p) \;=\; & \Bigg\{ \varepsilon^2 - g^2 + \varepsilon\eta - (\eta + \varepsilon)p^2 + (-1)^k \bigg[(\eta - \varepsilon)^2 p^4 \\
& + \; 2\Big(g^2(\eta + \varepsilon) - \varepsilon(\eta - \varepsilon)^2\Big)p^2 + (\varepsilon^2 - g^2 - \varepsilon\eta)^2 \bigg]^{1/2} \Bigg\} \Big/ 2\varepsilon, \quad (2.64)
\end{aligned}
$$

where $k = 1, 2$. For definiteness, the square roots in (2.63) and (2.64) are chosen to have a positive real part. As was done for branches of $n^2(\vartheta)$, we denote those of $p^2(q)$ as "ordinary" ($\alpha = $ o) and "extraordinary" ($\alpha = $ x). However, two branches of $q^2(p)$ will be named as branches " 1 " and " 2 ". Such a difference in nomenclature is justified by convenience of the later work with the functions (2.63) and (2.64).

It is easy to see that the determinant D may be written as

$$
D = -\eta\Big(p^2 - p_o^2(q)\Big)\Big(p^2 - p_x{}^2(q)\Big) \tag{2.65}
$$

or

$$
D = -\varepsilon\Big(q^2 - q_1^2(p)\Big)\Big(q^2 - q_2^2(p)\Big). \tag{2.66}
$$

Through a similar argument to that used for derivation of (2.60), we can now seek each field component $\mathcal{F}$ within a source-free region in the form

$$
\begin{aligned}
\mathcal{F}(\rho, \phi, z) \;=\; & \sum_{\alpha=o}^{x} \int_0^{\infty} dq \int_0^{2\pi} \Bigg[F_\alpha^{(+)}(q, \varphi) \exp\Big(-ik_0 p_\alpha(q)z \Big) \\
& + \; F_\alpha^{(-)}(q, \varphi) \exp\Big(ik_0 p_\alpha(q)\, z \Big) \Bigg] \exp\Big(-ik_0 q\rho \cos(\phi - \varphi) \Big) d\varphi \quad (2.67)
\end{aligned}
$$

or

$$
\begin{aligned}
\mathcal{F}(\rho, \phi, z) \;=\; & \sum_{k=1}^{2} \int_{-\infty}^{+\infty} dp \int_0^{2\pi} \Bigg\{ F_k^{(+)}(\varphi, p) \exp\Big[-ik_0 q_k(p)\, \rho \cos(\phi - \varphi) \Big] \\
& + \; F_k^{(-)}(\varphi, p) \exp\Big[ik_0 q_k(p)\, \rho \cos(\phi - \varphi) \Big] \Bigg\} \exp(-ik_0 pz)\, d\varphi. \quad (2.68)
\end{aligned}
$$

Here the ' k ' summation covers the relevant branches which are defined by (2.64).

Thus, there are various ways of describing the refractive indices and hence of representing the field. All the field representations, being entirely equivalent, differ in what coordinates in the n-space are taken as integration variables. A choice of the kind of a representation depends on whether it is convenient for solution of the particular problem in question. For much of the later work in this book, the representation (2.67) and consequently the formula (2.63) for description of the refractive index behavior will be the most suitable.

2.6. The refractive index surfaces in some frequency bands

An effective way of describing the qualitative behavior of wave propagation is by using refractive index surfaces for the "ordinary" and "extraordinary" waves. For each of these waves, take the refractive index and draw a line from the origin in the **n**-space parallel to the wavenormal and of length equal to the refractive index. If the wavenormal takes successively all possible directions, the locus of the end-point of this line is then the refractive index surface. Since the refractive indices $n_{o,x}(\vartheta)$ depend only on ϑ, and $n_{o,x}(\vartheta) = n_{o,x}(\pi - \vartheta)$, each surface is a surface of revolution about the n_z-axis, and the plane $\vartheta = \pi/2$ is a plane of symmetry. It is shown in the electromagnetic theory that if the spatial z, ρ coordinates are superimposed upon the p, q axes, respectively, the normal to the refractive index surface at the point $p_\alpha(q)$, q locates the average power flow direction in the plane wave with the wavenormal $\mathbf{n}_\alpha = \hat{\boldsymbol{\rho}}_0 q + \hat{\mathbf{z}}_0\, p_\alpha(q)$ (see, for example, Landau and Lifshitz, 1960, Chapter 11). In conventional terms of geometrical optics, this power flow direction may be called the *ray direction*. It is also known that for a collisionless, cold magnetoplasma, the angle between the wavenormal direction and the ray direction cannot exceed $\pi/2$ (see Felsen and Marcuvitz, 1973, Chapter 1, § 7).

The structure of refractive index surfaces in a magnetoplasma is discussed in many books and papers. A thorough description of these surfaces is given, for example, by Felsen and Marcuvitz (1973) for the case when plasma dispersion properties are unaffected by the ions. Here we give some examples of refractive index surfaces for a lossless magnetoplasma modeled on the earth's ionosphere and magnetosphere in the so-called whistler-frequency band. This may be defined thus:

$$\Omega_H \ll \omega < \omega_H \ll \omega_p, \tag{2.69}$$

and its lower part is related to the frequency intervals known as the extremely low frequency band (ELF, 3–3 000 Hz) and the very low frequency band (VLF, 3–30 kHz) in the theory of wave propagation. Band (2.69) allows for the existence of whistlers which are guided by ducts in geospace and laboratory plasmas. (This is discussed briefly in Chapter 1.) At frequencies (2.69), the only propagating electromagnetic mode is the "extraordinary" mode which is also called the *whistler mode*. The "ordinary" mode is evanescent in this case since its refractive index is purely imaginary for all wavenormal directions.

Typical whistler-mode refractive index surfaces (corresponding to $n_x(\vartheta)$ in (2.58)) are shown in Figures 2.1 to 2.3 for the following intervals of the band (2.69):

$$\Omega_H \ll \omega < \omega_{LH}, \tag{2.70}$$

$$\omega_{LH} < \omega < \omega_H/2, \tag{2.71}$$

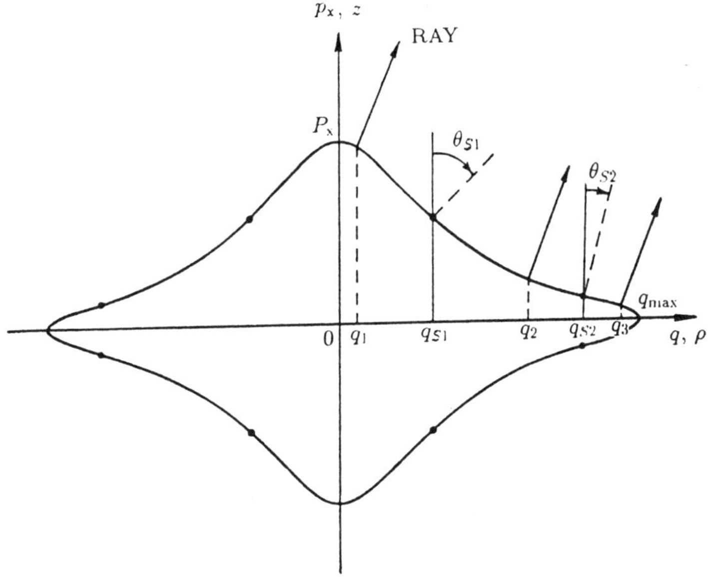

Figure 2.1. Typical whistler-mode refractive index surface when $\Omega_H \ll \omega < \omega_{LH}$.

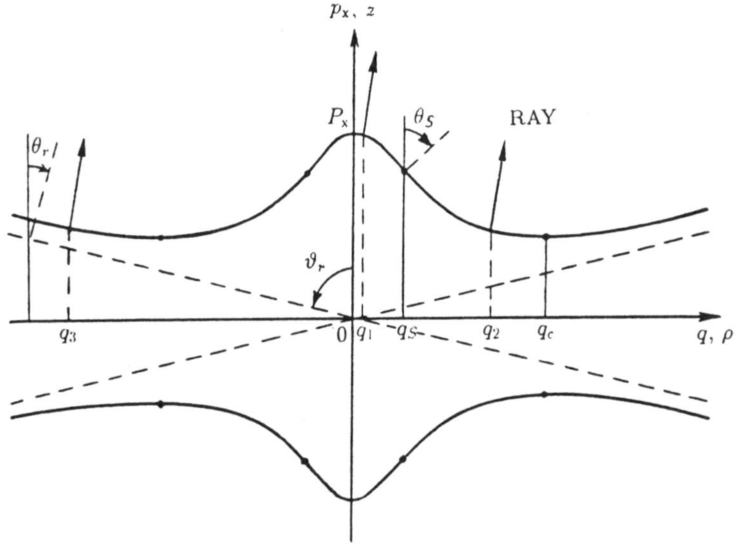

Figure 2.2. Typical whistler-mode refractive index surface when $\omega_{LH} < \omega < \omega_H/2$.

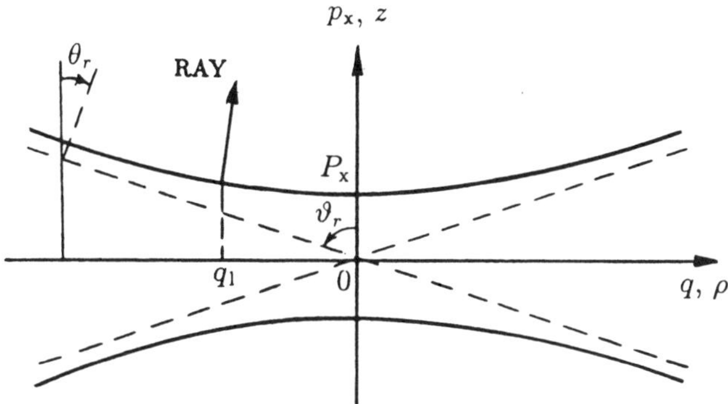

Figure 2.3. Typical whistler-mode refractive index surface when $\omega_H/2 < \omega < \omega_H$.

$$\omega_H/2 < \omega < \omega_H. \tag{2.72}$$

Looked upon as functions of q, all surfaces given in Figures 2.1 to 2.3 are described by $p = p_x(q)$ (see (2.63)). The quantity P_x denotes the refractive index for purely longitudinal propagation along the static magnetic-field direction on all plots. That is

$$P_x = n_x(0) = p_x(0) = (\varepsilon - g)^{1/2}.$$

It is evident that the surface corresponding to the interval (2.70) and shown in Figure 2.1 yields real values for the refractive index for all wavenormal directions, the refractive index reaching the value $q_{\max} = \left[(\varepsilon^2 - g^2)/\varepsilon\right]^{1/2}$ for purely transverse propagation. In the other intervals, (2.71) and (2.72), the refractive index is purely real only if the angle ϑ between the wavenormal and the static magnetic-field direction is less than the resonance-cone half angle ϑ_r defined by (2.54). As ϑ approaches ϑ_r, the refractive index surface extends theoretically to infinity (see Figures 2.2 and 2.3). The rays corresponding to the wavenormals at the angle ϑ_r form in the observation space the surface of a cone of half angle $\theta_r = \pi/2 - \vartheta_r$ whose axis is along the ambient static magnetic field. The angle θ_r is complementary to the wavenormal resonance-cone half angle ϑ_r. The cone of half angle θ_r is therefore called the *complementary cone* (Wang and Bell, 1972 a), or the *resonance cone in the observation space* (Fisher and Gould, 1971), and the direction given by $\theta = \theta_r$ is called the *resonance direction*.

It should be noted that the infinite value of the refractive index appears only within the collisionless cold plasma approximation, and this infinity is removed when collisions or warm plasma effects are included. In a collisionless magnetoplasma whose properties are unaffected by ions, the warm effects are negligible if

$$k_\perp V_{Te} \ll \omega_H, \quad k_\parallel V_{Te} \ll \omega, \quad k_\parallel V_{Te} \ll |\omega \pm \omega_H|, \tag{2.73}$$

where $V_{Te} = (T_e/m)^{1/2}$ is the electron thermal velocity, and $k_\perp$ and $k_\parallel$ are transverse and longitudinal components, with respect to the static magnetic-field direction, of the vector $\mathbf{k}$. When the inequalities (2.73) are violated, it is necessary to take into account the thermal motion of electrons that, in turn, leads to the spatial dispersion and changes in the refractive index surfaces. Since we use the cold plasma approximation, the conditions (2.73) are assumed throughout. This implies that a boundedness of the spatial spectrum is to be secured in the $\mathbf{n}$-space in some manner. This topic will be discussed in detail later.

In the frequency interval (2.71), the whistler-mode refractive index surface $p = p_x(q)$ shows that there is a point $q = q_c$ given by $p'_x(q_c) = 0$ ($q_c \neq 0$) where the prime indicates a derivative with respect to an argument. This point, for which $p(q_c) = P_c$, corresponds to the conical-refraction whistler-mode waves whose wavenormals make an angle $\vartheta_c = \arctan(q_c/P_c)$, known also as the *Gendrin angle*, with the static magnetic field, whereas the appropriate ray directions are parallel to that field. We notice that a conical refraction can be observed in biaxial crystals, and there may, in general, be external and internal conical refraction. In this case, we deal with the external conical refraction. Looking ahead, we note that this leads to the formation of the *axial caustic* along the static magnetic-field direction.

In the intervals (2.70) and (2.71) inflection points $q = q_S$ given by $p''_x(q_S) = 0$ (see Figures 2.1 and 2.2) may exist. For each inflection point wherein $p_x(q_S) = P_S$, the angle between the wavenormal and the magnetic field is $\vartheta_S = \arctan(q_S/P_S)$. The angle between the ray direction and the static-field direction then is $\theta_S = -\arctan p'_x(q_S)$. This angle θ_S is called the *Storey angle*, and a cone of half angle θ_S will be denoted as the *Storey cone*. The presence of inflection points on $p_x(q)$ results in the illumination of some portions of the observation space by multiple rays. For the frequency interval (2.70), there are three rays in the range $\theta_{S2} < \theta < \theta_{S1}$ where $\theta_{S1,2} = -\arctan p'_x(q_{S1,2})$ (see Figure 2.1), and only one ray in the ranges $0 \leq \theta \leq \theta_{S2}$ and $\theta_{S1} \leq \theta \leq \pi/2$. For the interval (2.71), there may be two possible cases: $\theta_r < \theta_S$ and $\theta_S < \theta_r$. In the former case, when $\theta_r < \theta_S$, the region $0 < \theta < \theta_r$ is illuminated by three rays (see Figure 2.2), whereas the region $\theta_r \leq \theta < \theta_S$ by two rays. In the other case, when $\theta_S < \theta_r$, three rays appear in the angular domain $0 < \theta < \theta_S$ and only one ray in the domain $\theta_S \leq \theta < \theta_r$. We notice that no rays exist for the part of space where $\max(\theta_S, \theta_r) < \theta \leq \pi/2$, and consequently there is no illumination of this region. At the directions $\theta = \theta_{S1,2}$ in the frequency interval (2.70) and at the direction $\theta = \theta_S$ in the frequency interval (2.71), the coalescence of two rays occurs. This leads to the formation of the *simple conical caustic*, which is a "shadow-boundary" for these rays. The pertinent complications associated with the presence of the caustic directions will be discussed in more detail later, in § 3.7. For the frequency interval (2.72), no inflection points exist on $p_x(q)$ and there are no multiple rays, and illumination is confined in the cone of half angle θ_r (see Figure 2.3). Note that the above consideration made for the half-space $0 \leq \theta \leq \pi/2$ may be extended without difficulty to the half-space $\pi/2 \leq \theta \leq \pi$ because of symmetry of refractive index surfaces about the plane $\theta = \pi/2$.

In the preceding analysis, we used the functions $n_x(\vartheta)$ and $p_x(q)$ to describe the whistler-mode refractive index surface. However, the functions $q_k(p)$ defined by (2.64) can also be used for this purpose. It is easily shown that in the interval (2.70) the whistler-mode refractive index surface is given by $q = q_1(p)$. In the interval (2.71) it is given by $q = q_1(p)$ when $0 \leq q \leq q_c$ ($P_c \leq p \leq P_x$), and by $q = q_2(p)$ when $q \geq q_c$ ($p \geq P_c$). At $p = P_c$, (2.64) yields $q_1(P_c) = q_2(P_c) = q_c$. For the interval (2.72), the refractive index surface is described by $q = q_2(p)$. Thus, a one-to-one correspondence between the branches of $p_\alpha^2(q)$ and $q_k^2(p)$ does not exist. That is why we avoid using the terms "ordinary" or "extraordinary" for branches of q_k^2.

For the following, it is useful to find the closed-form expressions for q_S and q_c. This is possible by making some simplifications in the general formulas for the refractive index.

In the interval (2.70), if the condition

$$|\eta| \gg \frac{g^2}{p^2 - \varepsilon} \tag{2.74}$$

is satisfied, we deduce from (2.64) that, to a good approximation,

$$q^2 = q_1^2(p) \approx \frac{g^2}{p^2 - \varepsilon} - p^2 + \varepsilon. \tag{2.75}$$

For all directions of propagation except those rather close to the transverse direction, the terms of the order ε/p^2 may be neglected. If this is used, the expression (2.75) becomes

$$q^2(p) = \frac{g^2}{p^2} - p^2. \tag{2.76}$$

This, for the range $0 \leq \vartheta < \pi/2$, leads to the formula

$$n_x^2 = p^2 + q^2 = \frac{|g|}{\cos \vartheta} \tag{2.77}$$

known as the quasi-longitudinal approximation to the refractive index of the "extraordinary" (whistler) mode (see Booker, 1935). Within this approximation, the equation determining the inflection point $q = q_{S1}$ is written as

$$\frac{\mathrm{d}^2 q}{\mathrm{d}p^2} = \frac{\mathrm{d}^2}{\mathrm{d}p^2} \left[\left(\frac{g^2}{p^2} - p^2 \right)^{1/2} \right] = 0.$$

This gives $q_{S1} = \dfrac{\sqrt{2}}{3^{1/4}} |g|^{1/2}$ whence one obtains the universal values $\vartheta_{S1} = \arctan \sqrt{2} = 54° 44'$ and $\theta_{S1} = \arctan \left(\dfrac{1}{2\sqrt{2}} \right) = 19° 28'$ (Storey, 1953). Unfortunately, the quasi-longitudinal approximation does not allow the determination of the quantity q_{S2} which corresponds to the wavenormal close to the transverse direction. However, for such a quasi-transverse propagation when $|g| \ll q^2 \lesssim q_{\max}^2$, the condition (2.74) holds again if $\omega \ll \omega_{\mathrm{LH}}$. The expression (2.75) can then be transformed into

$$p = \left(\frac{g^2}{q^2} - |\varepsilon| \right)^{1/2}.$$

If this is inserted in the equation $\mathrm{d}^2 p / \mathrm{d}q^2 = 0$ it gives $q_{S2} = \sqrt{\dfrac{2}{3}} \dfrac{|g|}{|\varepsilon|^{1/2}}$ whence $\vartheta_{S2} = \arctan \left(\dfrac{2}{\sqrt{3}} \dfrac{|g|}{|\varepsilon|} \right)$ and $\theta_{S2} = \arctan \left(\dfrac{3\sqrt{3}}{2} \dfrac{|\varepsilon|}{|g|} \right)$ (Bellyustin, 1978). Since in the interval (2.70) the inequality $|\varepsilon| \ll |g|$ is always satisfied, we infer that θ_{S1} exceeds appreciably ϑ_{S2}.

In the interval (2.71) there is some simplification for the whistler-mode refractive index if the effect of ion motion in the field of the wave is ignored. This is possible if $\omega_{\mathrm{LH}} \ll \omega$, in which case formulas (2.20) for the dielectric tensor elements may be written as

$$\varepsilon = \frac{X}{Y^2 - 1}, \quad g = -\frac{XY}{Y^2 - 1}, \quad \eta = -X. \tag{2.78}$$

Then (2.64) becomes

$$q_k^2(p) = p^2 \left(\frac{Y^2}{2} - 1 \right) - X + (-1)^k \frac{pY^2}{2} \left(p^2 - 4 \frac{X}{Y^2} \right)^{1/2}, \tag{2.79}$$

and this immediately leads to

$$P_c = 2X^{1/2}/Y, \quad q_c = X^{1/2}(1 - 4/Y^2)^{1/2}.$$

In the limiting case $p^2 \gg P_c^2$ the branches of (2.79) can be written, to a reasonable approximation, in the form

$$q_1^2 \approx \left(\frac{X}{pY} \right)^2 - p^2, \tag{2.80}$$

$$q_2^2 \approx p^2(Y^2 - 1) - 2X. \tag{2.81}$$

For the part of a whistler-mode refractive index surface whereon $q < q_c$, the inequality $p^2 \gg P_c^2$ becomes $Y \cos \vartheta \gg 4$, which, in effect, is the condition of validity of the quasi-longitudinal approximation. If this is used, the formula (2.80) can easily be reduced to the results coinciding in form with (2.76) and (2.77). Hence the inflection point $q = q_S$ and the corresponding angles ϑ_S and θ_S will now be given by the same formulas as those obtained respectively for q_{S1}, ϑ_{S1}, and θ_{S1} at the frequencies (2.70).

In the interval (2.72) the expressions (2.78) and (2.79) are valid again, and the refractive index surface is roughly described by (2.81) if $p^2 \gg p_x^2(0) = X/(Y-1)$.

It follows from (2.52) (see also (2.63) and (2.64)) that in a resonant magnetoplasma the refractive index surface which extends to infinity may be approximated, beginning with a certain large value of the refractive index n, by

$$\varepsilon q^2 + \eta p^2 = 0. \tag{2.82}$$

(The accurate criterion of validity of (2.82) must be found individually for each resonant frequency band.) Note that Equation (2.82) is readily deduced from (2.81) in the case $p^2 \gg 4X/Y^2$, $q \gg q_c$, when ω belongs to the interval (2.71), and in the case $p^2 \gg X/(Y-1)$, when ω belongs to the interval (2.72).

It can be shown that the electric field in a wave satisfying (2.82) may be represented through only one scalar potential, ψ, via $\mathbf{E} = -\nabla\psi$. Substituting this into the Maxwell equation

$$\nabla \cdot \mathbf{D} = 0$$

yields

$$\varepsilon \nabla_\perp^2 \psi + \eta \frac{\partial^2 \psi}{\partial z^2} = 0, \tag{2.83}$$

where the subscript '$\perp$' is used to indicate the component of ∇ transverse to the static magnetic-field direction. If we take the scalar potential ψ in the form $\psi(\mathbf{r}) = \psi_0 e^{-ik_0 \mathbf{n}\mathbf{r}}$, with ψ_0 being a constant, Equation (2.83) gives the approximate dispersion relation (2.82). The waves satisfying (2.82) are usually called the *electrostatic waves*. As mentioned earlier, they can exist only if $\varepsilon\eta < 0$, when Equation (2.83) is of the hyperbolic type. These waves, however, have a non-vanishing magnetic field which is defined by

$$\nabla \times \mathcal{H} = -ik_0 \hat{\varepsilon} \cdot \nabla\psi, \qquad \nabla \cdot \mathcal{H} = 0.$$

Therefore, the more preferable term for these waves is the *quasi-electrostatic waves*. The approach based on Equations (2.82) and (2.83) for representing the wave field in a resonant magnetoplasma is called the *quasi-static theory* (Balmain, 1964). For a detailed discussion of this approach and the possibilities of its application for solution of the concrete problems see Andronov and Chugunov (1975).

Problems

2.1. Consider the motion of individual electrons and ions, assuming that the force on these particles arises from harmonically varying electromagnetic field $\mathbf{E}$, $\mathbf{B}$ and the static magnetic field $\mathbf{B}_0$. Let the induction $\mathbf{B}_0$ of the static magnetic field be much greater than the magnetic induction $\mathbf{B}$ of the time-harmonic field. Then Newton's laws of motion give

$$m \frac{d\mathbf{v}_e}{dt} = -e(\mathbf{E} + \mathbf{v}_e \times \mathbf{B}_0) - m\nu_e \mathbf{v}_e,$$

$$M \frac{d\mathbf{v}_i}{dt} = e(\mathbf{E} + \mathbf{v}_i \times \mathbf{B}_0) - M\nu_i \mathbf{v}_i,$$

where $\mathbf{v}_e$ and $\mathbf{v}_i$ are velocities of an electron and an ion, respectively. The last term on the right side of each of the above equations shows that a particle is subjected to a retarding force arising from collisions. Find the velocities $\mathbf{v}_e$ and $\mathbf{v}_i$ which vary with time through the factor $e^{i\omega t}$, as does the time-harmonic field. Assuming that the average numbers of electrons and ions per unit volume are equal to each other, and denoting those quantities as N, determine the current density $\mathbf{j} = eN(\mathbf{v}_i - \mathbf{v}_e)$ and the electric polarization $\mathbf{P} = (i\omega)^{-1}\mathbf{j}$. Substitute $\mathbf{P}$ into $\mathbf{D} = \epsilon_0 \mathbf{E} + \mathbf{P}$ and bring the latter relationship to the form (2.17). Verify that the elements of the dielectric tensor are then given by expressions (2.28). Discuss the conditions for the validity of such a derivation of the relation (2.17).

2.2. Consider the refractive indices given by Equation (2.58) for the characteristic modes of a magnetoplasma, assuming that the elements of the dielectric tensor are given by formulas (2.20). Show that in the case $\omega_p \gg \omega_H$ the effect of ions on the refractive indices is negligible if $\omega \gg (\omega_H \Omega_H)^{1/2}$. Also, prove that when a wavenormal makes a small enough angle with the static magnetic-field direction, the refractive indices are unaffected by ions if $\omega \gg \Omega_H$.

2.3. Let a plane wave travel in a cold collisionless magnetoplasma. Prove that the angle between the wavenormal and the time-averaged power flow direction in the wave cannot exceed $\pi/2$.

2.4. Show that when a static magnetic field tends to infinity ($B_0 \to \infty$), the dielectric tensor in (2.19) takes the form

$$\hat{\varepsilon} = \begin{pmatrix} \varepsilon & 0 & 0 \\ 0 & \varepsilon & 0 \\ 0 & 0 & \eta \end{pmatrix},$$

characteristic of uniaxial anisotropy, with the optic axis directed along $\mathbf{B}_0$. In this limiting case, derive expressions for the refractive indices of the characteristic modes. Prove that when $\omega \ll \Omega_H$, the corresponding expressions will be approximately valid for a finite static magnetic field.

2.5. Assuming that $\omega_p > \omega_H$, discuss how the refractive index surfaces change (a) when ω approaches Ω_H but still remains greater than Ω_H, (b) when ω becomes less than Ω_H.

In the case (b) the refractive index surface for one of the characteristic modes will be open. This mode is called the *Alfvén wave*. What happens with the refractive index surface for the Alfvén wave in the limiting case $\omega \ll \Omega_H$? Sketch the form of the refractive index surface for the other characteristic mode in that case.

Chapter 3

Radiation from Given Sources in a Uniform Unbounded Magnetoplasma

3.1. Introduction

Before we apply the theory of excitation and propagation of electromagnetic waves in density ducts, it is necessary to consider some properties of radiation from given sources immersed in a uniform unbounded magnetoplasma. This is important for two reasons: (a) familiarity with this topic will facilitate understanding the problem of radiation from given sources in the presence of a duct; (b) the radiation characteristics obtained for the uniform magnetoplasma will be utilized for comparison with those in the case of a plasma containing density ducts.

3.2. General representation of source-excited fields

The expressions for electric and magnetic fields excited by given sources will now be deduced. In this section the Maxwell equations (2.32) to (2.37) are used. Elimination of $\mathcal{H}$ from those equations gives

$$\frac{\partial^2 E_x}{\partial y^2} + \frac{\partial^2 E_x}{\partial z^2} - \frac{\partial^2 E_y}{\partial x \partial y} - \frac{\partial^2 E_z}{\partial x \partial z} + k_0^2(\varepsilon E_x - ig E_y) = \mathrm{i}k_0 a_x,$$

$$-\frac{\partial^2 E_x}{\partial x \partial y} + \frac{\partial^2 E_y}{\partial x^2} + \frac{\partial^2 E_y}{\partial z^2} - \frac{\partial^2 E_z}{\partial y \partial z} + k_0^2(ig E_x + \varepsilon E_y) = \mathrm{i}k_0 a_y, \qquad (3.1)$$

$$-\frac{\partial^2 E_x}{\partial x \partial z} - \frac{\partial^2 E_y}{\partial y \partial z} + \frac{\partial^2 E_z}{\partial x^2} + \frac{\partial^2 E_z}{\partial y^2} + k_0^2 \eta E_z = \mathrm{i}k_0 a_z,$$

where

$$\mathbf{a} = \boldsymbol{\mathcal{J}}^e - \mathrm{i}k_0^{-1} \nabla \times \mathbf{J}^m. \qquad (3.2)$$

To obtain the solutions to Equations (3.1), we employ a Fourier transformation of (3.1) and (3.2), find the relations between the Fourier transforms of $\mathbf{E}$ and $\mathbf{a}$, and then make an inverse Fourier transformation of the corresponding relations. Let us first establish our notations and conventions. Our definition of the three-dimensional Fourier transform of any function $F(\mathbf{r})$ and its inverse is

$$F(\mathbf{n}) = \int F(\mathbf{r})\, e^{ik_0\mathbf{n}\cdot\mathbf{r}}\, d\mathbf{r},$$

$$F(\mathbf{r}) = \frac{k_0^3}{(2\pi)^3} \int F(\mathbf{n})\, e^{-ik_0\mathbf{n}\cdot\mathbf{r}}\, d\mathbf{n}. \tag{3.3}$$

The Fourier transforms of (3.1) and (3.2) are

$$\hat{\mathbf{T}} \cdot \mathbf{E}(\mathbf{n}) = -ik_0^{-1}\mathbf{a}(\mathbf{n}) \tag{3.4}$$

and

$$\mathbf{a}(\mathbf{n}) = \boldsymbol{\mathcal{J}}^e(\mathbf{n}) - \mathbf{n} \times \mathbf{J}^m(\mathbf{n}). \tag{3.5}$$

The elements T_{ij} $(i,j = x,y,z)$ of the tensor $\hat{\mathbf{T}}$ are defined by

$$T_{ij} = n^2\delta_{ij} - n_i n_j - \varepsilon_{ij}. \tag{3.6}$$

For given current sources, Equation (3.4) can be resolved formally for $\mathbf{E}(\mathbf{n})$ by multiplying both sides by $\hat{\mathbf{T}}^{-1}$, the inverse of $\hat{\mathbf{T}}$, yielding

$$\mathbf{E}(\mathbf{n}) = -ik_0^{-1}\hat{\mathbf{T}}^{-1} \cdot \mathbf{a}(\mathbf{n}). \tag{3.7}$$

Once $\mathbf{E}(\mathbf{n})$ is known, the Fourier transform of the magnetic field $\boldsymbol{\mathcal{H}}$ can readily be found from Equations (2.32) to (2.34) yielding

$$\boldsymbol{\mathcal{H}}(\mathbf{n}) = \mathbf{n} \times \mathbf{E}(\mathbf{n}) + ik_0^{-1}\mathbf{J}^m(\mathbf{n}). \tag{3.8}$$

Each of the field quantities can be transformed back to ordinary observation space using the inverse Fourier transform formula. Now we proceed to obtaining the integral representation of the field excited by given currents $\mathbf{J}^{e,m}(\mathbf{r})$.

We begin by studying the tensor $\hat{\mathbf{T}}^{-1}$, the individual elements of which are given by

$$T_{ij}^{-1} = C_{ij}/D, \tag{3.9}$$

where C_{ij} and D are the adjoint and determinant of the matrix T_{ij}. It is seen from comparison of (3.6) with (2.48) that the determinant D coincides with that in (2.48), and therefore it can be written in the form (2.53) or (2.65) or (2.66). The elements of the matrix C_{ij} are found to be

$$C_{xx} = n_x^2 n^2 - \varepsilon(n_x^2 + n_y^2) - \eta(n_x^2 + n_z^2) + \varepsilon\eta,$$

$$C_{xy} = n_x n_y n^2 - ig(n_x^2 + n_y^2) - \eta n_x n_y + ig\eta,$$

$$C_{xz} = n_x n_z n^2 - \varepsilon n_x n_z - ign_y n_z,$$

$$C_{yx} = n_x n_y n^2 + ig(n_x^2 + n_y^2) - \eta n_x n_y - ig\eta,$$

$$C_{yy} = n_y^2 n^2 - \varepsilon(n_x^2 + n_y^2) - \eta(n_y^2 + n_z^2) + \varepsilon\eta, \qquad (3.10)$$

$$C_{yz} = n_y n_z n^2 - \varepsilon n_y n_z + ign_x n_z,$$

$$C_{zx} = n_x n_z n^2 - \varepsilon n_x n_z + ign_y n_z,$$

$$C_{zy} = n_y n_z n^2 - \varepsilon n_y n_z - ign_x n_z,$$

$$C_{zz} = n_z^2 n^2 - \varepsilon(n^2 + n_z^2) + \varepsilon^2 - g^2.$$

It is clear that the behavior of the inverse Fourier transforms of the individual components of $\mathbf{E}(\mathbf{n})$, that is of the integrals

$$E_i(\mathbf{r}) = -\mathrm{i}\,\frac{k_0^2}{(2\pi)^3} \int \frac{1}{D} C_{ij}\, a_j(\mathbf{n})\, e^{-\mathrm{i}k_0 \mathbf{n}\cdot\mathbf{r}} \mathrm{d}\mathbf{n}, \qquad (3.11)$$

is governed by the zeros of D in the $\mathbf{n}$-space.* There are various ways of identifying zeros of D (see (2.53), (2.65) and (2.66)), depending on what kind of coordinates we use in the $\mathbf{n}$-space. As mentioned earlier, for the problems discussed in this book the cylindrical coordinate system defined by (2.61) will be the most suitable. When this system is used, it is convenient to perform first the integration over n_z in (3.11). For this purpose, we take D in the form (2.65) and rewrite (3.11), thus:

$$E_i(\mathbf{r}) = \mathrm{i}\,\frac{k_0^2}{(2\pi)^3\eta} \int \mathrm{d}\mathbf{n}_\perp \int \frac{C_{ij} a_j \exp(-\mathrm{i}k_0\mathbf{n}_\perp \cdot \mathbf{r}_\perp)}{\left[p^2 - p_\mathrm{o}^2(q)\right]\left[p^2 - p_\mathrm{x}^2(q)\right]} \exp(-\mathrm{i}k_0 pz)\, \mathrm{d}p, \qquad (3.12)$$

where the following notations are used: $\mathbf{r}_\perp = x\hat{\mathbf{x}}_0 + y\hat{\mathbf{y}}_0$, $\mathbf{n}_\perp = n_x\hat{\mathbf{x}}_0 + n_y\hat{\mathbf{y}}_0$, $\mathrm{d}\mathbf{n}_\perp = \mathrm{d}n_x\,\mathrm{d}n_y$, $q = (n_x^2 + n_y^2)^{1/2}$, $p = n_z$; the functions $p_{\mathrm{o,x}}(q)$ are given by (2.63). Looked upon as a function of p, the denominator in (3.12) has two pairs of zeros, which are poles of the integrand of (3.12):

$$p = \pm p_\mathrm{o}, \quad p = \pm p_\mathrm{x}. \qquad (3.13)$$

A close study of $p_{\mathrm{o,x}}$ shows that their locations in the complex p plane depend essentially on what frequency band is considered. For example, in a loss-free magnetoplasma in the whistler band (2.69) the quantity p_o is pure imaginary for all real values of q, while the quantity p_x may be real. When the poles $\pm p_\mathrm{x}$ lie exactly on the integration path, an uncertainty appears in performing the p integration. To avoid this uncertainty, we assume the medium to be slightly 'lossy'. This may always be ensured by including a minute correction in the dielectric tensor due

---

* Here, and throughout the book, the repeated subscript is summed over.

to collision processes. The poles $\pm p_x$ are then found to be slightly shifted off the real p axis, as shown in Figure 3.1. The p integration can then be performed without difficulty by using the technique of contour integration. After the integration the losses will be allowed to approach zero.

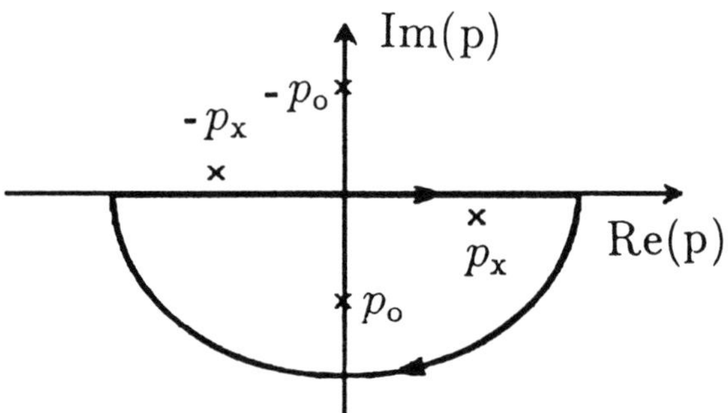

Figure 3.1. Path of contour integration of (3.12) for $z > 0$ and location of the poles $\pm p_o$, $\pm p_x$ in the complex p plane.

Having analyzed the behavior of singularities of the integrand, we are now ready to perform the integration over p in (3.12). This may be done by residues in the complex p plane if the integration contour is closed according to the criterion imposed by Jordan's lemma. Recall that according to this lemma, if f is a function of complex variable p, holomorphic everywhere in the lower half-plane of p, except for an isolated set of singular points, and if

$$\lim_{R \to \infty} \max_{p \in C_R^{(-)}} |f(p)| = 0, \tag{3.14}$$

where $C_R^{(-)}$ is the cemicircle defined by $|p| = R$ and $\mathrm{Im}\,(p) \leq 0$, then, for any $\zeta > 0$,

$$\lim_{R \to \infty} \int_{C_R^{(-)}} f(p)\, e^{-i\zeta p}\, \mathrm{d}p = 0. \tag{3.15}$$

The sense of the lemma consists in that the quantity $\max |f(p)|$ may vanish slowly when $p \in C_R^{(-)}$ and $R \to \infty$, so that the integral $\int_{C_R^{(-)}} f(p)\, \mathrm{d}p$ does not tend to zero, whereas the integral in (3.15) vanishes, as R approaches infinity, due to the presence of a decreasing exponential $\exp(-i\zeta p)$.* It is evident that a statement similar to (3.15) can be established for $\zeta < 0$ if we use in (3.14) and (3.15), instead of $C_R^{(-)}$, the cemicircle $C_R^{(+)}$ defined by $|p| = R$ and $\mathrm{Im}\,(p) \geq 0$, and take the function

---

* Sometimes, more generalized formulations of Jordan's lemma are used. These as well as their proofs may be found in books on the theory of complex variables.

$f(p)$ which satisfies the conditions imposed by Jordan's lemma in the upper half-plane of the complex variable p.

Now suppose that $z > 0$ and that the integrand of (3.12) satisfies the conditions imposed by Jordan's lemma in the lower half-plane of p. Then we can close the integration path in this half-plane, without changing the value of the integral, as shown in Figure 3.1. By virtue of Cauchy's theorem of residues it is found that only the poles p_o and p_x contribute to the integral value, and the following expression is obtained for the electric field:

$$E_i(\mathbf{r}) = \sum_{\alpha=o}^{x} \chi_\alpha \frac{k_0^2}{8\pi^2\eta} \int \frac{C_{ij}(\mathbf{n}_\alpha)\, a_j(\mathbf{n}_\alpha) \exp\left[-ik_0(\mathbf{n}_\perp \cdot \mathbf{r}_\perp + p_\alpha z)\right]}{p_\alpha(q)\left[p_x{}^2(q) - p_o^2(q)\right]}\, d\mathbf{n}_\perp, \quad (3.16)$$

where $\mathbf{n}_\alpha = \mathbf{n}_\perp + p_\alpha\hat{\mathbf{z}}_0$ and $\chi_o = -\chi_x = -1$. A similar integration can be performed for $z < 0$, if we close the contour in the upper half-plane, taking care of the fulfillment of the conditions required for validity of Jordan's lemma. It is worth noting that very often, for many current distributions encountered in practice, the quantities $C_{ij}a_j/D$ do not satisfy the requisite conditions, but can be split into a combination of separate terms so that each satisfies these conditions individually in either the lower or upper half-plane of the integration variable. Then the separate integrals are evaluated by an application of Jordan's lemma and Cauchy's theorem, and the results are added.

Sometimes it is convenient to specify a z-dependence of given sources $\mathbf{J}^{e,m}(\mathbf{r})$ by the Dirac delta-function $\delta(z)$. In this case all quantities $\mathcal{J}_j^e(\mathbf{n})$ and $J_j^m(\mathbf{n})$ are independent of p, and the foregoing method of evaluating (3.12) needs to be refined. This is due to the fact that some ratios $C_{ij}a_j/D$ will contain the term $p^2(n^2 - \varepsilon)$ in the numerator, and the integrand of (3.12) will not vanish when $p = n_z \rightarrow \infty$. In such a situation the p integration can be performed by the following procedure. We first divide the term $p^2(n^2 - \varepsilon)$ in the numerator $C_{ij}a_j$ by the denominator D, using the identity

$$\frac{p^2(n^2 - \varepsilon)}{\left[p^2 - p_o^2(q)\right]\left[p^2 - p_x^2(q)\right]} = 1 - \frac{(q^2 - \eta)\left[\varepsilon(n^2 - \varepsilon) + g^2\right]}{\eta\left[p^2 - p_o^2(q)\right]\left[p^2 - p_x^2(q)\right]}, \quad (3.17)$$

which can easily be examined, and then reduce each quantity $C_{ij}a_j/D$ to a sum of two terms. It is clear that one term will not possess any poles, and the integral containing this term can be evaluated in a straightforward fashion. The other term will have the poles $\pm p_{o,x}$, and is integrated by using Cauchy's theorem of residues.

To illustrate in more detail the procedure described above, consider the evaluation of the component $E_z(\mathbf{r})$. It follows from (3.7) and (3.10) that $E_z(\mathbf{n})$ may be written

$$E_z(\mathbf{n}) = \frac{\mathrm{i}}{k_0\eta\left[p^2 - p_\mathrm{o}^2(q)\right]\left[p^2 - p_\mathrm{x}{}^2(q)\right]}\left\{p\left[n_x(n^2 - \varepsilon) + ign_y\right]a_x(\mathbf{n})\right.$$

$$\left. + \; p\left[n_y(n^2 - \varepsilon) - ign_x\right]a_y(\mathbf{n}) + \left[p^2(n^2 - \varepsilon) - \varepsilon\left(n^2 + \frac{g^2 - \varepsilon^2}{\varepsilon}\right)\right]a_z(\mathbf{n})\right\}.$$

By rearranging the terms and using (3.17) the last formula is rewritten thus:

$$E_z(\mathbf{n}) = \frac{\mathrm{i}\mathcal{J}_z^e(\mathbf{n})}{k_0\eta} + \frac{\mathrm{i}}{k_0\eta\left[p^2 - p_\mathrm{o}^2(q)\right]\left[p^2 - p_\mathrm{x}{}^2(q)\right]}$$

$$\times \; \left\{p\left[n_x(n^2 - \varepsilon)\Big(a_x(\mathbf{n}) - pJ_y^m(\mathbf{n})\Big) + ign_y\,a_x(\mathbf{n})\right]\right.$$

$$+ \; p\left[n_y(n^2 - \varepsilon)\Big(a_y(\mathbf{n}) + pJ_x^m(\mathbf{n})\Big) - ign_x\,a_y(\mathbf{n})\right]$$

$$\left. - \; \varepsilon\left(n^2 + \frac{g^2 - \varepsilon^2}{\varepsilon}\right)\left[a_z(\mathbf{n}) + \frac{q^2 - \eta}{\eta}\,\mathcal{J}_z^e(\mathbf{n})\right]\right\}. \tag{3.18}$$

It is to be noted that such a rearrangement is usually very convenient even if the p integration can be done without a separation of the term $\mathrm{i}\mathcal{J}_z^e/(k_0\eta)$ from the initial representation.

If the first term in (3.18) is inserted in the inverse Fourier transform formula, it leads to

$$E_z^{(\mathrm{I})}(\mathbf{r}) = \frac{k_0^3}{(2\pi)^3} \int \frac{\mathrm{i}\mathcal{J}_z^e(\mathbf{n})}{k_0\eta}\,\exp(-\mathrm{i}k_0\mathbf{n}\cdot\mathbf{r})\,\mathrm{d}\mathbf{n} = \frac{\mathrm{i}\mathcal{J}_z^e(\mathbf{r})}{k_0\eta}. \tag{3.19}$$

We have now the p integration of the second term in (3.18) to perform. Assuming, for simplicity, that the properties of the integrand allow closing the integration contour in the lower half-plane for $z > 0$ and in the upper half-plane for $z < 0$, without changing the integral value, we perform the p integration to obtain

$$E_z^{(\mathrm{II})}(\mathbf{r}) = \sum_{\alpha=\mathrm{o}}^{\mathrm{x}} \int E_z^{(\alpha)}(\mathbf{n}_\alpha^{(\pm)})\exp\left[-\mathrm{i}k_0(\mathbf{n}_\perp\cdot\mathbf{r} + p_\alpha|z|)\right]\mathrm{d}\mathbf{n}_\perp, \tag{3.20}$$

where

$$E_z^{(\alpha)}(\mathbf{n}_\alpha^{(\pm)}) = \chi_\alpha\sigma\,\frac{k_0^2}{8\pi^2\eta}\,\frac{n_\alpha^2 - \varepsilon}{\left[\left(1 - \dfrac{\varepsilon}{\eta}\right)^2 q^4 - 4\dfrac{g^2}{\eta}q^2 + 4g^2\right]^{1/2}}$$

$$\times \; \left[W^e(\mathbf{n}_\alpha^{(\pm)}) + W^m(\mathbf{n}_\alpha^{(\pm)})\right],$$

$$
\begin{aligned}
W^e(\mathbf{n}_\alpha^{(\pm)}) &= \mathcal{J}_x^e(\mathbf{n}_\alpha^{(\pm)})\left(n_x + \frac{ign_y}{n_\alpha^2 - \varepsilon}\right) + \mathcal{J}_y^e(\mathbf{n}_\alpha^{(\pm)})\left(n_y - \frac{ign_x}{n_\alpha^2 - \varepsilon}\right) \\
&\quad + \mathcal{J}_z^e(\mathbf{n}_\alpha^{(\pm)})\frac{q^2}{q^2 - \eta}\,(\mathbf{n}_\alpha^{(\pm)}\cdot\hat{\mathbf{z}}_0),
\end{aligned}
\tag{3.21}
$$

$$
\begin{aligned}
W^m(\mathbf{n}_\alpha^{(\pm)}) &= J_x^m(\mathbf{n}_\alpha^{(\pm)})\left(\frac{\eta n_y}{q^2 - \eta} + \frac{ign_x}{n_\alpha^2 - \varepsilon}\right)(\mathbf{n}_\alpha^{(\pm)}\cdot\hat{\mathbf{z}}_0) \\
&\quad + J_y^m(\mathbf{n}_\alpha^{(\pm)})\left(-\frac{\eta n_x}{q^2 - \eta} + \frac{ign_y}{n_\alpha^2 - \varepsilon}\right)(\mathbf{n}_\alpha^{(\pm)}\cdot\hat{\mathbf{z}}_0) - J_z^m(\mathbf{n}_\alpha^{(\pm)})\frac{igq^2}{n_\alpha^2 - \varepsilon}\,,
\end{aligned}
$$

$$
\mathbf{n}_\alpha^{(\pm)} = \mathbf{n}_\perp \pm p_\alpha\hat{\mathbf{z}}_0,\ n_\alpha^2 = n_x^2 + n_y^2 + p_\alpha^2 = q^2 + p_\alpha^2,\ \sigma = \frac{(\mathbf{n}_\alpha^{(\pm)}\cdot\hat{\mathbf{z}}_0)}{p_\alpha}\,.
$$

In the above expressions, one must take $\mathbf{n}_\alpha^{(+)}$ for $z > 0$, and $\mathbf{n}_\alpha^{(-)}$ for $z < 0$. In obtaining (3.20) and (3.21) we made use of the relations

$$
p_\times^2(q) - p_\circ^2(q) = \left[\left(1 - \frac{\varepsilon}{\eta}\right)^2 q^4 - 4\frac{g^2}{\eta}q^2 + 4g^2\right]^{1/2},
$$

$$
p_\alpha^2(n_\alpha^2 - \varepsilon) = -\varepsilon\eta^{-1}(q^2 - \eta)\left(n_\alpha^2 + \frac{g^2 - \varepsilon^2}{\varepsilon}\right),
$$

$$
\tag{3.22}
$$

which can readily be deduced from (2.63) and (3.17), respectively. The final result for $E_z(\mathbf{r})$ is

$$
E_z(\mathbf{r}) = E_z^{(\mathrm{I})}(\mathbf{r}) + E_z^{(\mathrm{II})}(\mathbf{r}).
\tag{3.23}
$$

The remaining components $E_{x,y}(\mathbf{r})$ of the electric field and components of the magnetic field $\boldsymbol{\mathcal{H}}(\mathbf{r})$ may similarly be analyzed, and after some lengthy algebra they are found to be

$$
E_{x,y}(\mathbf{r}) = \sum_{\alpha=\circ}^{\times}\int E_{x,y}^{(\alpha)}(\mathbf{n}_\alpha^{(\pm)})\exp\left[-ik_0(\mathbf{n}_\perp\cdot\mathbf{r}_\perp + p_\alpha|z|)\right]d\mathbf{n}_\perp,
\tag{3.24}
$$

$$
\boldsymbol{\mathcal{H}}(\mathbf{r}) = ik_0^{-1}J_z^m(\mathbf{r})\hat{\mathbf{z}}_0 + \sum_{\alpha=\circ}^{\times}\int \boldsymbol{\mathcal{H}}^{(\alpha)}(\mathbf{n}_\alpha^{(\pm)})\exp\left[-ik_0(\mathbf{n}_\perp\cdot\mathbf{r}_\perp + p_\alpha|z|)\right]d\mathbf{n}_\perp,
\tag{3.25}
$$

where

$$
E_x^{(\alpha)}(\mathbf{n}_\alpha^{(\pm)}) = \sigma\frac{q^2 - \eta}{p_\alpha q^2}\left(n_x - \frac{ign_y}{n_\alpha^2 - \varepsilon}\right)E_z^{(\alpha)}(\mathbf{n}_\alpha^{(\pm)}),
$$

$$
E_y^{(\alpha)}(\mathbf{n}_\alpha^{(\pm)}) = \sigma\frac{q^2 - \eta}{p_\alpha q^2}\left(n_y + \frac{ign_x}{n_\alpha^2 - \varepsilon}\right)E_z^{(\alpha)}(\mathbf{n}_\alpha^{(\pm)})
\tag{3.26}
$$

and

$$
\mathcal{H}_x^{(\alpha)}(\mathbf{n}_\alpha^{(\pm)}) = \frac{q^2 - \eta}{q^2}\left(\frac{\eta n_y}{q^2 - \eta} - \frac{ign_x}{n_\alpha^2 - \varepsilon}\right)E_z^{(\alpha)}(\mathbf{n}_\alpha^{(\pm)}),
$$

$$\mathcal{H}_y^{(\alpha)}(\mathbf{n}_\alpha^{(\pm)}) = -\frac{q^2 - \eta}{q^2} \left(\frac{\eta n_x}{q^2 - \eta} + \frac{ign_y}{n_\alpha^2 - \varepsilon} \right) E_z^{(\alpha)}(\mathbf{n}_\alpha^{(\pm)}),$$

$$\mathcal{H}_z^{(\alpha)}(\mathbf{n}_\alpha^{(\pm)}) = \sigma \frac{ig}{p_\alpha} \frac{q^2 - \eta}{n_\alpha^2 - \varepsilon} E_z^{(\alpha)}(\mathbf{n}_\alpha^{(\pm)}). \tag{3.27}$$

Employing cylindrical coordinates in $\mathbf{n}$-space, any field quantity $\mathcal{F}$ in a source-free domain may now be represented by

$$\mathcal{F}(\rho, \phi, z) = \sum_{\alpha=o}^{x} \int_0^\infty \mathrm{d}q \int_0^{2\pi} F_\alpha(q, \varphi) \exp\left[-ik_0\Big(q\rho\cos(\phi-\varphi)+p_\alpha(q)\,|z| \Big) \right] \mathrm{d}\varphi. \tag{3.28}$$

It should be noted that the use of the proper branch of p_α, for which $\mathrm{Im}\,(p_\alpha) < 0$ when minute losses are included, ensures the fulfillment of the radiation condition at infinity, which requires that the energy transopt in the wave be outgoing.

Analogously, by applying the formula (2.53) for D and performing the n integration in (3.11) it is possible to obtain the field representation in another form, in spherical polar coordinates:

$$\mathcal{F}(r, \theta, \phi) = \sum_{\alpha=o}^{x} \int_0^\pi \mathrm{d}\vartheta \int_0^{2\pi} F_\alpha(\vartheta, \varphi)$$

$$\times \ \exp\left[-ik_0 n_\alpha(\vartheta)\, r\Big(\sin\theta\sin\vartheta\cos(\phi - \varphi) + \cos\theta\cos\vartheta \Big) \right] \mathrm{d}\varphi. \tag{3.29}$$

Again, when dealing with the D representation in the form (2.66), the q integration in (3.11) can be performed first to yield

$$\mathcal{F}(\rho, \phi, z) = \sum_{k=1}^{2} \int_{-\infty}^{+\infty} \mathrm{d}p \int_0^{2\pi} F_k(\varphi, p) \exp\left[-ik_0\Big(q_k(p)\,\rho\cos(\phi - \varphi) + pz \Big) \right] \mathrm{d}\varphi. \tag{3.30}$$

In derivation of (3.29) and (3.30), making use of the proper branches of n_α ($\mathrm{Im}\,(n_\alpha) < 0$) or q_k ($\mathrm{Im}\,(q_k) < 0$) excludes the appearance of terms contradicting the radiation condition at infinity (compare (3.29) and (3.30) with (2.60) and (2.68), respectively). For full mathematical details of the methods used for obtaining the fields of given sources in a magnetoplasma the reader should consult the works of Bunkin (1957), Kogelnik (1960), Kuehl (1962), Arbel and Felsen (1963), Mittra and Deschamps (1963), Bellyustin (1978), Al'pert *et al.* (1983), and Lukin *et al.* (1988).

Concluding this section, we note that in all the field representations given above, the φ integration can easily be performed. We shall show this using (3.28) as a case in point. If we utilize for the exponential $\exp[-ik_0q\rho\cos(\phi - \varphi)]$ the well-known series representation

$$\exp\left[-ik_0q\rho\cos(\phi - \varphi) \right] = \sum_{m=-\infty}^{\infty} (-i)^m J_m(k_0q\rho)\, e^{-im(\phi-\varphi)}, \tag{3.31}$$

where $J_m(\zeta)$ is the Bessel function of the first kind of order m (Abramowitz and Stegun, 1964), and expand $F_\alpha(q, \varphi)$ into its Fourier series

$$F_\alpha(q, \varphi) = \sum_{\ell=-\infty}^{\infty} F_\ell^{(\alpha)}(q) e^{-i\ell\varphi},$$

the φ integration in (3.28) then yields

$$\mathcal{F}(\rho, \phi, z) = \sum_{\alpha=o}^{x} \sum_{m=-\infty}^{\infty} 2\pi(-i)^m e^{-im\phi} \int_0^\infty F_m^{(\alpha)}(q)\, J_m(k_0 q\rho) \exp\left[-ik_0 p_\alpha(q)\,|z|\right] dq.$$

$$(3.32)$$

This formula represents the classical starting point and constitutes a representation of $\mathcal{F}$ by a sum of Fourier–Bessel integrals. Later (§ 3.4) it will be shown that in the far zone the integrals in (3.32) can be evaluated by the method of stationary phase.

3.3. Power radiated

It is now necessary to determine the power that can be radiated from given sources into a homogeneous, cold magnetoplasma. This has an important bearing on future comparisons with radiation from the same sources in the presence of a density duct. The problem of finding the radiation power of given currents immersed in a magnetoplasma has been treated by many authors (see, for example, Wang and Bell, 1969 a, 1970; Bell and Wang, 1971; Wang and Bell, 1972 b; Dokuchaev *et al.*, 1976; Bellyustin, 1978; Mareev and Chugunov, 1991). In this section, based partially on the works referenced above, we shall discuss the general procedure of evaluating the total radiation power of the electric-current and magnetic-current sources.

3.3.1. Integral formulation of the radiated power

It is well known that the time-averaged power radiated in a lossless medium by the electric-current distribution $\mathbf{J}^e(\mathbf{r})$ can be written

$$P_\Sigma^e = -\frac{1}{2}\operatorname{Re}\int \left(\mathbf{J}^e(\mathbf{r})\right)^* \cdot \mathbf{E}(\mathbf{r})\, d\mathbf{r} = -\frac{1}{2} Z_0^{-1} \operatorname{Re}\int \left(\boldsymbol{\mathcal{J}}^e(\mathbf{r})\right)^* \cdot \mathbf{E}(\mathbf{r})\, d\mathbf{r} \qquad (3.33)$$

where $\mathbf{E}(\mathbf{r})$ is the electric field excited by the source, and the star stands for complex conjugate. Using Parseval's theorem (Sneddon, 1951), we can express the radiated power via the integral in $\mathbf{n}$-space:

$$P_\Sigma^e = -\frac{k_0^3}{16\pi^3}\operatorname{Re}\int \left(\mathbf{J}^e(\mathbf{n})\right)^* \cdot \mathbf{E}(\mathbf{n})\, d\mathbf{n} = -\frac{k_0^3 Z_0^{-1}}{16\pi^3}\operatorname{Re}\int \left(\boldsymbol{\mathcal{J}}^e(\mathbf{n})\right)^* \cdot \mathbf{E}(\mathbf{n})\, d\mathbf{n}. \qquad (3.34)$$

It is clear from the preceding section that this integral can be represented in the form

$$P_\Sigma^e = \mathrm{Re} \int \frac{\mathrm{i}\,\Pi(\mathbf{n})}{(p^2 - p_\mathrm{o}^2)(p^2 - p_\mathrm{x}^2)}\, \mathrm{d}\mathbf{n}_\perp\, \mathrm{d}p, \qquad (3.35)$$

with $\Pi(\mathbf{n})$ being a real function. Although the method of contour integration outlined in §3.2 may generally be applied to evaluate integrals of the type (3.35), it is not entirely convenient in some particular cases. An alternative method of evaluating (3.35) makes use of Plemelj's formula which may be written

$$\lim_{\epsilon \to +0} \int_{-\infty}^{+\infty} \frac{f(x)\,\mathrm{d}x}{x - y \pm \mathrm{i}\epsilon} = \mp \pi \mathrm{i}\, f(y) + \fint_{-\infty}^{+\infty} \frac{f(x)\,\mathrm{d}x}{x - y}, \qquad (3.36)$$

where the bar on the integral sign indicates the principal value.

Since our main interest involves the VLF radiation characteristics when the source is immersed in a plasma modeled upon the near-Earth space, we shall focus our attention on the plasma parameters most appropriate to the ionosphere and inner magnetosphere and to frequencies in the band (2.69). As mentioned in §2.6, at these frequencies the only propagating electromagnetic mode is the whistler mode, that is, according to our nomenclature, the "extraordinary" mode. Bearing this in mind, we apply Plemelj's formula (3.36) to evaluate (3.35). The result is

$$P_\Sigma^e = \int \pi\, \frac{\Pi(\mathbf{n}_\perp + p_\mathrm{x}\hat{\mathbf{z}}_0) + \Pi(\mathbf{n}_\perp - p_\mathrm{x}\hat{\mathbf{z}}_0)}{2p_\mathrm{x}(p_\mathrm{x}^2 - p_\mathrm{o}^2)}\, \mathrm{d}\mathbf{n}_\perp. \qquad (3.37)$$

In obtaining (3.37) we made use of the fact that the poles $\pm p_\mathrm{x}$ are displaced slightly off the real p axis by introducing minute losses, as depicted in Figure 3.1. Taking into account expressions (3.21), (3.26) and (3.37), the general formula (3.34), in the lossless limit, becomes

$$P_\Sigma^e = -Z_0^{-1}\, \frac{k_0^2}{32\pi^2\eta} \int \frac{(n_\mathrm{x}^2 - \varepsilon)(q^2 - \eta)\left(\left| W^e\left(\mathbf{n}_\mathrm{x}^{(+)}\right)\right|^2 + \left| W^e\left(\mathbf{n}_\mathrm{x}^{(-)}\right)\right|^2 \right)}{p_\mathrm{x} q^2 \left[\left(1 - \dfrac{\varepsilon}{\eta}\right)^2 q^4 - 4\dfrac{g^2}{\eta} q^2 + 4g^2 \right]^{1/2}}\, \mathrm{d}\mathbf{n}_\perp, \qquad (3.38)$$

in which the integration is extended over the domain where p_x is purely real, and the notations are like those in (3.21).

The time-averaged power radiated in a lossless medium by a magnetic current $\mathbf{J}^m(\mathbf{r})$ can be written:

$$P_\Sigma^m = -\frac{1}{2}\,\mathrm{Re} \int \left(\mathbf{J}^m(\mathbf{r})\right)^* \cdot \mathbf{H}(\mathbf{r})\,\mathrm{d}\mathbf{r} = -\frac{1}{2} Z_0^{-1}\,\mathrm{Re} \int \left(\mathbf{J}^m(\mathbf{r})\right)^* \cdot \boldsymbol{\mathcal{H}}(\mathbf{r})\,\mathrm{d}\mathbf{r}, \qquad (3.39)$$

where $\mathbf{H}(\mathbf{r})$ is the magnetic field excited by the magnetic current. It is evident that (3.39) may be rewritten thus:

$$P_\Sigma^m = -\frac{k_0^3}{16\pi^3}\,\mathrm{Re}\int\left(\mathbf{J}^m(\mathbf{n})\right)^*\cdot\mathbf{H}(\mathbf{n})\,d\mathbf{n} = -\frac{k_0^3}{16\pi^3}\,Z_0^{-1}\,\mathrm{Re}\int\left(\mathbf{J}^m(\mathbf{n})\right)^*\cdot\boldsymbol{\mathcal{H}}(\mathbf{n})\,d\mathbf{n}.$$

$$(3.40)$$

Following the procedure similar to that used in obtaining (3.38), we have

$$P_\Sigma^m = -Z_0^{-1}\frac{k_0^2}{32\pi^2\eta}\int\frac{(n_x^2-\varepsilon)(q^2-\eta)\left(\left|W^m\left(\mathbf{n}_x^{(+)}\right)\right|^2+\left|W^m\left(\mathbf{n}_x^{(-)}\right)\right|^2\right)}{p_x q^2\left[\left(1-\dfrac{\varepsilon}{\eta}\right)^2 q^4 - 4\dfrac{g^2}{\eta}\,q^2 + 4g^2\right]^{1/2}}\,d\mathbf{n}_\perp,\quad(3.41)$$

where the notations of (3.21) are again understood.

The above expressions (3.38) and (3.41) relate to the case when only the "extraordinary" mode of the plasma is propagating. If both modes are propagating, the contribution from the "ordinary" mode should be added.

It should be noted that there is an alternative way of computing the value of radiated power. It is well known that the time-averaged flux of the electromagnetic energy in the harmonically varying field is given by the vector

$$\mathbf{S} = \frac{1}{2}\,\mathrm{Re}\left(\mathbf{E}\times\mathbf{H}^*\right),\qquad(3.42)$$

which is called the time-averaged Poynting vector. It is to be stressed that the total flux of energy in a medium coincides with that associated with the electromagnetic field and given by (3.42) only when the spatial dispersion effect may be neglected. This is ignored within the cold plasma approximation used in the present book. For full detailes of the definition of the total flux of energy in the presence of the spatial dispersion the reader should consult a general treatise such as Ginzburg (1970).

The rate of a total energy loss to radiation can be computed by integrating the Poynting vector $\mathbf{S}$ over an arbitrary closed surface Σ surrounding the source

$$P_\Sigma^{e,m} = \oint_\Sigma \mathbf{S}\cdot\hat{\mathbf{n}}\,d\sigma,\qquad(3.43)$$

where $\hat{\mathbf{n}}$ is an outward directed normal to the surface of integration. Once the electric and magnetic field intensities $\mathbf{E}$ and $\mathbf{H}$ and consequently the Poynting vector $\mathbf{S}$ are known, the integral in (3.43) immediately gives the value of radiated power. Therefore, the integral representation (3.43) for the radiation power is enirely equivalent to that given by the previous formulas (3.38) or (3.41). However, the power integral formulation using (3.38) and (3.41) can be more preferable in practical computations, especially if we wish to study the distribution of radiated power over the spatial spectrum of excited waves. This is due to the fact that the integrands in (3.38) and (3.41) immediately give this distribution in a desirable form.

3.3.2. Radiated power distribution over the spatial spectrum

In the following three sections (§§ 3.4–3.6), we apply the preceding formulation by computing the radiation characteristics of dipole and loop sources in the whistler band. A major aim will be to find the total radiated power and to study the distribution of the power over the spatial spectrum of quasi-plane waves excited in a surrounding plasma. A difficulty encountered in performing this study is that there are three frequency intervals (2.70), (2.71), and (2.72) in the whistler band (2.69), and the radiation characteristics should be analyzed individually for each interval. We do not present complete results for all of the frequencies here; instead, we focus on the interval (2.71). On one hand, this interval is of great importance for many applications, some of which were discussed in Chapter 1. On the other, it is this interval in which such fundamental singularities as the resonance cone, the simple caustic, and the axial caustic are encountered simultaneously. Therefore, having analyzed the radiation characteristics at the frequencies (2.71), one may show how this should be made, in principle, for the other frequencies. Accordingly, we give some preference to the interval (2.71) in various examples throughout this book when applying the general formulation to the particular problems.

As is seen from Figure 2.2, at the chosen frequencies, the quantity q may take all possible values from zero up to infinity. For further purposes, let us divide the interval $0 < q < \infty$ into the following ranges:

$$
\begin{aligned}
&\text{I)} \quad && 0 < q < q_S, \\
&\text{II)} \quad && q_c < q < \infty, \\
&\text{III)} \quad && q_S < q < q_c.
\end{aligned}
\qquad (3.44)
$$

The first range includes those whistler-mode waves whose refractive index is described by the quasi-longitudinal approximation (2.77). Whistlers of range I can penetrate into the Earth — ionosphere waveguide at middle and high latitudes and enable a VLF signal made up of these waves to be received on the surface of the Earth (see § 1.1). The second range contains so-called quasi-electrostatic whistler-mode waves, for which the quasi-static approximation (2.82) holds well. The third range involves the "intermediate" whistler-mode waves. It is useful to distinguish the "conical-refraction" whistler-mode waves which, according to convention, are those corresponding to a certain narrow interval of width $\Delta q \ll q_c$ with point $q = q_c$ as center. The partitioning (3.44) is reasonable and convenient for the study of the distribution of radiated power over the spatial spectrum (see Kondrat'ev *et al.*, 1992).

Including small electron collisions, one can easily find that if $|\omega - i\nu_e| \ll \omega_H$ the real and imaginary parts of the refractive index $n_x = n'_x - in''_x$ will satisfy the relation $n''_x/n'_x \simeq \nu_e/2\omega_H$ for quasi-longitudinal whistlers, and the relation $n''_x/n'_x \sim \nu_e/\omega$

for quasi-electrostatic waves. Hence the rate of the collision damping for waves from range I is very weak compared with that for waves from range II. Noting this dissimilarity in the behavior of the whistler-mode waves belonging to different parts of the spatial spectrum is important for the correct prediction of the VLF field pattern at long distances from the source in geospace.

As to the ducted whistler propagation, it will be shown later, in Chapter 4, that waves of range I and sometimes waves of range III can be trapped in ducts with enhanced density. By contrast, waves of range II cannot be trapped in such ducts. In the case of a duct with reduced ionization, the guided propagation may be observed for conical-refraction whistler-mode waves.

A knowledge of the radiated power distribution over the spatial spectrum of quasi-plane waves for various sources is crucial when it is necessary to determine the type of radiator to be used for the particular application. For instance, in experiments involving ducted whistler propagation, it is important to use a source whose radiation goes mainly to the whistler-mode waves which can be trapped in the duct and propagate along it with small attenuation. On the other hand, in a heating experiment, it is important to maximize the radiation in rapidly damping waves and the use of a source exciting the quasi-electrostatic waves well would seem in order.

3.4. Radiation from a linear electric current

An important contribution to the problem of electromagnetic radiation from linear dipole sources in a cold magnetized plasma in the whistler band was made by Wang and Bell (1969 a, 1970), Dokuchaev *et al.* (1976), and Bellyustin (1978), who derived closed-form expressions for the field and radiation power of such sources in various special cases. In this section, the radiation characteristics of linear electric currents will be examined in detail for the frequencies

$$\omega_{\text{LH}} \ll \omega \ll \omega_{\text{H}} \ll \omega_{\text{p}} \tag{3.45}$$

belonging to the interval (2.71). The limitations imposed by (3.45) are not essential for our purporses. However, when they are used, the computations are much more easily handled and it is possible to obtain the resulting expressions in a closed form.

At the frequencies (3.45), the elements of the dielectric tensor may be written

$$\varepsilon = X/Y^2, \quad g = -X/Y, \quad \eta = -X, \tag{3.46}$$

where $X \gg Y^2 \gg 1$, in which case they obey the following inequalities $|\eta| \gg |g| \gg \varepsilon \gg 1$. These will be used for the later work in this section.

We consider a linear source oriented either parallel or perpendicular to the static magnetic field. The current distribution on the source will be assumed to be triangular.[*]

3.4.1. Parallel orientation of a source

We take a linear dipole source oriented along the ambient magnetic field. The electric current distribution in the source is assumed to be

$$\mathbf{J}^e(\mathbf{r}) = \hat{\mathbf{z}}_0 I_0^e \, \delta(x) \, \delta(y) \left(1 - \frac{|z|}{L} \right) \left[U(z+L) - U(z-L) \right], \qquad (3.47)$$

where L is the source half length and $U(\zeta)$ is the Heaviside step function defined as

$$U(\zeta) = \begin{cases} 1 & \text{if } \zeta > 0, \\ 0 & \text{if } \zeta < 0. \end{cases} \qquad (3.48)$$

The Fourier transform of (3.47) is

$$\mathbf{J}^e(\mathbf{n}) = \hat{\mathbf{z}}_0 \, \frac{I_0^e}{L} \, \frac{\sin^2(k_0 L \, n_z/2)}{(k_0 n_z/2)^2}. \qquad (3.49)$$

We substitute this into the expression for W^e in (3.21) and then utilize the integral field representation which was obtained in § 3.2 (see formulas (3.23) to (3.25)). The resulting expression for the field excited by the source (3.47) in the zone $|z| > L$ is readily found to be

$$\mathcal{F}_\ell(\mathbf{r}) = \sum_{\alpha=o}^{x} \int_0^\infty \chi_\alpha F_\ell^{(\alpha)}(q) \, J_n(k_0 q \rho) \exp\left[-ik_0 p_\alpha(q) \, |z| \right] dq, \qquad (3.50)$$

where the quantities $\mathcal{F}_\ell$ ($\ell = 1, 2, \ldots, 6$) denote the field components E_ρ, E_ϕ, E_z, H_ρ, H_ϕ, H_z, respectively; and $n = 0$ for $\ell = 3, 6$, while $n = 1$ for $\ell = 1, 2, 4, 5$. In the preceding,

$$F_\ell^{(\alpha)}(q) = \frac{I_0^e}{\pi \eta L} \, \frac{q^2 f_\ell^{(\alpha)} \sin^2\left(k_0 L \, p_\alpha(q)/2 \right)}{p_\alpha^2(q) \left[q^4 \left(1 - \frac{\varepsilon}{\eta} \right)^2 - 4q^2 \frac{g^2}{\eta} + 4g^2 \right]^{1/2}}, \qquad (3.51)$$

---

[*] Finding the actual current distribution on a wire antenna excited by a given voltage is the basic problem of the theory of thin antennas. The problem of a cylindrical dipole antenna immersed in a *uniaxial* resonant magnetoplasma, in which $g \to 0$, has been treated in considerable detail by Lee (1969). However, an accurate solution to the current distribution on the antenna in a *gyrotropic* resonant magnetoplasma is yet to be found. Nevertheless, there is some evidence based on the experimental data and reasonable theoretical estimates that a triangular distribution can be a satisfactory approximation for the current behavior if the antenna length is much less than the maximum wavelength of the whistler mode.

where

$$f_1^{(\alpha)} = -iZ_0(n_\alpha^2 - \varepsilon)\,\mathrm{sgn}\,z, \qquad f_2^{(\alpha)} = Z_0 g\,\mathrm{sgn}\,z,$$

$$f_3^{(\alpha)} = -Z_0\varepsilon q\left(n_\alpha^2 + \frac{g^2}{\varepsilon} - \varepsilon\right)\Big/\eta p_\alpha(q), \tag{3.52}$$

$$f_4^{(\alpha)} = -g p_\alpha(q), \quad f_5^{(\alpha)} = -i\varepsilon\left(n_\alpha^2 + \frac{g^2}{\varepsilon} - \varepsilon\right)\Big/p_\alpha(q), \quad f_6^{(\alpha)} = igq\,\mathrm{sgn}\,z.$$

In obtaining expressions (3.50) to (3.52) we made use of the identities given by (3.22). Here, we omit the expressions for the field components in the zone $|z| < L$. Their derivation is left as an exercise for the reader.

We now have only the integration over q to perform. This may be facilitated by extending the q integration along the entire q axis. For this, we first make use of the fact that the Bessel function $J_n(\zeta)$ may be represented

$$J_n(\zeta) = \frac{1}{2}\left[H_n^{(1)}(\zeta) + H_n^{(2)}(\zeta)\right], \tag{3.53}$$

where $H_n^{(1)}(\zeta)$ and $H_n^{(2)}(\zeta)$ are the nth-order Hankel functions of the first and second kind, respectively. Using the circuital relations for the cylindrical functions, viz.,

$$H_n^{(1)}(e^{i\pi}\zeta) = -e^{-in\pi}H_n^{(2)}(\zeta),$$
$$H_n^{(2)}(e^{-i\pi}\zeta) = -e^{in\pi}H_n^{(1)}(\zeta) \tag{3.54}$$

(Korn and Korn, 1968), and taking into account the relations

$$F_\ell^{(\alpha)}(q) = F_\ell^{(\alpha)}(-q), \quad \ell = 1, 2, 4, 5,$$
$$F_\ell^{(\alpha)}(q) = -F_\ell^{(\alpha)}(-q), \quad \ell = 3, 6, \tag{3.55}$$

it is possible to rewrite (3.50) in the two alternative forms:

$$\mathcal{F}_\ell(\mathbf{r}) = \sum_{\alpha=o}^{x} \frac{1}{2}\int_{\Gamma_1} \chi_\alpha F_\ell^{(\alpha)}(q)\, H_n^{(1)}(k_0 q\rho)\exp\left[-ik_0\, p_\alpha(q)\,|z|\right]\mathrm{d}q \tag{3.56}$$

$$= \sum_{\alpha=o}^{x} \frac{1}{2}\int_{\Gamma_2} \chi_\alpha F_\ell^{(\alpha)}(q)\, H_n^{(2)}(k_0 q\rho)\exp\left[-ik_0\, p_\alpha(q)\,|z|\right]\mathrm{d}q. \tag{3.57}$$

The integration contours $\Gamma_{1,2}$ used in (3.56) and (3.57) are shown in Figure 3.2, where the branch cut originating at the branch point $q = 0$ of the Hankel functions $H_n^{(1,2)}(k_0 q\rho)$ is also indicated. Further, we shall employ the field representation in the form (3.57).

Since we are interested in the behavior of the electromagnetic field far from the source, only the "extraordinary" mode for which $p_\mathrm{x}(q)$ is real is important, as the "ordinary" mode for which $p_\mathrm{o}(q)$ is pure imaginary will decay exponentially

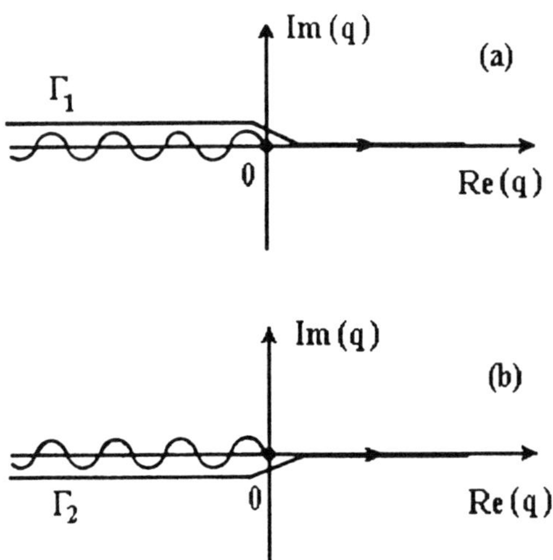

Figure 3.2. Contours Γ_1 and Γ_2 in the complex q plane.

with distance from the source. Bearing this in mind, we retain in (3.57) only the contribution from the "extraordinary" mode, for which $\chi_x = 1$, and then expand the Hankel function $H_n^{(2)}(k_0 q \rho)$ into its large-argument approximation

$$H_n^{(2)}(k_0 q \rho) \approx \left(\frac{2}{\pi k_0 q \rho} \right)^{1/2} \exp \left[-i k_0 q \rho + i \frac{\pi}{4} (2n + 1) \right], \quad k_0 q \rho \gg n \qquad (3.58)$$

(Korn and Korn, 1968). If this is inserted in the integral with 'α' = 'x' in (3.57) it gives

$$\mathcal{F}_\ell(\mathbf{r}) = \int_{\Gamma_2} \left(\frac{1}{2\pi k_0 q \rho} \right)^{1/2} F_\ell^{(x)}(q) \exp \left[-i k_0 \left(q \rho + p_x(q) |z| \right) + i \frac{\pi}{4} (2n + 1) \right] dq. \quad (3.59)$$

The final integration in (3.59) may then be done by the method of stationary phase.

Recall that this is an approximate method which is used to evaluate integrals of the form

$$I = \int_{-\infty}^{+\infty} f(x) \exp \left[i \Lambda g(x) \right] dx, \qquad (3.60)$$

where Λ is large and positive and $g(x)$ is real function of x. It can be shown that the principal contribution to an integral of this type arises from the neighborhoods of the points where $dg/dx = 0$. Such points are called the *saddle points*. The method of stationary phase consists first in evaluating the contributions to the integral from the vicinity of each saddle point and then adding the results. It is well known that the contribution from the neighborhood of one saddle point $x = x_i$ may be found as follows:

$$I \sim \sqrt{\frac{2\pi}{\Lambda\,|g''(x_i)|}}\, f(x_i)\, e^{\mathrm{i}\Lambda g(x_i)\pm\mathrm{i}\pi/4}, \qquad g''(x_i) \gtrless 0 \tag{3.61}$$

(see, for details, Felsen and Marcuvitz, 1973, Chapter 4). Here the primes indicate a derivative with respect to an argument.

Evaluating the integral in (3.59) using the foregoing method of stationary phase yields

$$\begin{aligned}
\mathcal{F}_\ell(\mathbf{r}) &= \sum_i \mathcal{F}_{\ell,i}(\mathbf{r}) = \sum_i \frac{F_\ell^{(\mathrm{x})}(q_i)}{k_0 r\,\sqrt{q_i\, p_{\mathrm{x}}''(q_i)\,\sin\theta\,\cos\theta}} \\
&\quad \times\ \exp\left[-\mathrm{i}k_0 q_i \rho - \mathrm{i}k_0\, p_{\mathrm{x}}(q_i)\,|z| + \mathrm{i}\,\frac{\pi n}{2}\right],
\end{aligned} \tag{3.62}$$

where $r = (\rho^2 + z^2)^{1/2}$ and the 'i' summation covers all the saddle points $q = q_i$ determined by

$$p_{\mathrm{x}}'(q) = -\rho/|z|. \tag{3.63}$$

Further, we shall confine our analysis to the angle domain $0 < \theta < \pi/2$ since the field pattern is symmetric about a plane $\theta = \pi/2$. In that domain, Equation (3.63) is rewritten

$$p_{\mathrm{x}}'(q) = -\tan\theta. \tag{3.64}$$

Equation (3.64) has a simple geometrical interpretation. Clearly each saddle point q_i locates the point $(q_i, p_{\mathrm{x}}(q_i))$ on the refractive index surface at which the normal to the surface makes an angle θ with the static magnetic-field direction (see Figures 2.1 to 2.3). Since the normal to the refractive index surface locates the ray direction, the saddle point equation implies that energy reaches the distant observation point (r, θ) via rays individually traveling far from the source along the radial direction from the source point to the observation point.

Consider in more detail the location of the saddle points and the behavior of propagating rays for the frequency range (3.45), which is of special interest to us. In that range, $\omega \ll \omega_{\mathrm{H}}$ whence $\theta_r < \theta_S$. It follows from the analysis presented in § 2.6 (see Figure 2.2) that, for a given direction θ, there are three saddle points q_i $(i = 1, 2, 3)$ if $0 < \theta < \theta_r$, and two saddle points q_i $(i = 1, 2)$ if $\theta_r \le \theta < \theta_S$. The saddle points lie in the following intervals of q:

$$\begin{aligned}
0 < \ & q_1 \ < q_S, \\
q_S < \ & q_2 \ < q_c, \\
-\infty < \ & q_3 \ < -q_c, \quad \left(q_3 = e^{-\mathrm{i}\pi}|q_3|\right).
\end{aligned} \tag{3.65}$$

Thus, we have seen that the radiation field far from the source is approximated by contributions of the form (3.62) that corresponds in reality to the standard geometrical optics approximation. This implies that the field is represented by a sum of propagating ray contributions whose number for a given observation point can be immediately inferred from the refractive index surface. In the case under consideration this approximation fails as $\theta \to \theta_S$ and $\theta \to 0$. A difficulty arises since at the shadow boundary when $\theta \to \theta_S$, the two rays corresponding to saddle points $q = q_1$ and $q = q_2$ tend to merge, thereby forming a simple caustic surface known as the Storey cone. Along the static magnetic field, $\theta = 0$, an infinite number of rays appropriate to conical refraction (i. e. to waves with the wavenormal angle $\vartheta = \vartheta_c$) tend to merge. These merged rays form an axial caustic along the static magnetic field $\mathbf{B}_0$. An asymptotic approximation for the integral (3.59) in the vicinity of caustics cannot be derived via an isolated first-order saddle point procedure; some caution is also necessary in the use of that procedure near the resonance direction $\theta = \theta_r$ (Arbel and Felsen, 1963; Brodskii $et\ al.$, 1969, Al'pert $et\ al.$, 1983). These considerations will be discussed in more detail in § 3.7, applied to calculations of the radiation pattern.

Now consider the time-averaged power radiated by the current distribution (3.47). The foregoing analysis shows that evaluation of the total radiated power using the Poynting-flux formulation (3.43) needs an understanding of the radiation field everywhere far from the source and turns out to be considerably complicated. Fortunately, the problem becomes more tractable when it is attacked by employing a power integral formulation via (3.38).

To obtain the radiated power within the framework of the power integral method, we first substitute (3.49) into the expression for W^e in (3.21) and then utilize the general formula (3.38). Employing cylindrical coordinates q and φ in the $\mathbf{n}$-space and taking into account that $\mathrm{d}\mathbf{n}_\perp = q\,\mathrm{d}q\,\mathrm{d}\varphi$, we perform the trivial φ integration in (3.38). Then the resulting formula for the total radiated power is

$$
\begin{aligned}
P_\Sigma^e \;=\;& |I_0^e|^2 Z_0 \,\frac{2}{\pi(k_0 L)^2 |\eta|} \int_0^\infty \frac{q^3}{p_\mathrm{x}^3(q)} \frac{q^2 + p_\mathrm{x}^2(q) - \varepsilon}{q^2 + |\eta|} \\
&\times \frac{\sin^4(k_0 L\,p_\mathrm{x}(q)/2)\,\mathrm{d}q}{\left[q^4\left(1 + \dfrac{\varepsilon}{|\eta|}\right)^2 + 4q^2\,\dfrac{g^2}{|\eta|} + 4g^2 \right]^{1/2}} \, .
\end{aligned}
\tag{3.66}
$$

Use is made of the fact that in a lossless magnetoplasma, η is purely real and negative for all the frequencies from the whistler band, so that $\eta = -|\eta|$.

The spectral representation (3.66) of the radiated power allows us to investigate the distribution of this power over the spectrum of quasi-plane waves excited in the surrounding plasma. We perform the integral (3.66) using as the integration intervals those defined by (3.44) to find the "partial" powers P_W^e, P_Q^e, and P_I^e going to waves of ranges I, II, and III, respectively.

Firstly, we shall determine the power P_W^e that goes to whistler-mode waves obeying the quasi-longitudinal approximation (2.77). The quantity P_W^e is obtainable by setting the upper limit of integration in (3.66) equal to $q = q_S$. As mentioned earlier, we assume that $k_0 L(\varepsilon - g)^{1/2} \approx k_0 L |g|^{1/2} \ll 1$, so that the sine function in the integral can be replaced by its argument, i.e., $\sin(k_0 L\, p_x(q)/2) \approx k_0 L\, p_x(q)/2$. As a result, we obtain

$$P_W^e = |I_0^e|^2 Z_0 \frac{(k_0 L)^2}{8\pi\eta^2} \int_0^{q_S} p_x(q)\, q^3\, \frac{q^2 + p_x^2(q)}{(q^4 + 4g^2)^{1/2}}\, dq. \tag{3.67}$$

By making use of the fact that within the quasi-longitudinal approximation it is possible to write

$$q = \sqrt{\frac{|g|}{\cos\vartheta}}\, \sin\vartheta, \quad p_x = \sqrt{\frac{|g|}{\cos\vartheta}}\, \cos\vartheta, \tag{3.68}$$

the integral (3.67) reduces to

$$P_W^e = |I_0^e|^2 Z_0 \frac{(k_0 L)^2}{16\pi} \frac{|g|^{5/2}}{\eta^2} \int_0^{\vartheta_S} \frac{\sin^3\vartheta}{\cos^{5/2}\vartheta}\, d\vartheta, \tag{3.69}$$

where $\vartheta_S = \arctan\sqrt{2} = \arccos\left(\dfrac{1}{\sqrt{3}}\right)$. The integral in (3.69) can be evaluated in a straightforward fashion; the result follows

$$P_W^e = |I_0^e|^2 Z_0 \frac{(k_0 L)^2}{48\pi} \frac{|g|^{5/2}}{\eta^2}\, \alpha^e, \tag{3.70}$$

where $\alpha^e = 4 \times (3^{3/4} - 2) \approx 1.12$.

Secondly, we shall evaluate the power P_Q^e that goes to quasi-electrostatic whistler-mode waves. For this, we take $q = q_c$ as a lower integration limit in (3.66). It follows from (2.79) and (2.81) that, at the frequencies (3.45), $q_c \simeq X^{1/2} \simeq |\eta|^{1/2}$, $n_x^2 + X = p_x n_x Y$ and, for $q \geq q_c$, $p_x^2/q^2 \simeq Y^{-2} \simeq \varepsilon/|\eta|$. The power P_Q^e will then be given by

$$P_Q^e = |I_0^e|^2 Z_0 \frac{1}{\pi k_0 L\varepsilon} \int_{\frac{k_0 L\varepsilon^{1/2}}{2}}^{\infty} \frac{\sin^4\xi}{\xi^2}\, d\xi. \tag{3.71}$$

Taking into account that

$$\int_0^{\infty} \frac{\sin^4\xi}{\xi^2}\, d\xi = \frac{\pi}{4}, \tag{3.72}$$

we find that the expression in (3.71) reduces to

$$P_Q^e = |I_0^e|^2 Z_0 \frac{1}{4k_0 L\varepsilon} \left(1 - \frac{4}{\pi} \int_0^{\frac{k_0 L\varepsilon^{1/2}}{2}} \frac{\sin^4\xi}{\xi^2}\, d\xi\right). \tag{3.73}$$

In the limiting case $k_0 L \varepsilon^{1/2} \ll 1$, (3.73) becomes

$$P_Q^e \simeq |I_0^e|^2 Z_0 \frac{1}{4 k_0 L \varepsilon} . \qquad (3.74)$$

Thirdly, it remains to evaluate the power P_I^e radiated in the whistler-mode waves of the intermediate range $q_S < q < q_c$. This power may be written

$$P_I^e \simeq |I_0^e|^2 Z_0 \frac{2}{\pi (k_0 L)^2 |\eta|} \int_{q_S}^{q_c} \frac{q^3}{p_x^3(q)} \frac{q^2 + p_x^2(q) - \varepsilon}{q^2 + |\eta|} \frac{\sin^4(k_0 L \, p_x(q)/2)}{(q^4 + 4g^2)^{1/2}} \, dq. \qquad (3.75)$$

Since a derivation of the closed-form expression for P_I^e needs cumbersome calculations, we shall only make an estimate of (3.75) in the limiting case $k_0 L \, |g|^{1/2} \ll 1$. Taking into account the last identity in (3.22), one has

$$P_I^e \simeq |I_0^e|^2 Z_0 \frac{(k_0 L)^2}{8\pi} \frac{\varepsilon}{\eta^2} \int_{q_S}^{q_c} \frac{q^3 \left(q^2 + p_x^2(q) + \dfrac{g^2 - \varepsilon^2}{\varepsilon} \right)}{p_x(q)(q^4 + 4g^2)^{1/2}} \, dq. \qquad (3.76)$$

We change an integration variable in (3.76) through the use of (3.68) and then perform integration over a new variable ϑ between limits $\vartheta = \vartheta_S$ and $\vartheta = \vartheta_c = \arctan(q_c/P_c) \approx \arctan(Y/2)$. This leads to the following form:

$$P_I^e \simeq |I_0^e|^2 Z_0 \frac{(k_0 L)^2 \varepsilon^{1/2}}{32 \sqrt{2} \, \pi} . \qquad (3.77)$$

Since formulas (3.68) used in obtaining (3.77) fail to approximate the behavior of q and $p = p_x(q)$ in the near vicinity of $q = q_c$ (i.e., $\vartheta = \vartheta_c$), the last expression should be considered as a rough, but acceptable, estimate of P_I^e.

Note that the chosen partial powers P_W^e, P_I^e, and P_Q^e completely overlap all of the spatial spectrum of excited quasi-plane waves, so that, due to power orthogonality, we have $P_\Sigma^e = P_W^e + P_I^e + P_Q^e$.

By comparing (3.70), (3.74), and (3.77) one can readily deduce that in the case $k_0 L \varepsilon^{1/2} \ll 1$, almost all of the radiated power goes to quasi-electrostatic whistler-mode waves

$$P_W^e \ll P_I^e \ll P_Q^e \simeq P_\Sigma^e. \qquad (3.78)$$

In obtaining this chain of inequalities we made use of the fact that the expressions (3.70) and (3.77) derived under the condition $k_0 L \, |g|^{1/2} \ll 1$ give the majorizing estimates of P_W^e and P_I^e in the case when $k_0 L \varepsilon^{1/2} \ll 1$ but $k_0 L \, |g|^{1/2}$ is not small. It is of importance that the last inequality is usually satisfied for the standard satellite-based dipole antennas operating in the VLF band. This means that such sources are not efficient in producing the radiation in whistler-mode waves belonging to ranges I and III.

3.4.2. Perpendicular orientation of a source

For the case of a source oriented perpendicular to the static magnetic field, it is necessary to introduce a finite radius for the cross-sectional extent of the current distribution to secure the convergence (in the resonant bands) of the integral giving the total radiated power. Therefore we take the current density in the form

$$\mathbf{J}^e(\mathbf{r}) = \hat{\mathbf{x}}_0 I_0^e \left(1 - \frac{|x|}{L}\right) \left[U(x+L) - U(x-L)\right] \delta\left(\sqrt{y^2 + z^2} - d\right), \qquad (3.79)$$

where d is the radius of the current distribution. The Fourier-transformed current density then is

$$\mathbf{J}^e(\mathbf{n}) = \hat{\mathbf{x}}_0 \frac{I_0^e}{L} \frac{\sin^2(k_0 L n_x/2)}{(k_0 n_x/2)^2} J_0\left(k_0 d \sqrt{n_y^2 + n_z^2}\right). \qquad (3.80)$$

The radiation field and the power that radiates from the perpendicular current (3.79) may be obtained by using the procedures similar to those utilized in the preceding consideration for the parallel source orientation. To avoid repetition, we shall confine our analysis only to presentation of the resulting integral for the total radiated power P_Σ^e and evaluation of the partial powers P_W^e, P_I^e, P_Q^e in the most important cases.

The power radiated can now be written

$$P_\Sigma^e = |I_0^e|^2 Z_0 \frac{4}{\pi^2 (k_0 L)^2 |\eta|} \int_0^\infty dn_x \int_0^\infty \frac{(q^2 + p_x^2(q) - \varepsilon)(q^2 + |\eta|)}{q^2 p_x(q)\left[q^4\left(1 + \frac{\varepsilon}{|\eta|}\right)^2 + 4q^2 \frac{g^2}{|\eta|} + 4g^2\right]^{1/2}}$$

$$\times \left[n_x^2 + \frac{g^2 n_y^2}{(q^2 + p_x^2(q) - \varepsilon)^2}\right] \frac{\sin^4(k_0 L n_x/2)}{n_x^4} J_0^2\left(k_0 d \sqrt{n_y^2 + p_x^2(q)}\right) dn_y, \qquad (3.81)$$

where the notation $q = (n_x^2 + n_y^2)^{1/2}$ is again used. Now consider the distribution of radiated power over the spatial spectrum. This may be investigated similarly as applied to a parallel orientation of the electric current.

The power P_W^e that goes to waves of range I (see (3.44)) in the case when $k_0 L |g|^{1/2} \ll 1$ and $k_0 d |g|^{1/2} \ll 1$, reduces to

$$P_W^e = |I_0^e|^2 Z_0 \frac{(k_0 L)^2}{16\pi} \int_0^{q_s} \frac{q\left[q^2 + p_x^2(q)\right]}{p_x(q)\left(q^4 + 4g^2\right)^{1/2}} \left[1 + \frac{g^2}{(q^2 + p_x^2(q))^2}\right] dq. \qquad (3.82)$$

In obtaining (3.82) we made use of cylindrical coordinates q and φ in the $\mathbf{n}$-space and performed the φ integration. By using a change of variable given by (3.68), the integral in (3.82) is evaluated in a straightforward fashion. The result is

$$P_W^e = |I_0^e|^2 Z_0 \frac{(k_0 L)^2 |g|^{1/2}}{24\pi}.$$

(3.83)

The power P_Q^e radiated in quasi-electrostatic whistler-mode waves under the conditions $k_0 L |\eta|^{1/2} \ll 1$ and $d \ll L$ becomes

$$
\begin{aligned}
P_Q^e &= |I_0^e|^2 Z_0 \frac{4}{\pi^2 (k_0 L)^2 (\varepsilon |\eta|)^{1/2}} \int_0^\infty dn_x \frac{\sin^4(k_0 L n_x/2)}{n_x^2} \\
&\times \int_0^\infty \frac{J_0^2 \left(k_0 d \sqrt{n_y^2 + p_x^2(q)} \right)}{\sqrt{n_x^2 + n_y^2}} \, dn_y.
\end{aligned}
$$

(3.84)

It can easily be verified from the behavior of the integrand of (3.84) that the contributions to the partial power P_Q^e from the ranges of integration $n_x \gg (k_0 L)^{-1}$ and $n_y \gg (k_0 d)^{-1}$ are insignificant. Therefore, by choosing the ratio d/L small enough, it is possible to find a large number μ such that

$$1 \ll \mu \ll L/d,$$

(3.85)

and then rewrite the integral with respect to n_y in (3.84) thus:

$$\int_0^\infty \frac{J_0^2 \left(k_0 d \sqrt{n_y^2 + p_x^2(q)} \right)}{\sqrt{n_x^2 + n_y^2}} \, dn_y \approx \int_0^{\mu n_x} \frac{dn_y}{\sqrt{n_x^2 + n_y^2}} + \int_{\mu n_x}^\infty \frac{J_0^2(k_0 \, dn_y)}{n_y} \, dn_y.$$

(3.86)

In the derivation of (3.86) use is made of the fact that for the range of n_x whose contribution to the integral in (3.84) is significant, $\mu k_0 dn_x \ll 1$. In the light of this, the last integral in the right-hand side of (3.86) is evaluated thus:

$$\int_{\mu n_x}^\infty \frac{J_0^2(k_0 \, dn_y)}{n_y} \, dn_y \simeq -\ln(\mu k_0 dn_x).$$

(3.87)

The remaining integral is known

$$\int_0^{\mu n_x} \frac{dn_y}{\sqrt{n_x^2 + n_y^2}} = \ln\left(\mu + \sqrt{\mu^2 + 1}\right) \approx \ln(2\mu).$$

(3.88)

As a result, one arrives at

$$P_Q^e \simeq |I_0^e|^2 Z_0 \frac{4}{\pi^2 (k_0 L)^2 (\varepsilon |\eta|)^{1/2}} \int_0^\infty \left(-\ln n_x + \ln \frac{2}{k_0 d} \right) \frac{\sin^4(k_0 L n_x/2)}{n_x^2} \, dn_x.$$

(3.89)

Taking into account that

$$\int_0^\infty \ln \xi \, \frac{\sin^4(a\xi)}{\xi^2} \, d\xi = \frac{\pi a}{4} (1 - \gamma - \ln a),$$

(3.90)

with $\gamma = 0.5772\ldots$ being Euler's constant, and using (3.72), we find that the expression in (3.89) reduces to

$$P_Q^e \simeq |I_0^e|^2 Z_0 \, \frac{1}{2\pi k_0 L(\varepsilon\,|\eta|)^{1/2}} \left[\ln\frac{2L}{d} - 1 + \beta^e\right], \qquad (3.91)$$

where $\beta^e = \gamma - \ln 2 \approx -0.12$ is a negligibly small term.

The remaining partial power P_I^e can be estimated roughly for $k_0 L\,|\eta|^{1/2} \ll 1$ and $k_0 d\,|\eta|^{1/2} \ll 1$ by

$$P_I^e \simeq |I_0^e|^2 Z_0 \, \frac{(k_0 L)^2}{16\pi\,|\eta|} \int_{q_S}^{q_c} \frac{q\,(q^2 + |\eta|)\,[q^2 + p_{\mathrm{x}}^2(q)]}{p_{\mathrm{x}}(q)\,(q^2 + 4g^2)^{1/2}}\,dq. \qquad (3.92)$$

Making use of a change of variable in accordance with (3.68) and integrating over ϑ between ϑ_S and ϑ_c gives

$$P_I^e \simeq |I_0^e|^2 Z_0 \, \frac{(k_0 L)^2}{80\sqrt{2}\,\pi} \, \frac{|\eta|^{3/2}}{|g|}. \qquad (3.93)$$

It can now be found that when $k_0 L\,|\eta|^{1/2} \ll 1$ and $d \ll L$, the main contribution to the total radiated power $P_\Sigma^e = P_W^e + P_I^e + P_Q^e$ is due to the partial power P_Q^e, so that $P_\Sigma^e \simeq P_Q^e$. This may be valid even when the condition $k_0 L\,|\eta|^{1/2} \ll 1$ is violated, provided that the quantity $\ln(2L/d)$ is large enough. It is also important to note that by comparing the corresponding partial radiation powers for two particular orientations of a linear current, it is possible to deduce that a perpendicular current will radiate a larger fraction of its energy into ranges I and III than will a parallel current.

Thus, for resonant bands, it is necessary to take the current distribution to be more regular than the delta-function $\delta(\mathbf{r})$. The use of infinitesimally small sources described by the delta-function leads to divergence of the integral giving the total radiated power. Fuzzifying the linear electric current over its cross-section limits the spatial spectrum of the source and hence the spectrum of excited quasi-plane waves, thereby securing boundedness of the total radiated power. An exception to this can only occur if the current symmetry axis lies precisely along $\mathbf{B}_0$ (see (3.47)), in which case the total radiated power of a linear source is finite without the cross-sectional spreading.

Aside from fuzzifuing the current distribution, the presence of collisional losses or the spatial dispersion would remove divergence of the power integral. However, when the spatial spectrum decay associated with a distributed character of the current becomes apparent before the collisional or warm-plasma effects appear, it is possible to employ the model of a collisionless cold magnetoplasma, as made in the foregoing analysis.

3.5. Radiation from a ring electric current

Let us now consider the power that can be radiated from a ring electric current immersed in a magnetoplasma. This type of a current distribution is realized on the loop antenna having the form of a wire ring. Only one particular orientation of the ring, when its axis is along the static magnetic-field direction, will be discussed. Other possible orientations may be left as an exercise for the reader. As in § 3.4, the frequency region (3.45) will be taken when presenting the concrete results.

3.5.1. Uniform current distribution

We first consider the current which is uniform round the ring. Then the current density may be written

$$\mathbf{J}^e(\mathbf{r}) = \hat{\boldsymbol{\phi}}_0\, I_0^e\, \delta(\rho - b)\, \delta(z), \tag{3.94}$$

where b is the ring radius.

To obtain the field excited by this current, we make use of the fact that the quantities $n_x J_x^e(\mathbf{n}) + n_y J_y^e(\mathbf{n})$ and $n_y J_x^e(\mathbf{n}) - n_x J_y^e(\mathbf{n})$ appearing in the expression for $W^e(\mathbf{n})$ in (3.21) can be found to be

$$\begin{aligned}
n_x J_x^e(\mathbf{n}) + n_y J_y^e(\mathbf{n}) &= 0, \\
n_y J_x^e(\mathbf{n}) - n_x J_y^e(\mathbf{n}) &= -\mathrm{i}2\pi b I_0^e q\, J_1(k_0 bq).
\end{aligned} \tag{3.95}$$

From (3.19) to (3.27), we then find that the field expressions would coincide in form with (3.50); it is only necessary to take the following expression for $F_\ell^{(\alpha)}(q)$ ($\ell = 1,\ldots,6$; $\alpha = $ 'o', 'x'):

$$F_\ell^{(\alpha)}(q) = I_0^e\, \frac{k_0^2 b}{2}\, \frac{g}{\eta}\, \frac{q f_\ell^{(\alpha)} J_1(k_0 bq)}{\left[q^4 \left(1 - \dfrac{\varepsilon}{\eta}\right)^2 - 4q^2 \dfrac{g^2}{\eta} + 4g^2 \right]^{1/2}}, \tag{3.96}$$

where

$$f_1^{(\alpha)} = -\mathrm{i}Z_0\, \frac{q^2 - \eta}{p_\alpha(q)}, \quad f_2^{(\alpha)} = Z_0\, g\, \frac{q^2 - \eta}{p_\alpha(q)(n_\alpha^2 - \varepsilon)}, \quad f_3^{(\alpha)} = Z_0 q\, \mathrm{sgn}\, z,$$

$$f_4^{(\alpha)} = -g\, \frac{q^2 - \eta}{n_\alpha^2 - \varepsilon}\, \mathrm{sgn}\, z, \quad f_5^{(\alpha)} = \mathrm{i}\eta\, \mathrm{sgn}\, z, \quad f_6^{(\alpha)} = \mathrm{i}g\, \frac{q(q^2 - \eta)}{p_\alpha(q)(n_\alpha^2 - \varepsilon)}.$$

The general formula (3.38) for the radiated power now reduces to

$$
\begin{aligned}
P_\Sigma^e &= |I_0^e|^2 Z_0 \, \frac{\pi}{2} \, (k_0 b)^2 \, \frac{g^2}{|\eta|} \int_0^\infty \frac{q}{p_{\mathrm{x}}(q)} \, \frac{q^2 + |\eta|}{q^2 + p_{\mathrm{x}}^2(q) - \varepsilon} \\
&\quad \times \frac{J_1^2(k_0 b q)\, \mathrm{d}q}{\left[q^4 \left(1 + \dfrac{\varepsilon}{|\eta|} \right)^2 + 4q^2 \dfrac{g^2}{|\eta|} + 4g^2 \right]^{1/2}} \, .
\end{aligned}
\tag{3.97}
$$

As in § 3.4, to obtain the power P_W^e radiated in quasi-longitudinal whistlers, we perform the integral in (3.97) between $q = 0$ and $q = q_S$. After some obvious simplifications we arrive at

$$
P_W^e = |I_0^e|^2 Z_0 \, \frac{\pi}{2} \, (k_0 b g)^2 \int_0^{q_S} \frac{q}{p_{\mathrm{x}}(q)\,[q^2 + p_{\mathrm{x}}^2(q)]} \, \frac{J_1^2(k_0 b q)\, \mathrm{d}q}{(q^4 + 4g^2)^{1/2}} \, .
\tag{3.98}
$$

In the limiting case $k_0 b\, |g|^{1/2} \ll 1$, the Bessel function in (3.98) is replaced by its small-argument approximation, i.e. $J_1(k_0 b q) \simeq k_0 b q / 2$. Using a change of variable such as given by (3.68) and performing the integration similarly to that as was made for the electric dipole in § 3.4, we obtain

$$
P_W^e = |I_0^e|^2 Z_0 \, \frac{\pi}{48} \, (k_0 b)^4 |g|^{3/2} \alpha^e,
\tag{3.99}
$$

where $\alpha^e \approx 0.77$. For arbitrary values of the source radius, (3.98) may be rewritten approximately, thus:

$$
P_W^e \simeq |I_0^e|^2 Z_0 \, \frac{\pi}{4} \, (k_0 b)^2 |g|^{-1/2} \int_0^{q_S} q\, J_1^2(k_0 b q)\, \mathrm{d}q.
\tag{3.100}
$$

Evaluating the integral in (3.100), we come to the formula

$$
P_W^e \simeq |I_0^e|^2 \, Z_0 \, \frac{\pi}{8} \, |g|^{-1/2} \, G_W(\xi_S),
\tag{3.101}
$$

where

$$
G_W(\xi_S) = 2 \int_0^{\xi_S} \xi\, J_1^2(\xi)\, \mathrm{d}\xi = \xi_S^2 \left[J_0^2(\xi_S) + J_1^2(\xi_S) \right] - 2\xi_S J_0(\xi_S)\, J_1(\xi_S),
\tag{3.102}
$$

$$
\xi_S = k_0 b q_S = \frac{2^{1/2}}{3^{1/4}} \, k_0 b\, |g|^{1/2}.
$$

It can readily be found that

$$
G_W(\xi) \approx \frac{\xi^4}{8}, \quad \text{for } \xi \ll 1,
\tag{3.103}
$$

$$
G_W(\xi) \approx \frac{2\xi}{\pi} \left(1 + \frac{\cos 2\xi}{2\xi} \right), \quad \text{for } \xi \gg 1.
\tag{3.104}
$$

If (3.103) is inserted in (3.101) it gives the result differing from (3.99) only by a factor of order α^e. In the opposite limiting case, when $k_0 b \, |g|^{1/2} \gg 1$, substituting (3.104) into (3.101) yields

$$P_W^e \simeq |I_0^e|^2 Z_0 \, \frac{\xi_S}{4|g|^{1/2}} \left(1 + \frac{\cos 2\xi_S}{2\xi_S} \right). \tag{3.105}$$

Next, we calculate the power P_Q^e launched into quasi-electrostatic whistler-mode waves. It follows from (2.79) that in the band (3.45), for $q \geq |\eta|^{1/2}$, we can put $p_x \approx (q^2 + |\eta|)/Y n_x = (q^2 + |\eta|)(\varepsilon/|\eta|)^{1/2}/n_x$. Making use of this fact and taking into account the inequalities $n_x^2 = q^2 + p_x^2(q) \gg |g| \gg \varepsilon$ valid for quasi-electrostatic waves, we obtain

$$P_Q^e = |I_0^e|^2 Z_0 \, \frac{\pi}{2} \, (k_0 b)^2 \, \frac{g^2}{(\varepsilon \, |\eta|)^{1/2}} \int_{q_c}^{\infty} \frac{J_1^2(k_0 b q)}{q^2} \, \mathrm{d}q. \tag{3.106}$$

The integral on the right is known

$$\int_{\xi_c}^{\infty} \frac{J_1^2(\xi)}{\xi^2} \, \mathrm{d}\xi = \frac{4}{3\pi} \left[1 - G_Q(\xi_c) \right], \tag{3.107}$$

where

$$G_Q(\xi) = \frac{\pi}{2\xi} \left\{ \left[\xi \, J_0(\xi) - \frac{1}{2} \, J_1(\xi) \right]^2 + \left(\xi^2 - \frac{3}{4} \right) J_1^2(\xi) \right\}. \tag{3.108}$$

With the help of (3.107), one has

$$P_Q^e = |I_0^e|^2 Z_0 \, \frac{2}{3} \, (k_0 b)^3 \, \frac{g^2}{(\varepsilon \, |\eta|)^{1/2}} \left[1 - G_Q(\xi_c) \right], \tag{3.109}$$

where $\xi_c = k_0 b q_c \approx k_0 b \, |\eta|^{1/2}$. For electrically small radiator, when $k_0 b \, |\eta|^{1/2} \ll 1$, using the approximation

$$G_Q(\xi) \approx \frac{3\pi}{16} \, \xi, \quad \xi \ll 1,$$

the power P_Q^e becomes

$$P_Q^e = |I_0^e|^2 Z_0 \, \frac{2}{3} \, (k_0 b)^3 \, \frac{g^2}{(\varepsilon \, |\eta|)^{1/2}} \left(1 - \frac{3\pi}{16} \, k_0 b \, |\eta|^{1/2} \right). \tag{3.110}$$

Finally, it remains to evaluate the residual part P_I^e of the radiated power, which corresponds to the intermediate region of values of integration variable: $q_S < q < q_c$. A comparatively simple formula for P_I^e can only be found in the limiting case $k_0 b \, |\eta|^{1/2} \ll 1$. That is

$$P_I^e \simeq |I_0^e|^2 Z_0 \, \frac{\pi}{8 \sqrt{2}} \, (k_0 b)^4 \, \frac{g^2}{\varepsilon^{1/2}}. \tag{3.111}$$

In obtaining this result we used the same approximations as for our previous derivation of (3.77).

Recall that adding the partial powers P_W^e, P_I^e, and P_Q^e yields the total radiated power P_Σ^e.

For the ring electric current, it is also of interest to evaluate the power P_C^e going to conical-refraction whistler-mode waves which we specify as those relating to the range $q_c - q_S/2 \leq q \leq q_c + q_S/2$. It is not difficult to find that

$$P_C^e = |I_0^e|^2 Z_0 \frac{\pi}{4} (k_0 b)^2 \frac{|g|}{|\eta|^{1/2}} \sum_{k=0}^{1} \left[J_k^2 \left(\xi_c - \frac{1}{2} \xi_S \right) - J_k^2 \left(\xi_c + \frac{1}{2} \xi_S \right) \right]. \quad (3.112)$$

In the case when $k_0 b \, |g|^{1/2} \ll 1$, for $k_0 b q_c \neq \mu_m^{(1)}$ ($\mu_m^{(1)}$ are zeros of the Bessel function $J_1(\xi)$), the formula (3.112) simplifies to

$$P_C^e \simeq |I_0^e|^2 Z_0 \frac{\pi}{2} (k_0 b)^2 \frac{|g|^{3/2}}{|\eta|} J_1^2 \left(k_0 b \, |\eta|^{1/2} \right). \quad (3.113)$$

Now it is easy to see that, for an electrically small radiator ($k_0 b \, |\eta|^{1/2} \ll 1$), we have

$$P_W^e \ll P_C^e, \quad P_W^e \ll P_I^e \ll P_Q^e \simeq P_\Sigma^e, \quad (3.114)$$

so that practically all of the radiated power goes to quasi-electrostatic waves, while for larger values of b ($k_0 b \, |g|^{1/2} \gg 1$),

$$P_Q^e \ll P_W^e \sim P_\Sigma^e, \quad (3.115)$$

and the most of the radiated power goes to whistler-mode waves of range I (i. e. $0 \leq q \leq q_S$).

More specific and refined conclusions can be drawn based on the results of numerical computations. The corresponding results are shown in Figure 3.3 as plots of total, $R_\Sigma^e = 2 P_\Sigma^e / |I_0^e|^2$, and partial, $R_{W,I,Q}^e = 2 P_{W,I,Q}^e / |I_0^e|^2$, radiation resistances versus the source radius b for the given frequency $\omega = 1.88 \times 10^5\,\mathrm{s}^{-1}$ and the plasma parameters typical of the earth's ionosphere: $N = 10^6\,\mathrm{cm}^{-3}$ and $B_0 = 0.5\,\mathrm{G}$. With these values the plasma has $\omega_\mathrm{p} = 5.64 \times 10^7\,\mathrm{s}^{-1}$ and $\omega_\mathrm{H} = 8.78 \times 10^6\,\mathrm{s}^{-1}$ (give also the values $|\eta|^{1/2} = 299.1$ and $|g|^{1/2} = 43.8$). Note that the chosen frequency ω is sufficiently above the lower-hybrid frequency ω_LH so that ion effects are negligible. It is evident that partial radiation resistances R_W^e, R_Q^e, and R_I^e correspond to the ranges I, II, and III, respectively, and $R_\Sigma^e = R_W^e + R_I^e + R_Q^e$. Besides, the left part of Figure 3.3 shows the dependence of the radiation resistance in conical-refraction waves $R_C^e = 2 P_C^e / |I_0^e|^2$ on the radius b.

One sees from Figure 3.3 that for sufficiently small values of b, the major portion of the radiated power goes to quasi-electrostatic whistler-mode waves, as was predicted from the analytical evaluation of P_W^e, P_I^e, and P_Q^e. As the radius b increases,

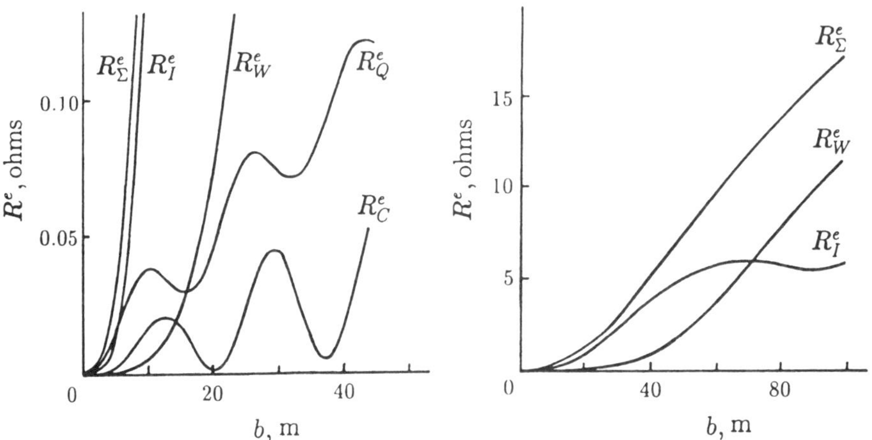

Figure 3.3. Total and partial radiation resistances for the uniform ring electric current. $\omega = 1.88 \times 10^5\,\mathrm{s}^{-1}$, $\omega_\mathrm{H} = 8.78 \times 10^6\,\mathrm{s}^{-1}$ $(B_0 = 0.5\,\mathrm{G})$, $\omega_\mathrm{p} = 5.64 \times 10^7\,\mathrm{s}^{-1}$ $(N = 10^6\,\mathrm{cm}^{-3})$.

the power P_I^e accordingly increases, so that for $b = b_1 = 5.5\,\mathrm{m}$ the quantities R_I^e and R_Q^e become equal: $R_Q^e(b_1) = R_I^e(b_1)$. Then, in a rather extensive range of b, up to $b = b_2 = 71\,\mathrm{m}$, where $R_W^e(b_2) = R_I^e(b_2)$, the contribution from the intermediate range III dominates. Only for comparatively large values of b when $b > b_2$, does the major portion of the radiated power go to whistlers of range I, that is to the long-wavelength part of the spatial spectrum. We also notice that the power R_C^e is of pronounced oscillation character. This fact would result essentially in the behavior of the radiation pattern of a ring current along the static magnetic-field direction.

A notable feature of a uniform current distribution along the ring is that any change of the orientation of the ring axis does not affect appreciably the total radiation resistance of the source (Wang and Bell, 1972 b). It should be mentioned, however, that a relative fraction of energy radiated in whistlers of range I is normally much larger for the case of a perpendicular orientation of the axis than for its parallel orientation. Also, this fraction is larger for a ring with a uniform current distribution than for a linear dipole source of length commensurate with the ring radius.

3.5.2. Nonuniform current distribution

In reality, some nonuniformity always exists in the current distribution on a loop antenna. It is clear that as the radius of the ring increases, the nonuniformity of the current distribution may become apparent, and the purely uniform distribution cannot be used. To investigate the effect of a current nonuniformity on the radiation characteristics of a ring source, we take the current in the form

$$\mathbf{J}^e(\mathbf{r}) = \hat{\phi}_0\, I^e(\phi)\, \delta(\rho - b)\, \frac{d}{\pi(z^2 + d^2)} \, . \tag{3.116}$$

The need to fuzzify the current over its cross-section arises from the fact that the total radiated power tends to infinity in the limit $d \to 0$, i. e. when

$$\frac{d}{\pi(z^2 + d^2)} \to \delta(z).$$

Note that employing the current distribution in the model form chosen above simplifies the algebra but does not lead to the loss of any important features of the radiation characteristics.

The quantity $I^e(\phi)$ in (3.116) may be expanded into a Fourier series

$$I^e(\phi) = I_0^e + \sum_{m=1}^{\infty} (I_{c,m} \cos m\phi + I_{s,m} \sin m\phi). \tag{3.117}$$

In the case of a nonuniform current distribution, with $I^e(\phi)$ being given by (3.117), the general formula (3.38) for the total radiated power takes the following form:

$$P_\Sigma^e = P_0^e + \sum_{m=1}^{\infty} P_m^e \tag{3.118}$$

$$= Z_0\, \frac{\pi}{2}\, (k_0 b)^2\, \frac{g^2}{|\eta|} \int_0^{\infty} \frac{q}{p_\mathsf{x}(q)}\, \frac{q^2 + |\eta|}{q^2 + p_\mathsf{x}^2(q) - \varepsilon}\, \frac{\exp[-2k_0 d\, p_\mathsf{x}(q)]}{\left[q^4 \left(1 + \dfrac{\varepsilon}{|\eta|} \right)^2 + 4q^2\, \dfrac{g^2}{|\eta|} + 4g^2 \right]^{1/2}}$$

$$\times \left\{ |I_0^e|^2 J_1^2(k_0 bq) + \sum_{m=1}^{\infty} \frac{|I_m^e|^2}{2} \left[J_m'^2(k_0 bq) + \left(\frac{(q^2 + p_\mathsf{x}^2(q) - \varepsilon)\, m\, J_m(k_0 bq)}{k_0 bgq} \right)^2 \right] \right\}\, dq,$$

where $|I_m^e|^2 = |I_{c,m}^e|^2 + |I_{s,m}^e|^2$, and the prime indicates a derivative with respect to an argument. We shall examine, for comparison, the same particular cases that were considered previously as applied to a uniform ring current, following the similar procedures. Therefore, the discussion of mathematical steps involved in obtaining some formulas will be much briefer (for more details, see Zaboronkova $et\ al.$, 1991, and Kondrat'ev $et\ al.$, 1992). To simplify the mathematics, we shall give the results relating only to the case when

$$I_0^e \neq 0, \quad I_{c,1}^e \neq 0, \quad I_{s,1}^e \neq 0, \quad I_{c,m}^e = I_{s,m}^e = 0, \quad m = 2, 3, \ldots\, . \tag{3.119}$$

The power P_W^e radiated now in whistlers belonging to range I appears, for $k_0 d\, |g|^{1/2} \ll 1$, as

$$P_W^e = P_{0,W}^e + P_{1,W}^e \simeq Z_0\, \frac{\pi}{4}\, |g|^{-1/2} \left[|I_0^e|^2 \hat{I}_1 + |I_1^e|^2 \left(\frac{1}{2}\, \hat{I}_2 - \hat{I}_3 + \hat{I}_4 \right) \right], \tag{3.120}$$

where

$$\hat{I}_1 = \int_0^{\xi_S} \xi\, J_1^2(\xi)\, d\xi = \frac{1}{2}\, G_W(\xi_S),$$

$$\hat{I}_2 = \int_0^{\xi_S} \xi\, J_0^2(\xi)\, d\xi = \frac{1}{2}\, \xi_S^2 \left[J_0^2(\xi_S) + J_1^2(\xi_S) \right],$$

$$\hat{I}_3 = \int_0^{\xi_S} J_0(\xi)\, J_1(\xi)\, d\xi = \frac{1}{2} \left[1 - J_0^2(\xi_S) \right], \tag{3.121}$$

$$\hat{I}_4 = \int_0^{\xi_S} \frac{J_1^2(\xi)}{\xi}\, d\xi = \frac{1}{2} \left[1 - J_0^2(\xi_S) - J_1^2(\xi_S) \right].$$

Other notations in (3.121) are like those in (3.102). On substituting the results given by (3.121) into (3.120), we get

$$P_W^e \simeq Z_0\, \frac{\pi}{8}\, |g|^{-1/2} \left\{ |I_0^e|^2 G_W(\xi_S) + \frac{1}{2}\, |I_1^e|^2 \left[\xi_S^2\, J_0^2(\xi_S) + (\xi_S^2 - 2)\, J_1^2(\xi_S) \right] \right\}. \tag{3.122}$$

The expression (3.122) may be simplified to the following forms:

$$P_W^e \simeq |I_0^e|^2 Z_0\, \frac{\pi}{48}\, (k_0 b)^4 |g|^{3/2} \alpha^e + |I_1^e|^2 Z_0\, \frac{\pi}{16\sqrt{3}}\, (k_0 b)^2 |g|^{1/2} \beta^e \tag{3.123}$$

if $k_0 b\, |g|^{1/2} \ll 1$, $d \ll b$, and

$$P_W \simeq |I_0^e|^2 Z_0\, \frac{\xi_S}{4\, |g|^{1/2}} \left(1 + \frac{\cos 2\xi_S}{2\xi_S} \right) + |I_1^e|^2 Z_0\, \frac{\xi_S}{8\, |g|^{1/2}} \tag{3.124}$$

if $k_0 b\, |g|^{1/2} \gg 1$, $k_0 d\, |g|^{1/2} \ll 1$ (α^e and β^e are numerical coefficients of the order unity). Within the quasi-longitudinal approximation (2.77), $\alpha^e = 0.77$ and $\beta^e = 1.15$.

Power P_Q^e, which goes to quasi-electrostatic whistler-mode waves, is described under conditions $k_0 d \varepsilon^{1/2} \ll 1$ and $\varepsilon^{1/2} d \ll |\eta|^{1/2} b$ by the expression

$$\begin{aligned}
P_Q^e \;=\; & P_{0,Q}^e + P_{1,Q}^e = Z_0\, \frac{\pi}{2}\, (k_0 b)^3\, \frac{g^2}{(\varepsilon\, |\eta|)^{1/2}} \\[2mm]
& \times\ \left\{ |I_0^e|^2\, \hat{J}_1 + \frac{1}{2}\, |I_1^e|^2 \left[\hat{J}_2 - 2\hat{J}_3 + \hat{J}_4 + \frac{1}{(k_0 b)^4 g^2}\, (\hat{J}_5 - \hat{J}_6) \right] \right\}, \tag{3.125}
\end{aligned}$$

where

$$\hat{J}_1 = \int_{\xi_c}^{\infty} \frac{J_1^2(\xi)}{\xi^2}\, d\xi = \frac{4}{3\pi} \left[1 - G_Q(\xi_c) \right],$$

$$\hat{J}_2 = \int_{\xi_c}^{\infty} \frac{J_0^2(\xi)}{\xi^2}\, d\xi = -\frac{4}{\pi} \left[1 - Q_Q(\xi_c) \right],$$

$$\hat{J}_3 = \int_{\xi_c}^{\infty} \frac{J_0(\xi)\, J_1(\xi)}{\xi^3}\, d\xi = \frac{1}{3} \left\{ \frac{J_0(\xi_c)\, J_1(\xi_c)}{\xi_c^2} - \frac{4}{3\pi} \left[1 - G_Q(\xi_c) \right] - \frac{4}{\pi} \left[1 - Q_Q(\xi_c) \right] \right\},$$

$$\hat{J}_4 = \int_{\xi_c}^{\infty} \frac{J_1^2(\xi)}{\xi^4}\, d\xi = \frac{1}{5}\, \frac{J_1^2(\xi_c)}{\xi_c^3} + \frac{2}{15}\, \frac{J_0(\xi_c)\, J_1(\xi_c)}{\xi_c^2} + \frac{8}{45\pi} \left[G_Q(\xi_c) + 3Q_Q(\xi_c) - 4 \right],$$

$$\hat{J}_5 = \int_0^{\infty} J_1^2(\xi) \exp\left(-2\, \frac{\varepsilon^{1/2} d}{|\eta|^{1/2} b}\, \xi \right) d\xi = \frac{1}{\pi \Lambda} \left[(2 - \Lambda^2)\, K(\Lambda) - 2E(\Lambda) \right], \tag{3.126}$$

$$\hat{J}_6 = \int_0^{\xi_c} J_1^2(\xi)\, d\xi = \frac{\xi_c^3}{12}\, {}_3F_4\left(\frac{3}{2}, 2, \frac{3}{2};\, 2, 2, 3, \frac{5}{2};\, -\xi_c^2 \right),$$

and

$$Q_Q(\xi) = \frac{\pi}{2\xi} \left\{ \left[\xi\, J_1(\xi) - \frac{1}{2}\, J_0(\xi) \right]^2 + \left(\xi^2 + \frac{1}{4} \right) J_0^2(\xi) \right\}.$$

In the preceding, $_3F_4$ is the generalized hypergeometric series (see Bateman Manuscript Project, 1953, vol. 1, Chapter 4; Gradshteyn and Ryzhik, 1965, §9.14); $K(\Lambda)$ and $E(\Lambda)$ are the complete elliptic integrals of the first and second kinds, respectively, of modulus Λ, where $\Lambda = (1 + d^2\varepsilon/b^2|\eta|)^{-1/2}$; other notations are like those in (3.108).

Because of the inequality $\varepsilon^{1/2}d \ll |\eta|^{1/2}b$, the value of Λ is close to unity. Thence the complete elliptic integrals $K(\Lambda)$ and $E(\Lambda)$ may be well approximated by

$$K(\Lambda) \approx \ln \frac{4}{\sqrt{1-\Lambda^2}} \approx \ln \left(\frac{4b}{d} \sqrt{\frac{|\eta|}{\varepsilon}} \right), \tag{3.127}$$

$$E(\Lambda) \approx 1.$$

From (3.125) and (3.126), with the help of (3.127), one obtains the following expression for the partial power P_Q^e in the limiting case $k_0 b |\eta|^{1/2} \ll 1$:

$$\begin{aligned}
P_Q^e &= |I_0^e|^2 Z_0 \frac{2}{3} (k_0 b)^3 \frac{g^2}{(\varepsilon|\eta|)^{1/2}} \\
&\quad + |I_1^e|^2 Z_0 \frac{1}{4k_0 b (\varepsilon|\eta|)^{1/2}} \left[\ln \left(\frac{4b}{d} \sqrt{\frac{|\eta|}{\varepsilon}} \right) - 2 \right].
\end{aligned} \tag{3.128}$$

In the last formula, the second term corresponds, in fact, to the appearance of an effective perpendicular (with respect to $\mathbf{B}_0$) electric dipole moment making the major contribution to the power P_Q^e if $|I_1^e|^2/|I_0^e|^2 > \frac{8}{3} (k_0 b)^4 g^2 / \ln \left(\frac{4b}{d} \sqrt{\frac{|\eta|}{\varepsilon}} \right)$. This dipole moment is associated with the nonuniform current distribution along the ring. As is clear from the continuity equation (2.11), a current nonuniformity would provide some charge distribution along the ring that, in turn, leads to the appearance of the electric dipole moment. It is of interest that the "dipole" term in (3.128) is nearly similar to that observed in the expression (3.91) for the power P_Q^e of an electric dipole with a perpendicular orientation. It can now be seen that, for vanishing d, when the z dependence of a current in (3.116) reduces to the delta-function case, the integral in (3.118) diverges, in contrast to the case of a uniform ring current. From the mathematical viewpoint, the difference between the two currents manifests itself in the behavior of the spatial spectrum of excited whistler-mode waves: a uniform distribution, for $k_0 b|\eta|^{1/2} \ll 1$, most-effectively excites quasi-electrostatic waves with the characteristic scale of the order $2\pi/(k_0 q) \sim b$, whereas a nonuniform distribution excites the waves with much smaller scales, down to $2\pi/(k_0 q) \sim d \ll b$. Ultimately, this results in the unboundedness of the partial power P_Q^e and hence the total radiated power P_Σ^e when d goes to zero. The unboundedness of the total power can also be deduced from the behavior of the radiation field. It is not difficult to understand the physical nature of such behavior. It is entirely associated with the fact that electron charges, which are absent for a uniform (symmetric) ring current but appear at the break of symmetry, excite effectively a "quasi-electrostatic part" of the total field. Thus, we can assert that, for resonant frequency bands, a purely uniform current distribution in a loop antenna is, in a sense, a somewhat degenerate case, and it is therefore important to know the actual form of this distribution.

In view of the above, it is necessary to remember that for sufficiently small values of d, the more realistic plasma models including the electron collisions and the spatial dispersion should be employed. Thus, in a collisionless magnetoplasma, when d becomes smaller than the electron thermal gyroradius $\rho_{\mathrm{He}} = V_{\mathrm{Te}}/\omega_{\mathrm{H}}$, the boundedness of the total radiated power is secured due to the warm-plasma effects rather than fuzzifying the current distribution. It can be verified, however, that for the VLF band and the plasma parameters typical of the earth's ionosphere, the partial radiation power P_Q^e of a nonuniform ring current with reasonable dimensions d and b is satisfactorily described by the formula obtained within the collisionless cold-plasma model.

The remainder of P_Σ^e, that is the power P_I^e radiated in waves of the intermediate range III, can be found in a closed form only for $k_0 d \, |g|^{1/2} \ll 1$ and $k_0 b \, |\eta|^{1/2} \ll 1$. The result is

$$P_I^e = P_{0,I}^e + P_{1,I}^e \simeq |I_0^e|^2 Z_0 \, \frac{\pi}{8\sqrt{2}} \, (k_0 b)^4 \, \frac{g^2}{\varepsilon^{1/2}} + |I_1^e|^2 Z_0 \, \frac{\pi}{48} \, (k_0 b)^2 \, \frac{|\eta|^{3/2}}{|g|}. \tag{3.129}$$

Thus, for a ring radiator of small size ($k_0 d \, |g|^{1/2} \ll k_0 b \, |\eta|^{1/2} \ll 1$) with nonuniform current distribution, almost all of the radiated power goes to quasi-electrostatic whistler-mode waves. With an increase of the loop radius b, the radiated power is redistributed over the spatial spectrum of excited waves similarly to that observed for a uniform circular current. An important difference, however, consists in the fact that the values of b_1 and b_2, at which the corresponding partial powers become equal to each other ($P_Q^e(b_1) = P_I^e(b_1)$ and $P_I^e(b_2) = P_W^e(b_2)$), are shifted to the side of larger values in the case of a nonuniform current. This is clearly illustrated by Table 2.1 which corresponds to the nonuniform distribution (3.116) with $|I_{0,1}^e| \neq 0$, $I_m^e = 0$ ($m = 2, 3, \ldots$), and $d = 0.01\,\mathrm{m}$ and gives values of b_1 and b_2 for different ratios $|I_1^e| / |I_0^e|$ and the parameters ω, N, and B_0 chosen earlier for plotting the curves of Figure 3.3.

Table 2.1. Values of b_1 and b_2 for nonuniform ring current; $d = 0.01\,\mathrm{m}$,
$\omega = 1.88 \times 10^5\,\mathrm{s}^{-1}$, $\omega_{\mathrm{H}} = 8.78 \times 10^6\,\mathrm{s}^{-1}$, $\omega_{\mathrm{p}} = 5.64 \times 10^7\,\mathrm{s}^{-1}$.

$|I_1^e| / |I_0^e|$	b_1, m	b_2, m
0	5.5	71.0
0.01	7.2	71.8
0.05	12.2	72.0
0.10	16.7	72.1
0.50	48.0	75.0
∞ ($|I_0^e| = 0$)	177.0	126.0

The above consideration shows that when the refractive index surface is open, a nonuniformity in the distribution of a ring electric current leads to an appreciable increase in the relative fraction of energy going to quasi-electrostatic waves as well as the total radiated power. It is worth noting that although a current nonuniformity causes some decrease in the ratios P_W^e/P_Σ^e and P_I^e/P_Σ^e, it can assist in increasing the magnitudes of P_W^e and P_I^e compared with those in the case of a uniform distribution. On the other hand, when current nonuniformity is rather weak, a considerable part of the power radiated from a loop source of reasonable radius $(k_0 b |\eta|^{1/2} \geq 1)$ goes to whistlers of ranges I and III, which, in turn, can be trapped in a density duct.

Unfortunately, a detailed full-wave theoretical study of an actual current distribution in loop antennas has not yet been made for the case of a resonant magnetoplasma, as far as the authors are aware. It can be inferred, however, from some experiments and preliminary theoretical estimates, that a current remains quasi-uniform, provided that the loop radius is small enough compared with the maximum wavelength of the whistler mode (Duff and Mittra, 1970). Otherwise, the current becomes nonuniform and the loop develops first an electric dipole moment, and then higher multipole moments (Ohnuki *et al.*, 1986). Based on recent experimental work by Zaboronkova *et al.* (1996 a), there is good reason to believe that satisfactory uniformity of the current distribution can be ensured by covering the antenna wire by a special dielectric shell of sufficient thickness.

In the preceding examples, we have restricted ourselves only to the situation where the loop axis was oriented along the static magnetic field. It was mentioned earlier for a uniform ring current that a change of its orientation does not appreciably influence the value of the total radiated power P_Σ^e. In contrast to this, if a current nonuniformity occurs, the radiation characteristics will essentially depend on the source orientation.

3.6. Radiation from a ring magnetic current

The general formulas (3.26) and (3.27) derived in § 3.2 allows comparison of the magnitudes of different components in a progressive plane wave. From those it follows that in whistler waves obeying the quasi-longitudinal approximation the magnetic component dominates, so that $|\mathcal{H}| \gg |\mathbf{E}|$, and, specifically, $|\mathcal{H}_{\rho,\phi}| \gg |\mathcal{H}_z|$. This fact raises the possibility of using ring magnetic currents to excite those waves.[*] In this section we shall examine only a ring magnetic current with the simplest uniform distribution[†]

$$\mathbf{J}^m(\mathbf{r}) = \hat{\phi}_0\, I_0^m\, \delta(\rho - b)\, \frac{d}{\pi(z^2 + d^2)}\,. \tag{3.130}$$

The need for "spreading" the magnetic current results from the fact that in this case (unlike that of a ring electric current) even a uniform distribution leads to unboundedness of the total radiated power when $d \to 0$.

---

[*] The whistler-band radiation of magnetic dipoles and magnetic quadrupoles has been investigated by Dokuchaev *et al.* (1976) and Bellyustin (1978).

[†] Such a source may be modeled, for example, by a toroidal coil with an opposing winding, which is widely used for diagnostic purposes and commonly known as Rogowski coil (Huddlestone and Leonard, 1965, Chapter 2, § 1).

The integral representation of the total radiation power P_Σ^m (see formula (3.41) in §3.3) that corresponds to current (3.130) in the frequency band (3.45) is as follows:

$$P_\Sigma^m = |I_0^m|^2 Z_0^{-1} \frac{\pi}{2} (k_0 b)^2 \varepsilon \int_0^\infty \frac{q}{p_x(q)} \frac{q^2 + p_x^2(q) + \dfrac{g^2 - \varepsilon^2}{\varepsilon}}{\left[q^4 \left(1 + \dfrac{\varepsilon}{|\eta|} \right)^2 + 4q^2 \dfrac{g^2}{|\eta|} + 4g^2 \right]^{1/2}} J_1^2(k_0 b q)$$

$$\times \quad \exp\left[- 2k_0 d\, p_x(q) \right] \mathrm{d}q \tag{3.131}$$

(Zaboronkova *et al.*, 1992 b). As in §§ 3.4 and 3.5, we divide P_Σ^m into the partial powers for quasi-longitudinal whistlers (P_W^m), quasi-electrostatic waves (P_Q^m), and intermediate waves (P_I^m) and evaluate all these quantities in the special cases that permit analytic description. We shall employ the previous partitioning (3.44) to define different parts of the spatial spectrum and the same approximations and restrictions as in § 3.5.

The power P_W^m converted to whistlers of range I in the case $k_0 d\, |g|^{1/2} \ll 1$ can be written

$$P_W^m = |I_0^m|^2 Z_0^{-1} \frac{\pi}{8} |g|^{1/2} G_W(\xi_S), \tag{3.132}$$

the notations of (3.102) being understood. For small and large radii b, we arrive, respectively, at formulas

$$P_W^m = |I_0^m|^2 Z_0^{-1} \frac{\pi}{48} (k_0 b)^4 |g|^{5/2} \alpha^m, \quad \alpha^m = 1.12 \quad \text{if} \quad k_0 b\, |g|^{1/2} \ll 1, \tag{3.133}$$

and

$$P_W^m \simeq |I_0^m|^2 Z_0^{-1} \frac{|g|^{1/2}}{4} \xi_S \left(1 + \frac{\cos 2\xi_S}{2\xi_S} \right) \quad \text{if} \quad k_0 b\, |g|^{1/2} \gg 1. \tag{3.134}$$

The power P_Q^m carried away by quasi-electrostatic whistler-mode waves of range II can be represented via integrals $\hat{J}_5$ and $\hat{J}_6$ of (3.126), when the additional condition $k_0 d\varepsilon^{1/2} \ll 1$ is satisfied. The result follows:

$$P_Q^m = |I_0^m|^2 Z_0^{-1} \frac{1}{2} k_0 b\, (\varepsilon\, |\eta|)^{1/2} \left\{ \left[(2 - \Lambda^2)\, K(\Lambda) - 2E(\Lambda) \right] \Lambda^{-1} \right.$$

$$\left. - \frac{\pi}{12} \xi_c^3 \, {}_3F_4 \left(\frac{3}{2}, 2, \frac{3}{2};\, 2, 2, 3;\, \frac{5}{2}, -\xi_c^2 \right) \right\}, \tag{3.135}$$

where the notations are like those in (3.126). Formula (3.135) may be simplified thus:

$$P_Q^m = |I_0^m|^2 Z_0^{-1} \frac{1}{2} k_0 b\, (\varepsilon\, |\eta|)^{1/2} \left[\ln \left(\frac{4b\, |\eta|^{1/2}}{d\varepsilon^{1/2}} \right) - 2 \right] \quad \text{if} \quad k_0 b|\eta|^{1/2} \ll 1, \tag{3.136}$$

and

$$P_Q^m \simeq |I_0^m|^2 Z_0^{-1} \frac{1}{2} k_0 b\, (\varepsilon\, |\eta|)^{1/2} \left[\ln \left(\frac{4}{k_0 d\varepsilon^{1/2}} \right) - 2 \right] \quad \text{if} \quad k_0 b|\eta|^{1/2} \gg 1. \tag{3.137}$$

In obtaining (3.136) and (3.137) we used the approximations given by (3.127). Formulas (3.136) and (3.137) clearly attest to the above-mentioned divergence of the radiation power integral for $d \to 0$. Note that this divergence has the same logarithmic character as in the case of a nonuniform ring electric current.

For the power P_I^m that goes to intermediate whistler-mode waves, the algebra in the general case is too complicated to give a useful result. While a source of small electrical size $(k_0 d \, |g|^{1/2} \ll 1, \ k_0 b \, |\eta|^{1/2} \ll 1)$ still permits a simple analytic description

$$P_I^m \simeq |I_0^m|^2 Z_0^{-1} \frac{\pi}{24} (k_0 b)^4 \, \varepsilon^{1/2} \, \eta^2, \tag{3.138}$$

for a source of large electrical size $(k_0 b \, |g|^{1/2} \gg 1, \ k_0 d \, |g|^{1/2} \ll 1)$ we can indicate only a limiting inequality

$$P_I^m \leq |I_0^m|^2 Z_0^{-1} \frac{1}{2} k_0 b |g| \ln \left(|\eta|^{1/2} / |g|^{1/2} \right). \tag{3.139}$$

It follows from the formulas obtained that for a ring magnetic current of small size $(k_0 d \, |g|^{1/2} \ll k_0 b \, |\eta|^{1/2} \ll 1)$, almost all of the radiation power goes to quasi-electrostatic waves

$$P_W^m \ll P_I^m \ll P_Q^m \simeq P_\Sigma^m. \tag{3.140}$$

For a radiator of large size $(k_0 b \, |g|^{1/2} \gg 1, \ k_0 d \, |g|^{1/2} \ll 1)$, in view of the relation

$$P_Q^m - P_I^m > |I_0^m|^2 Z_0^{-1} \frac{1}{2} k_0 b |g| \left[\ln \left(\frac{4}{k_0 d \, |g|^{1/2}} \right) - 2 \right], \tag{3.141}$$

which follows from (3.137) and (3.139), the character of inequalities (3.140) is preserved:

$$P_W^m < P_I^m < P_Q^m, \tag{3.142}$$

though they become less rigid. This means that with an increase in radiator radius b, despite an increase in the relative part of energy radiated in whistlers of ranges I and III, most of the radiation power goes to quasi-electrostatic whistler-mode waves. This conclusion is confirmed by the numerical calculations of Zaboronkova *et al.* (1992 b). Thus, the character of the redistribution of the radiated power as a function of radiator size for magnetic current (3.130) differs radically from the case of a ring electric current. This is due to differences in the behavior of the electric and magnetic components of the fields excited by ring electric and magnetic currents and can be traced directly from general formulas of § 3.2.

It can be shown that allowance for the nonuniformity of the magnetic current along the ring is not so important as it is for the electric current and does not result in a considerable increase in either the total radiation power or the relative share of power that goes to quasi-electrostatic waves (Zaboronkova *et al.*, 1992 b). At the same time, current nonuniformity can, under certain conditions, contribute to an appreciable increase in the power that radiates in the whistler-mode waves belonging to ranges I and III.

Summarizing, the most important conclusion to be drawn from the examination is that in frequency band (3.45) practically all of the power radiated by ring magnetic currents (under the reasonable restriction $k_0 d \, |g|^{1/2} \ll 1$) goes to quasi-electrostatic whistler-mode waves, regardless of the radiator radius.

3.7. On the definition of the radiation pattern

The *radiation pattern* $\mathcal{D}(\theta, \phi)$ in a loss-free medium is defined as the dependence of the mean radiated power flux per unit solid angle on the direction (θ, ϕ) in the observation space. That is

$$\mathcal{D}(\theta, \phi) = \lim_{\Delta\Omega \to 0} \frac{1}{\Delta\Omega} \lim_{r \to \infty} \int_{\Delta\Omega} S_r(r, \theta, \phi)\, r^2 \, d\Omega, \qquad (3.143)$$

where S_r is the radial component of the Poynting vector $\mathbf{S}$ and $d\Omega = \sin\theta\, d\theta\, d\phi$ is the element of the solid angle. In the "ordinary" medium, which does not have resonant and caustic directions, formula (3.143) is obviously reduces to the usual form

$$\mathcal{D}(\theta, \phi) = \lim_{r \to \infty} S_r(r, \theta, \phi)\, r^2.$$

Under the above-mentioned complicating circumstances relevant to the resonant magnetoplasma, one should employ the general definition (3.143) to avoid any inaccuracies in describing the radiation pattern. When dealing with it, some difficulties arise which need special discussion. This will be made in the present section for the whistler band. To render our consideration more concrete, we shall use as the source a ring uniform current (3.94).

Take the whistler-band field representation in the far zone (see (3.62)) in the conventional form

$$\mathcal{F}_\ell(\mathbf{r}) = \sum_i \mathcal{F}_{\ell,i}(\mathbf{r}) = \sum_i \frac{F_\ell(q_i)}{k_0 r \sqrt{q_i\, p_x''(q_i) \sin\theta \cos\theta}}$$

$$\times \quad \exp\left[-ik_0 q_i \rho - ik_0\, p_x(q_i)\, |z| + i\frac{\pi n}{2}\right],$$

where the 'i' summation covers all saddle points which are roots of Equation (3.64), and the quantities $F_\ell = F_\ell^{(x)}$ are obtainable from expressions (3.96). Substituting for $\mathcal{F}_\ell(\mathbf{r})$ into (3.143) gives

$$\mathcal{D}(\theta, \phi) \equiv \mathcal{D}(\theta) = \sum_i \mathcal{D}_i(\theta) = \sum_i \frac{\mathrm{Re}\left[F_1(q_i)\, F_5^*(q_i) - F_2(q_i)\, F_4^*(q_i)\right]}{2k_0^2 \sin\theta \cos^2\theta\, |q_i\, p_x''(q_i)|}, \qquad (3.144)$$

where a star denotes a complex conjugate. In obtaining (3.144) we made use of the fact that the field radiated from the chosen source is independent of the azimuth about $\mathbf{B}_0$. With the help of formulas (3.96) for quantities F_ℓ, the numerator in (3.144) can be written

$$\mathrm{Re}\left[F_1(q_i)\, F_5^*(q_i) - F_2(q_i)\, F_4^*(q_i)\right] = |I_0^e|^2 Z_0\, \frac{1}{4}\left(k_0^2 b\right)^2 \frac{g^2}{(-\eta)}$$

$$\times \frac{q_i^2 - \eta}{q_i^2 + p_x^2(q_i) - \varepsilon} \frac{q_i^2 \, J_1^2(k_0 b q_i)}{p_x(q_i) \left[q_i^4 \left(1 - \frac{\varepsilon}{\eta} \right)^2 - 4q_i^2 \frac{g^2}{\eta} + 4g^2 \right]}$$

$$\times \left[q_i^2 + p_x^2(q_i) - \varepsilon - \frac{g^2}{\eta} \frac{q_i^2 - \eta}{q_i^2 + p_x^2(q_i) - \varepsilon} \right]. \tag{3.145}$$

Making use of the fact that

$$q^2 + p_x^2(q) - \varepsilon - \frac{g^2}{\eta} \frac{q^2 - \eta}{q^2 + p_x^2(q) - \varepsilon} = \left[q^4 \left(1 - \frac{\varepsilon}{\eta} \right)^2 - 4q^2 \frac{g^2}{\eta} + 4g^2 \right]^{1/2}$$

and substituting (3.145) in (3.144), we get

$$\mathcal{D}(\theta) = \sum_i \mathcal{D}_i(\theta) = \sum_i |I_0^e|^2 Z_0 \frac{(k_0 b)^2}{8 \sin \theta \cos^2 \theta} \frac{g^2}{(-\eta)} \frac{q_i^2 - \eta}{q_i^2 + p_x^2(q_i) - \varepsilon}$$

$$\times \left| \frac{q_i}{p_x''(q_i)} \right| \frac{J_1^2(k_0 b q_i)}{p_x(q_i) \left[q_i^4 \left(1 - \frac{\varepsilon}{\eta} \right)^2 - 4q_i^2 \frac{g^2}{\eta} + 4g^2 \right]^{1/2}}. \tag{3.146}$$

This is the resulting expression for the whistler-band radiation pattern of the source with the current density (3.94).

It is interesting to note that the radiation pattern $\mathcal{D}$ involves a sum of individual propagating ray contributions but does not contain the interference terms. Although such terms appear in the expression for S_r (Arbel and Felsen, 1963; GiaRusso and Bergeson, 1970; Al'pert et $al.$, 1983), they vanish after performing integration and making passage to the limits in accordance with (3.143). This fact indicates the presence of the power orthogonality for propagating waves.

Since our main concern is to investigate the distinctive features of the radiation pattern under circumstances where resonance and caustic directions exist, we will consider its behavior in detail for the frequency range (3.45). Recall that at the chosen frequencies the radiation is confined within the Storey cone of half angle $\theta = \theta_S$ (see § 2.6 and Figure 2.2 therein). Note that we consider the behavior of the radiation pattern only in half-space $0 \le \theta \le \pi/2$ because of symmetry of $\mathcal{D}(\theta)$ about a plane $\theta = \pi/2$. As explained earlier, in §§ 2.6 and 3.4, in the angle domain $0 < \theta < \theta_r$ three rays exist for a given direction (θ, ϕ), so that three saddle points q_i can be found, while in the domain $\theta_r \le \theta < \theta_S$ two rays exist, and hence two saddle points q_i appear. Location of the saddle points q_i was discussed in § 3.4.

The far-field asymptotic expression taken in the form (3.62) to obtain the radiation pattern corresponds to the usual geometrical optics approximation and cannot be applicable near the caustic directions $\theta = 0$ and $\theta = \theta_S$. A more accurate evaluation of the field shows that the far field drops off as $r^{-5/6}$ along the simple caustic $\theta = \theta_S$ and as $r^{-1/2}$ along the axial caustic $\theta = 0$ instead of the conventional r^{-1} variation (Arbel and Felsen, 1963; Brodskii *et al.*, 1969; Wang and Bell, 1969 b). From a mathematical viewpoint, when $\theta \to \theta_S$, and two rays coalesce, the saddle points q_1 and q_2 approach one another in the vicinity of the inflection point $q = q_S$ (see Figure 2.2). In this case an isolated first-order saddle point approximation fails to represent the field, and an asymptotic approximation should be derived via a double saddle point procedure. This gives the uniform asymptotic expansion for the far field near the simple caustic at $\theta = \theta_S$ in terms of the Airy integral function $\mathrm{Ai}\,(\xi)$ and its first derivative $\mathrm{Ai}'(\xi)$ (Felsen and Marcuvitz, 1973). In the vicinity of the axial caustic at $\theta = 0$, the far-field uniform asymptotic expansion derived by Moiseyev (1985) may be used (see also Al'pert *et al.* (1983)). A close examination of both expansions shows that in the limiting case $r \to \infty$ the angular width of each near-caustic region tends eventually to zero, so that the geometrical optics approximation (i.e. the approximation of isolated first-order saddle points) may nevertheless be used to describe the radiation pattern near the caustic directions. Proof of this may be found, for example, in the work of Zaboronkova *et al.* (1993 a).

Now let us discuss the behavior of the radiation pattern in the vicinity of the caustic directions in some detail. As $\theta \to 0$, the formula for the radiation pattern immediately gives

$$\lim_{\theta \to 0} \mathcal{D}(\theta) = \mathcal{D}_c / \sin \theta, \tag{3.147}$$

where $\mathcal{D}_c$ is a certain constant; for $\theta \to \theta_S \to 0$, one has

$$\lim_{\theta \to \theta_S - 0} \mathcal{D}(\theta) = \mathcal{D}_S (\theta_S - \theta)^{-1/2}, \tag{3.148}$$

with $\mathcal{D}_S$ being a constant (Kondrat'ev *et al.*, 1992). In obtaining (3.148) we made use of the fact that $p_x''(q_S) = 0$ and that, for $\theta \to \theta_S - 0$,

$$\left| p_x''(q_{1,2}) \right| \approx \left[2 p'''(q_S) \left(\tan \theta_S - \tan \theta \right) \right]^{1/2}. \tag{3.149}$$

It follows from (3.147) and (3.148) that the radiation pattern tends to infinity as θ^{-1} when $\theta \to 0$, and as $(\theta_S - \theta)^{-1/2}$ when $\theta \to \theta_S - 0$. It is easily verified that both of these singularities of the radiation pattern are integrable, so that the power radiated into any given solid angle $\Delta \Omega$ does take the finite value. It is to be stressed that the near-caustic singularities of the radiation pattern depend only on the properties of the whistler-mode refractive index surface. Therefore, the behavior of these singularities described by (3.147) and (3.148) would remain in the case of an arbitrary source. Exception to this can only occur if the source does not excite the waves with wavenormals at $\vartheta = \vartheta_c$ or at $\vartheta = \vartheta_S$; in these cases the singularities respectively at $\theta = 0$ or $\theta = \theta_S$ would disappear. Note that for a ring current (3.94) such a situation occurs if $k_0 b q_c = \mu_m^{(1)}$ or $k_0 b q_S = \mu_m^{(1)}$, where $\mu_m^{(1)}$ is the mth zero of the first-order Bessel function $J_1(\zeta)$. We also note that formula (3.146) correctly describes the behavior of the radiation pattern in the vicinity of the resonance direction $\theta = \theta_r$. It is easy to verify that in the limiting case $\theta \to \theta_r - 0$ the radiation pattern $\mathcal{D}(\theta)$ defined by (3.146) remains finite. Its fraction $\mathcal{D}_3(\theta)$ has an oscillatory character, the frequency of oscillations increasing up to infinity as θ tends to the resonance-direction angle $\theta = \theta_r$. It can be demonstrated that although the details of the radiation pattern near the resonance direction depend strongly on the configuration of the source function, the radiation pattern is always regular and integrable there when reasonable descriptions of the source functions are used, which provide the boundedness of the total radiation power.

As an example Figure 3.4 shows the plot of the normalized radiation pattern $\mathcal{D}(\theta) / |I_0^e|^2$ of a ring current (3.94) with radius $b = 10\,\mathrm{m}$ for the following values of parameters: $\omega = 1.88 \times 10^5\,\mathrm{s}^{-1}$, $\omega_{\mathrm{H}} = 8.78 \times 10^6\,\mathrm{s}^{-1}$, and $\omega_{\mathrm{p}} = 5.64 \times 10^7\,\mathrm{s}^{-1}$ (curve 1). Curves 2 and 3 show the behavior of auxiliary functions $\tilde{\mathcal{D}}_c(\theta) = 2\pi \mathcal{D}(\theta) \sin\theta$ and $\tilde{\mathcal{D}}_S(\theta) = \mathcal{D}(\theta) (\theta_S - \theta)^{1/2} \sin\theta$, respectively. It is evident that $\tilde{\mathcal{D}}_c(0) = 2\pi \mathcal{D}_c$ and $\tilde{\mathcal{D}}_S(\theta_S) = \mathcal{D}_S \sin\theta_S$. Notice that the radiation pattern becomes unbounded at $\theta = 0$ and $\theta = \theta_S$, although the radial component S_r of the Poynting vector at any finite distance r is equal to zero when $\theta = 0$, and remains finite when $\theta = \theta_S$. This fact may be explained by analyzing the behavior of $S_r\, r^2$ near the caustic directions. As is seen from Figure 3.5, showing this quantity in the vicinity of both caustic directions for various distances r, as r increases, the angle position of the near-caustic maximum of $S_r r^2$ shifts to the caustic direction and accordingly increases in value. Ultimately, this leads to singularities in the radiation pattern, as expressed by (3.147) and (3.148). Some authors (see, for example, Wang and Bell, 1972 a) suggested defining the radiation pattern in the vicinity of the caustic directions through the near-caustic section of the quantity $S_r r^2$. However, in the near-caustic regions the angular behavior of $S_r r^2$ depends essentially on distance r and does not coincide with the radiated power flux per unit solid angle. We find it more relevant to use the usual integral definition (3.143) everywhere, with isolation of the character of singularity in $\mathcal{D}(\theta, \phi)$ along the caustic directions.

Finally, note that the features of the radiation pattern listed above will be encountered when we study the angular distribution of power radiated from a density duct into the surrounding medium.

To summarize: In this chapter we have discussed a procedure of finding the full-wave solutions to Maxwell's equations with source terms, in a cold uniform magnetoplasma. Integral representations of the source-excited fields have been de-

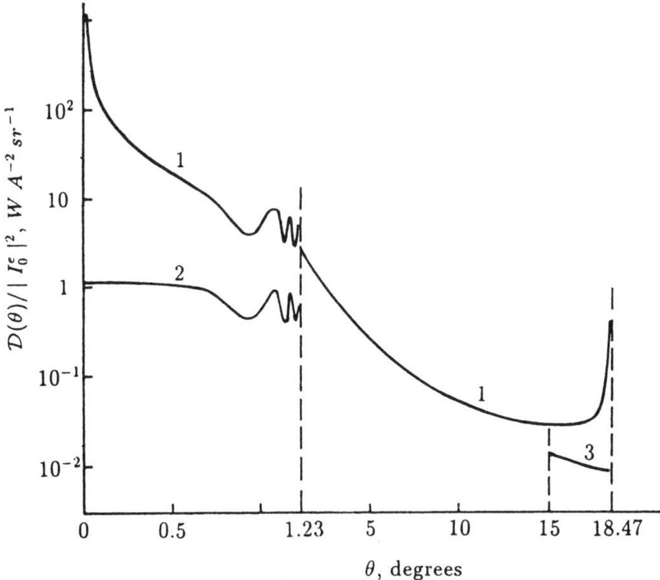

Figure 3.4. Normalized radiation pattern $\mathcal{D}$ (curve 1) and auxiliary functions $\tilde{\mathcal{D}}_c$ (curve 2) and $\tilde{\mathcal{D}}_S$ (curve 3) for the uniform ring electric current of radius $b = 10\,\mathrm{m}$, for $\omega = 1.88 \times 10^5\,\mathrm{s}^{-1}$, $\omega_\mathrm{H} = 8.78 \times 10^6\,\mathrm{s}^{-1}$, and $\omega_\mathrm{p} = 5.64 \times 10^7\,\mathrm{s}^{-1}$. Note that different scales are used for the ranges $\theta < \theta_r$ and $\theta_r < \theta < \theta_S$ ($\theta_r = 1.23°$ and $\theta_S = 18.47°$).

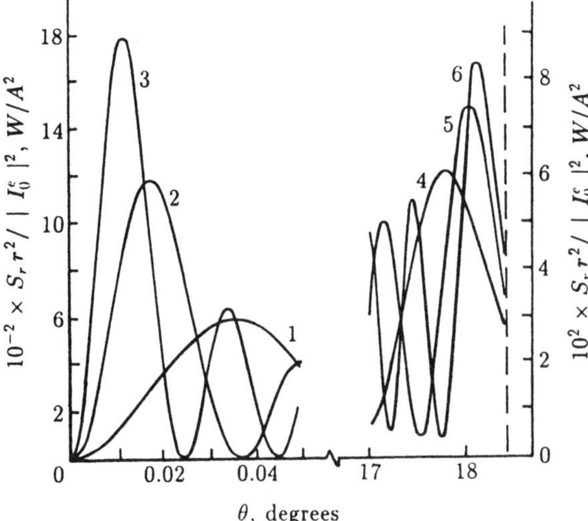

Figure 3.5. Angular dependence of $S_r r^2$ for the uniform ring electric current near the caustic directions at $k_0 r = 10$ (curves 1 and 4), $k_0 r = 20$ (curves 2 and 5) and $k_0 r = 30$ (curves 3 and 6). The values of b, ω, ω_H and ω_p are the same as in Figure 3.4.

rived for the simplest given current distributions. It has been discussed how the integral representations may be asymptotically evaluated for far field calculations.

Also, radiation losses of linear and ring sources have been considered in some detail. It was pointed out, when examining the usefulness of a particular source for performing any practical tasks, it is insufficient to have only the total radiation power and a knowledge of the distribution of power over the spatial spectrum of excited waves is very desirable. The definition and some properties of the radiation pattern have also been studied.

It is to be stressed that although we are able to determine the radiation characteristics when a distribution of currents is specified, the problem of determining the actual current distribution in a given radiator when it is excited by a voltage source at its input terminals still remains. There is much to be done to obtain a satisfactory solution of this problem for a resonant magnetoplasma, in which the refractive index surface of one of the characteristic modes is open. In this chapter we, however, restricted ourselves to given, but reasonable, current distributions in the conditions where they are known to be applicable.

Problems

3.1. Prove the identity given by (3.17).

Hint: Use the quartic equation (2.62) together with the formula (2.65).

3.2. Consider a dipole source possessing an electric current on a cylindrical surface of finite radius, oriented at an arbitrary angle with respect to the static magnetic field. Let the current density vary linearly from a maximum at the center to zero at both ends of the source. For such a current distribution, derive the integral representations for the field components and the total radiation power. Discuss the radiation power distribution over the spatial spectrum of excited waves in the frequency interval (3.45) and compare the results for an arbitrary dipole orientation with those obtained in § 3.4 for particular orientations parallel to and perpendicular to the static magnetic field. Show that when the source makes the angle $\theta_r = \arctan\left(-\varepsilon/\eta\right)^{1/2}$ with the static magnetic-field direction, the partial power going to quasi-electrostatic waves reaches a maximum value.

3.3. Consider a ring carrying a uniform electric current, oriented at an arbitrary angle with respect to the static magnetic field. Derive the integral representations for the field components and the total radiation power of such a source. Elucidate how the radiation power distribution over the spatial spectrum of excited waves and the total radiated power in the frequency interval (3.45) depend on the source orientation.

3.4. It is explained in § 3.4 for the frequency interval (3.45) that when a linear dipole source possessing a current density $\mathbf{J}^e$ is short enough, the major portion of its radiation power goes to quasi-electrostatic waves. Show that in this case the electric field can be found approximately by using the quasi-static theory. Note that within this theory the electric field is expressible via a scalar potential ψ, $\mathbf{E} = -\nabla\psi$, and the equation for ψ is

$$\varepsilon\nabla^2_\perp\psi + \eta\frac{\partial^2\psi}{\partial z^2} = -\frac{\rho^e}{\epsilon_0},$$

where $\nabla^2_\perp = \nabla^2 - \partial^2/\partial z^2$ and the charge density ρ^e is found from the equation of continuity, yielding

$$\rho^e = i\omega^{-1}\nabla\cdot\mathbf{J}^e.$$

Assuming the current density to be specified by (3.47), apply spatial Fourier transforms to solve the equation for ψ. Determine the electric field $\mathbf{E}$ and then find the radiation power for this source by the use of the general formula (3.34).

3.5. Within the scope of the quasi-static theory, find the electric field excited in the frequency interval (3.45) by a ring with the current density $\mathbf{J}^e$ given by (3.94). Noting for this case that $\rho^e = 0$, show that the electric field must now be expressed in terms of a scalar potential ψ and a vector potential $\mathbf{A}$, thus:

$$\mathbf{E} = -\nabla\psi - i\omega\mathbf{A}.$$

Prove that the equation for the scalar potential ψ is then

$$\varepsilon\nabla^2_\perp\psi + \eta\frac{\partial^2\psi}{\partial z^2} = -i\omega\nabla\cdot\hat{\varepsilon}\cdot\mathbf{A},$$

where the vector potential $\mathbf{A}$ can be found from the magnetostatic equation. Using the relationship $\mathbf{B} = \nabla\times\mathbf{A}$ and the Coulomb gauge condition $\nabla\cdot\mathbf{A} = 0$, verify that the equation for $\mathbf{A}$ may be written

$$\nabla^2\mathbf{A} = -\mu_0\mathbf{J}^e.$$

3.6. Derive the integral representations for the radiation power of sources with the current densities given by (3.47), (3.79) and (3.94) for the frequency interval (2.70). Assuming the sources to be small enough, obtain the closed-form expressions for the radiation power.

3.7. Find an expression for the radiation pattern $\mathcal{D}(\theta)$ of a ring source with the current density (3.94) for the frequency interval (2.70). Verify that the radiation pattern behaves as $(\theta_{S1} - \theta)^{-1/2}$ when $\theta \to \theta_{S1} - 0$, and as $(\theta - \theta_{S2})^{-1/2}$ when $\theta \to \theta_{S2} + 0$. (Angles θ_{S1} and θ_{S2} are defined in § 2.6 and presented in Figure 2.1.) Show that at angles $\theta = 0$ and $\theta = \pi/2$ the radiation pattern becomes

$$\mathcal{D}(0) = 0,$$

$$\mathcal{D}\left(\frac{\pi}{2}\right) = |I_0^e|^2\, Z_0\,\frac{(k_0 b)^2}{8}\,\frac{\left(\varepsilon\eta + g^2 - \varepsilon^2\right)\left[\left(g^2 - \varepsilon^2\right)|\varepsilon|\right]^{1/2} J_1^2\left(k_0 b\sqrt{(g^2 - \varepsilon^2)/|\varepsilon|}\right)}{\left\{(g^2 - \varepsilon^2)\left[g^2(\eta + \varepsilon)^2 - \varepsilon^2(\eta - \varepsilon)^2\right] + 4\varepsilon^2 g^2\eta^2\right\}^{1/2}}.$$

Compare these results with those obtained in § 3.7 for the frequency interval (3.45).

Chapter 4

Modes in Axially Uniform Ducts

4.1. Introduction

In this chapter we discuss the propagation of electromagnetic waves as waveguide modes along axially uniform ducts. One of the most important applications of this branch of the electromagnetic theory is the propagation of very low frequency waves, known as whistlers, in magnetospheric ducts. Another application is the guidance of electromagnetic waves along artificial ducts. These are formed in active ionospheric experiments as well as laboratory experiments conducted in various plasma devices. The "mode theory" and the full wave approach for electromagnetic waves in ducts received much careful study, and there are many accounts of them (see, for example, Adachi, 1965, 1966; Scarabucci and Smith, 1971; Walker, 1971, 1972; Laird and Nunn, 1975; Washimi, 1976; Karpman and Kaufman, 1982 a, 1982 b, 1984; Laird, 1992; Zaboronkova *et al.*, 1992 a, 1993 b). In the present chapter, no attempt is made to give all the material relating to the topics discussed in the above references. Instead, the essential basic theory is described and some important cases encountered in actual practice are given.

An important distinctive feature of the ducts is that they all are *open* guiding systems. For any open system capable of sustaining waveguide modes, the field propagating away from a given finite source consists of two portions. One portion travels without attenuation along an ideal, i.e. nonabsorbing, duct and can be expressed as a sum of the fields in individual modes analogous to the *bound* modes of ordinary dielectric waveguides. The remaining portion radiates from the duct.

Among the waves radiating outwards, there may be some waves which travel comparatively long distances along the duct prior to their eventual escape, or leakage, to the surrounding medium. This portion of the radiation field may also be represented in the modal form if one introduces (as for dielectric waveguides) a class of modes called the *leaky* modes. Both the bound- and leaky-mode fields are solutions of the source-free Maxwell equations and satisfy appropriate boundary conditions, but behave differently at infinity. In this chapter we consider what modes are possible in density ducts at the frequencies lying in the whistler band.

We leave aside the question of how such modes are set up by a given source. This problem is discussed in Chapters 5 and 6. Much of the present chapter deals with collisionless, i. e. loss-free, cylindrical ducts of circular cross-section.

A representative duct is illustrated in Figure 4.1 (a). The duct is here taken to be an infinitely long cylinder of radius a. The duct is uniform along its axis which is parallel to the external magnetic field $\mathbf{B}_0$. As in Chapters 2 and 3, the cylindrical coordinate system (ρ, ϕ, z) is used, with the z-axis chosen to be in the direction of $\mathbf{B}_0$. The plasma density is a function, $N(\rho)$, only of distance ρ from the z-axis in a core $\rho < a$. It is a constant, N_a, in an outer region $\rho > a$. The maximum value of the density in the core will be denoted by N_m. One of the possible density profiles is shown in Figure 4.1 (b).

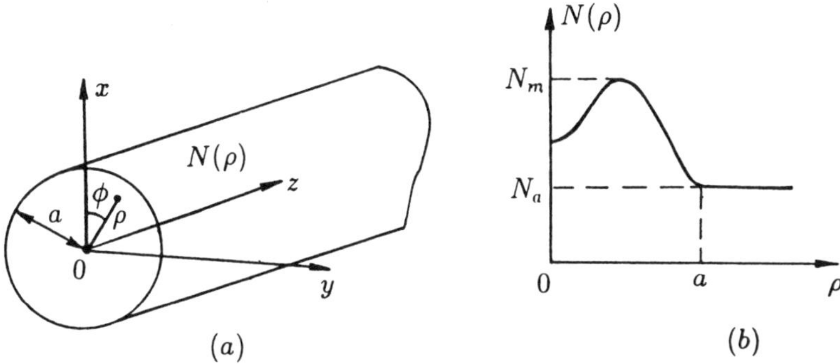

Figure 4.1. (a) Nomenclature and coordinates for describing cylindrical ducts, and (b) a representative density profile $N(\rho)$ which varies in the core and is uniform in the outer region.

For the plasma with density $N(\rho)$, the previous definitions (2.19) will be used to denote the dielectric tensor $\hat{\varepsilon}(\rho)$ and its elements $\varepsilon(\rho)$, $g(\rho)$, and $\eta(\rho)$. For the uniform ambient plasma in the outer region, we henceforth define the dielectric tensor $\hat{\varepsilon}_a$ with its elements ε_a, g_a, and η_a. In the outer region, where $N(\rho) = N_a$, one may evidently write $\hat{\varepsilon}(\rho) = \hat{\varepsilon}_a$.

4.2. The basic equations for modes of ducts

4.2.1. The field equations

A thorough discussion of the field equations for a general cylindrical geometry is given by Allis *et al.* (1963) and Felsen and Marcuvitz (1973). The treatment given here is restricted to what is necessary for the problem at hand.

It is clear from the symmetry of our problem (see Figure 4.1) that complex amplitudes of the field components in the duct whose parameters are allowed to be varying only with ρ, can depend on the azimuthal angle ϕ and the longitudinal coordinate z through the factor $\exp(-im\phi - ihz)$ where m is an integer and h is a propagation constant. In cylindrical coordinates, the field may therefore be written

$$\mathbf{E} = \left[\hat{\boldsymbol{\rho}}_0 \, E_\rho(\rho) + \hat{\boldsymbol{\phi}}_0 \, E_\phi(\rho) + \hat{\mathbf{z}}_0 \, E_z(\rho)\right] \exp(-im\phi - ik_0 pz),$$

$$\mathcal{H} = \left[\hat{\boldsymbol{\rho}}_0 \, \mathcal{H}_\rho(\rho) + \hat{\boldsymbol{\phi}}_0 \, \mathcal{H}_\phi(\rho) + \hat{\mathbf{z}}_0 \, \mathcal{H}_z(\rho)\right] \exp(-im\phi - ik_0 pz), \qquad (4.1)$$

where $p = h/k_0$ is the normalized (dimensionless) propagation constant, called also the axial wavenumber, and the time factor $\exp(i\omega t)$ is omitted. If this is used, in the Maxwell equations (2.38) to (2.43) we may write symbolically: $\partial/\partial\phi = -im$ and $\partial/\partial z = -ik_0 p$. Then the terms which remain are functions of ρ only and the partial differentiation sign $\partial/\partial\rho$ may be replaced by the total differentiation sign $\mathrm{d}/\mathrm{d}\rho$. Then Equations (2.38) to (2.43) in a source-free region become

$$\frac{m}{k_0\rho} E_z - pE_\phi = \mathcal{H}_\rho, \qquad (4.2)$$

$$pE_\rho - \frac{i}{k_0}\frac{\mathrm{d}E_z}{\mathrm{d}\rho} = \mathcal{H}_\phi, \qquad (4.3)$$

$$\frac{i}{k_0\rho}\frac{\mathrm{d}}{\mathrm{d}\rho}(\rho E_\phi) - \frac{m}{k_0\rho} E_\rho = \mathcal{H}_z, \qquad (4.4)$$

$$-\frac{m}{k_0\rho}\mathcal{H}_z + p\mathcal{H}_\phi = \varepsilon E_\rho - ig E_\phi, \qquad (4.5)$$

$$-p\mathcal{H}_\rho + \frac{i}{k_0}\frac{\mathrm{d}\mathcal{H}_z}{\mathrm{d}\rho} = ig E_\rho + \varepsilon E_\phi, \qquad (4.6)$$

$$-\frac{i}{k_0\rho}\frac{\mathrm{d}}{\mathrm{d}\rho}(\rho\mathcal{H}_\phi) + \frac{m}{k_0\rho}\mathcal{H}_\rho = \eta E_z. \qquad (4.7)$$

Equations (4.2), (4.3), (4.5), and (4.6) may be written more concisely in matrix notation. In the matrix form they become

$$\begin{pmatrix} \varepsilon & -ig & 0 & -p \\ ig & \varepsilon & p & 0 \\ 0 & p & 1 & 0 \\ p & 0 & 0 & -1 \end{pmatrix} \begin{pmatrix} E_\rho \\ E_\phi \\ \mathcal{H}_\rho \\ \mathcal{H}_\phi \end{pmatrix} = \begin{pmatrix} -\dfrac{m}{k_0\rho}\mathcal{H}_z \\[2ex] \dfrac{i}{k_0}\dfrac{\mathrm{d}\mathcal{H}_z}{\mathrm{d}\rho} \\[2ex] \dfrac{m}{k_0\rho}E_z \\[2ex] \dfrac{i}{k_0}\dfrac{\mathrm{d}E_z}{\mathrm{d}\rho} \end{pmatrix}. \qquad (4.8)$$

Let $\hat{W}$ denote the matrix in the left side of (4.8). The determinant of $\hat{W}$ is

$$\det \hat{W} = g^2 - \left(p^2 - \varepsilon\right)^2. \tag{4.9}$$

From the matrix equation (4.8) we can formally express the transverse components E_ρ, E_ϕ, $\mathcal{H}_\rho$, $\mathcal{H}_\phi$ via E_z and $\mathcal{H}_z$, which gives:

$$
\begin{aligned}
E_\rho \;=\; & \frac{1}{k_0 \left[g^2 - \left(p^2 - \varepsilon\right)^2\right]} \\
& \times \; \left\{ ipg \, \frac{m}{\rho} E_z + ip(\varepsilon - p^2) \frac{\mathrm{d}E_z}{\mathrm{d}\rho} + (\varepsilon - p^2) \frac{m}{\rho} \mathcal{H}_z + g \frac{\mathrm{d}\mathcal{H}_z}{\mathrm{d}\rho} \right\},
\end{aligned} \tag{4.10}
$$

$$
\begin{aligned}
E_\phi \;=\; & \frac{1}{k_0 \left[g^2 - \left(p^2 - \varepsilon\right)^2\right]} \\
& \times \; \left\{ p(\varepsilon - p^2) \frac{m}{\rho} E_z + pg \frac{\mathrm{d}E_z}{\mathrm{d}\rho} - ig \frac{m}{\rho} \mathcal{H}_z - i(\varepsilon - p^2) \frac{\mathrm{d}\mathcal{H}_z}{\mathrm{d}\rho} \right\},
\end{aligned} \tag{4.11}
$$

$$
\begin{aligned}
\mathcal{H}_\rho = & \frac{1}{k_0 \left[g^2 - \left(p^2 - \varepsilon\right)^2\right]} \\
& \times \left\{ \left[g^2 - \varepsilon\left(\varepsilon - p^2\right)\right] \frac{m}{\rho} E_z - p^2 g \frac{\mathrm{d}E_z}{\mathrm{d}\rho} + ipg \frac{m}{\rho} \mathcal{H}_z + ip\left(\varepsilon - p^2\right) \frac{\mathrm{d}\mathcal{H}_z}{\mathrm{d}\rho} \right\},
\end{aligned} \tag{4.12}
$$

$$
\begin{aligned}
\mathcal{H}_\phi = & \frac{1}{k_0 \left[g^2 - \left(p^2 - \varepsilon\right)^2\right]} \\
& \times \left\{ ip^2 g \frac{m}{\rho} E_z - i\left[g^2 - \varepsilon\left(\varepsilon - p^2\right)\right] \frac{\mathrm{d}E_z}{\mathrm{d}\rho} + p(\varepsilon - p^2) \frac{m}{\rho} \mathcal{H}_z + pg \frac{\mathrm{d}\mathcal{H}_z}{\mathrm{d}\rho} \right\}.
\end{aligned} \tag{4.13}
$$

If (4.10) and (4.11) are substituted into Equation (4.4), then:

$$
\begin{aligned}
\hat{L}_m \mathcal{H}_z \;+\; & k_0^2 \left(\frac{g^2}{p^2 - \varepsilon} - p^2 + \varepsilon \right) \mathcal{H}_z = \frac{ipg}{p^2 - \varepsilon} \hat{L}_m E_z \\
+\; & \left(\frac{g^2}{p^2 - \varepsilon} - p^2 + \varepsilon \right) \left\{ \left(ip \frac{\mathrm{d}E_z}{\mathrm{d}\rho} + \frac{m}{\rho} \mathcal{H}_z \right) \frac{\mathrm{d}}{\mathrm{d}\rho} \left[\frac{g}{g^2 - \left(p^2 - \varepsilon\right)^2} \right] \right. \\
-\; & \left. \left(ip \frac{m}{\rho} E_z + \frac{\mathrm{d}\mathcal{H}_z}{\mathrm{d}\rho} \right) \frac{\mathrm{d}}{\mathrm{d}\rho} \left[\frac{p^2 - \varepsilon}{g^2 - \left(p^2 - \varepsilon\right)^2} \right] \right\},
\end{aligned} \tag{4.14}
$$

where

$$\hat{L}_m = \frac{\mathrm{d}^2}{\mathrm{d}\rho^2} + \frac{1}{\rho} \frac{\mathrm{d}}{\mathrm{d}\rho} - \frac{m^2}{\rho^2}.$$

The substitution of (4.12) and (4.13) into Equation (4.7) gives:

$$\left[g^2 - \varepsilon(\varepsilon - p^2)\right]\hat{L}_m E_z + k_0^2\eta\left[g^2 - (p^2 - \varepsilon)^2\right]E_z = -\mathrm{i}pg\hat{L}_m\mathcal{H}_z$$

$$- \left[g^2 - (p^2 - \varepsilon)^2\right]\left\{\mathrm{i}p\left(\mathrm{i}p\,\frac{m}{\rho}\,E_z + \frac{\mathrm{d}\mathcal{H}_z}{\mathrm{d}\rho}\right)\frac{\mathrm{d}}{\mathrm{d}\rho}\left[\frac{g}{g^2 - (p^2 - \varepsilon)^2}\right]\right.$$

$$+ \frac{\mathrm{d}E_z}{\mathrm{d}\rho}\,\frac{\mathrm{d}}{\mathrm{d}\rho}\left[\frac{g^2 - \varepsilon(\varepsilon - p^2)}{g^2 - (p^2 - \varepsilon)^2}\right] + \mathrm{i}p\,\frac{m}{\rho}\,\mathcal{H}_z\,\frac{\mathrm{d}}{\mathrm{d}\rho}\left[\frac{\varepsilon - p^2}{g^2 - (p^2 - \varepsilon)^2}\right]\right\}. \qquad (4.15)$$

The coupled differential equations (4.14) and (4.15) are the starting point for much of the later work in this chapter. It will be clear from what follows that in the limiting case $p^2 \to \varepsilon \pm g$, when the determinant (4.9) tends to zero, the field components remain finite.

It is worth noting that there is a special case when the coupled equations (4.14) and (4.15) become independent. This occurs for axially symmetric modes (i. e. when $m = 0$) in a plasma with $g \to 0$ that corresponds to an infinite external magnetic field. In this limiting case, Equation (4.14) will govern the field of a mode of type TE in which the electron intensity $\mathbf{E}$ is everywhere transverse to the direction of propagation, whereas Equation (4.15) will govern the field of a mode of type TM in which the magnetic intensity $\mathcal{H}$ is everywhere transverse to that direction. Our main concern, however, involves the case when the plasma is essentially gyrotropic such that $g \neq 0$.

The second-order coupled equations (4.14) and (4.15) are very complicated for the general case. There is a case, however, where the field equations can be expressed in a simpler form. This is when the field is independent of the azimuth about $\mathbf{B}_0$. In that case, $m = 0$, and Equations (4.2) to (4.4) give

$$\mathcal{H}_\rho = -pE_\phi,$$

$$\mathcal{H}_\phi = pE_\rho - \frac{\mathrm{i}}{k_0}\,\frac{\mathrm{d}E_z}{\mathrm{d}\rho}, \qquad (4.16)$$

$$\mathcal{H}_z = \frac{\mathrm{i}}{k_0\rho}\,\frac{\mathrm{d}}{\mathrm{d}\rho}\left(\rho E_\phi\right).$$

Now $\mathcal{H}_\phi$ may be eliminated from Equation (4.5). Hence Equation (4.5) becomes

$$E_\rho = \frac{\mathrm{i}}{p^2 - \varepsilon}\left(\frac{p}{k_0}\,\frac{\mathrm{d}E_z}{\mathrm{d}\rho} - gE_\phi\right). \qquad (4.17)$$

If Equations (4.16) are substituted into (4.6) and (4.7), and if (4.17) is used to eliminate E_ρ, then

$$\frac{\mathrm{d}^2 E_\phi}{\mathrm{d}\rho^2} + \frac{1}{\rho}\,\frac{\mathrm{d}E_\phi}{\mathrm{d}\rho} - \frac{E_\phi}{\rho^2} + k_0^2\left(\frac{g^2}{p^2 - \varepsilon} - p^2 + \varepsilon\right)E_\phi = \frac{k_0pg}{p^2 - \varepsilon}\,\frac{\mathrm{d}E_z}{\mathrm{d}\rho}, \qquad (4.18)$$

$$\frac{\mathrm{d}^2 E_z}{\mathrm{d}\rho^2} + \frac{1}{\rho}\frac{\mathrm{d}E_z}{\mathrm{d}\rho} + \frac{p^2}{(p^2 - \varepsilon)\,\varepsilon}\frac{\mathrm{d}\varepsilon}{\mathrm{d}\rho}\frac{\mathrm{d}E_z}{\mathrm{d}\rho} + k_0^2\frac{\eta}{\varepsilon}\left(\varepsilon - p^2\right)E_z$$

$$= k_0\frac{p}{\varepsilon}\left(p^2 - \varepsilon\right)\frac{1}{\rho}\frac{\mathrm{d}}{\mathrm{d}\rho}\left(\frac{\rho g E_\phi}{p^2 - \varepsilon}\right). \tag{4.19}$$

These are entirely equivalent to Equations (4.14) and (4.15) when $m = 0$.

A real duct in the magnetosphere near the equatorial plane has an elongated oval cross-section with its broad dimension horizontal and its narrow dimension normal to the L-shell on which it lies (see § 1.1). The central portion of such an oval duct may be approximated by a plane stratified duct shown in Figure 4.2 (a). Therefore it is useful to give some general results for the case of a plane duct. Here Cartesian coordinates are used with the z-axis directed along the external magnetic field. The plasma density is assumed to be varying with x only, as shown in Figure 4.2 (b). The axis of a real duct is curved because of the curvature of the geomagnetic field line. The effect of this curvature is neglected in the type of duct discussed here. The conditions when this is possible are given, for example, in the work of Walker (1971).

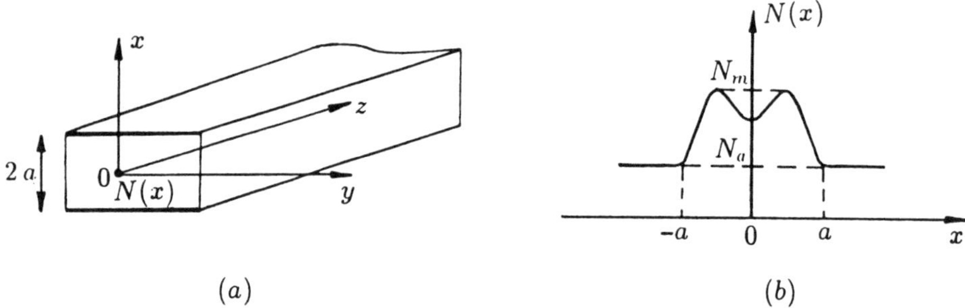

(a)   (b)

Figure 4.2. (a) Section of a plane duct, which is unbounded in the y- and z-directions, and (b) a representative density profile $N(x)$. The z-axis coincides with the duct axis midway between interfaces.

Consider a progressive wave with its propagation direction along the z-axis and its field independent of y. Then the field may be written

$$\mathbf{E} = \left[\hat{\mathbf{x}}_0 E_x(x) + \hat{\mathbf{y}}_0 E_y(x) + \hat{\mathbf{z}}_0 E_z(x)\right]\exp(-\mathrm{i}k_0 pz),$$

$$\mathcal{H} = \left[\hat{\mathbf{x}}_0 \mathcal{H}_x(x) + \hat{\mathbf{y}}_0 \mathcal{H}_y(x) + \hat{\mathbf{z}}_0 \mathcal{H}_z(x)\right]\exp(-\mathrm{i}k_0 pz). \tag{4.20}$$

If this is used, the Maxwell equations (2.32) to (2.37), in the absence of sources, give

$$\frac{\mathrm{d}^2 E_y}{\mathrm{d}x^2} + k_0^2\left(\frac{g^2}{p^2 - \varepsilon} - p^2 + \varepsilon\right)E_y = \frac{k_0 pg}{p^2 - \varepsilon}\frac{\mathrm{d}E_z}{\mathrm{d}x}, \tag{4.21}$$

$$\frac{\mathrm{d}^2 E_z}{\mathrm{d}x^2} + \frac{p^2}{(p^2 - \varepsilon)\varepsilon}\frac{\mathrm{d}\varepsilon}{\mathrm{d}x}\frac{\mathrm{d}E_z}{\mathrm{d}x} + k_0^2\frac{\eta}{\varepsilon}(\varepsilon - p^2)E_z = k_0\frac{p}{\varepsilon}(p^2 - \varepsilon)\frac{\mathrm{d}}{\mathrm{d}x}\left(\frac{g E_y}{p^2 - \varepsilon}\right), \tag{4.22}$$

and

$$E_x = \frac{\mathrm{i}}{p^2 - \varepsilon} \left(\frac{p}{k_0} \frac{\mathrm{d}E_z}{\mathrm{d}x} - g E_y \right), \tag{4.23}$$

$$\mathcal{H}_x = -p E_y, \qquad \mathcal{H}_y = p E_x - \frac{\mathrm{i}}{k_0} \frac{\mathrm{d}E_z}{\mathrm{d}x}, \qquad \mathcal{H}_z = \frac{\mathrm{i}}{k_0} \frac{\mathrm{d}E_y}{\mathrm{d}x}. \tag{4.24}$$

The algebra here is similar to that used in obtaining Equations (4.16) to (4.19) and the details need not be given.

4.2.2. The field solutions for a uniform plasma

The equations (4.14) and (4.15) are two linear differential equations of the second order, with two dependent variables E_z and $\mathcal{H}_z$. One of these could be eliminated to give a single linear differential equation of the fourth order. Such an equation can, in general, be solved only numerically if the density profile is prescribed. There is, however, a special case when the field solution may be found in the analytic form. This occurs if the plasma is uniform, so that the quantities ε, g, and η are constants independent of ρ. In this case, there is no need to form the fourth-order equation and it is more convenient to deal with the coupled equations (4.14) and (4.15). A method of solving these equations is to set the variables equal to $Z_m(\lambda\rho)$, where Z_m is the cylindrical function, say the Bessel function, of order m, and λ is a constant to be found. Clearly each of the variables E_z and $\mathcal{H}_z$ is then a multiple of $Z_m(\lambda\rho)$ for each solution, and

$$\hat{L}_m Z_m(\lambda\rho) = -\lambda^2 Z_m(\lambda\rho),$$

so that, symbolically, $\hat{L}_m \equiv -\lambda^2$. It is here convenient to take $\lambda = k_0 q$ that leads to $\hat{L}_m \equiv -(k_0 q)^2$. If this is substituted in (4.14) and (4.15) and the variables E_z and $\mathcal{H}_z$ are eliminated, we obtain the determinantal equation for q:

$$\begin{vmatrix} -q^2 + \dfrac{g^2}{p^2 - \varepsilon} - p^2 + \varepsilon & \dfrac{ipg}{p^2 - \varepsilon} q^2 \\[2ex] -ipg q^2 & -\left[g^2 - \varepsilon(\varepsilon - p^2) \right] q^2 + \eta \left[g^2 - (p^2 - \varepsilon)^2 \right] \end{vmatrix} = 0. \tag{4.25}$$

This is the Booker quartic, which may therefore be regarded as the characteristic equation of our differential equations when the plasma is uniform. Four roots of the quartic (see (2.64)) give four solutions. A typical solution is

$$E_z = \mathcal{A}_k Z_m(k_0 q_k \rho), \qquad \mathcal{H}_z = \mathcal{B}_k Z_m(k_0 q_k \rho), \tag{4.26}$$

where q_k is one root of the quartic. The quantities $\mathcal{A}_k$ and $\mathcal{B}_k$ are constants whose ratios $\mathcal{A}_k/\mathcal{B}_k$ can be found by substituting (4.26) in either (4.14) or (4.15). If this is used, then

$$\frac{\mathcal{A}_k}{\mathcal{B}_k} = \frac{\left(p^2 - \varepsilon\right)^2 + \left(p^2 - \varepsilon\right)q_k^2 - g^2}{ipgq_k^2}.$$

This, with the help of the Booker quartic (2.62), reduces to

$$\frac{\mathcal{A}_k}{\mathcal{B}_k} = -\frac{i}{\eta}\, n_k, \tag{4.27}$$

where

$$n_k = -\frac{\varepsilon}{pg}\left(p^2 + q_k^2 + \frac{g^2}{\varepsilon} - \varepsilon\right). \tag{4.28}$$

Hence the solutions (4.26) become

$$E_z = \frac{i}{\eta}\, A_k n_k q_k Z_m(k_0 q_k \rho),$$

$$\mathcal{H}_z = -A_k q_k Z_m(k_0 q_k \rho). \tag{4.29}$$

Here we have introduced new constants $A_k = \mathcal{A}_k \eta/(i n_k q_k)$ which are more convenient for the later work.

The transverse components E_ρ, E_ϕ, $\mathcal{H}_\rho$, $\mathcal{H}_\phi$ in each solution can then be expressed in terms of E_z and $\mathcal{H}_z$ by means of Equations (4.10) to (4.13). The resulting formulas are readily found to be

$$E_\rho = \frac{1}{2\eta} A_k n_k q_k^2 \left[\frac{p - \eta n_k^{-1}}{p^2 - \varepsilon - g} Z_{m+1}(k_0 q_k \rho) - \frac{p + \eta n_k^{-1}}{p^2 - \varepsilon + g} Z_{m-1}(k_0 q_k \rho) \right], \quad (4.30)$$

$$E_\phi = \frac{i}{2\eta} A_k n_k q_k^2 \left[\frac{p - \eta n_k^{-1}}{p^2 - \varepsilon - g} Z_{m+1}(k_0 q_k \rho) + \frac{p + \eta n_k^{-1}}{p^2 - \varepsilon + g} Z_{m-1}(k_0 q_k \rho) \right], \quad (4.31)$$

$$\mathcal{H}_\rho = \frac{i}{2} A_k q_k^2 \left[\frac{p - (\varepsilon + g)\,\eta^{-1} n_k}{p^2 - \varepsilon - g} Z_{m+1}(k_0 q_k \rho) \right.$$
$$\left. - \frac{p + (\varepsilon - g)\,\eta^{-1} n_k}{p^2 - \varepsilon + g} Z_{m-1}(k_0 q_k \rho) \right], \tag{4.32}$$

$$\mathcal{H}_\phi = -\frac{1}{2} A_k q_k^2 \left[\frac{p - (\varepsilon + g)\,\eta^{-1} n_k}{p^2 - \varepsilon - g} Z_{m+1}(k_0 q_k \rho) \right.$$
$$\left. + \frac{p + (\varepsilon - g)\,\eta^{-1} n_k}{p^2 - \varepsilon + g} Z_{m-1}(k_0 q_k \rho) \right]. \tag{4.33}$$

Although the denominators in these expressions tend to zero as $p^2 \to \varepsilon \pm g$, it can be verified that the field components are not singular at $p^2 - \varepsilon \pm g = 0$. In the whistler band (2.69), for a loss-free medium, the quantity $P_{\mathrm{x}} = (\varepsilon - g)^{1/2}$ is real, whereas the quantity $P_{\mathrm{o}} = (\varepsilon + g)^{1/2}$ is imaginary. For the later work, it is more convenient to rearrange the terms in (4.30) to (4.33) so that only the expression $p^2 - \varepsilon + g$ remains in the denominators. This can be made by applying the following formula valid for the cylindrical functions:

$$Z_{m-1}(\zeta) = \frac{2m}{\zeta} Z_m(\zeta) - Z_{m+1}(\zeta). \tag{4.34}$$

If (4.34) is inserted in (4.30) to (4.33) it gives the alternative form for E_ρ, E_ϕ, $\mathcal{H}_\rho$, $\mathcal{H}_\phi$:

$$E_\rho = -A_k \left[\frac{n_k p + g}{\varepsilon} Z_{m+1}(k_0 q_k \rho) + q_k \frac{n_k p + \eta}{\eta \left(p^2 - P_x^2\right)} \frac{m}{k_0 \rho} Z_m(k_0 q_k \rho) \right], \tag{4.35}$$

$$E_\phi = \mathrm{i} A_k \left[Z_{m+1}(k_0 q_k \rho) + q_k \frac{n_k p + \eta}{\eta \left(p^2 - P_x^2\right)} \frac{m}{k_0 \rho} Z_m(k_0 q_k \rho) \right], \tag{4.36}$$

$$\mathcal{H}_\rho = -\mathrm{i} A_k \left[p Z_{m+1}(k_0 q_k \rho) + q_k \frac{p\eta + n_k P_x^2}{\eta \left(p^2 - P_x^2\right)} \frac{m}{k_0 \rho} Z_m(k_0 q_k \rho) \right], \tag{4.37}$$

$$\mathcal{H}_\phi = -A_k \left[n_k Z_{m+1}(k_0 q_k \rho) + q_k \frac{p\eta + n_k P_x^2}{\eta \left(p^2 - P_x^2\right)} \frac{m}{k_0 \rho} Z_m(k_0 q_k \rho) \right]. \tag{4.38}$$

4.2.3. The field solutions for a uniform duct

Now we make an application of the preceding formulation by considering the modal fields in the uniform step-profile duct. Let the density profile be

$$N(\rho) = \begin{cases} \tilde{N} & \text{if } \rho < a, \\ N_a & \text{if } \rho > a, \end{cases} \tag{4.39}$$

where $\tilde{N}$ and N_a are constants and a is the duct radius. Separable solutions which are bounded everywhere must vary as $Z_m = J_m(k_0 \tilde{q}_k \rho)$ in the core, and as $Z_m = H_m^{(2)}(k_0 q_k \rho)$, with $\mathrm{Im}\,(q_k) < 0$, in the outer region, where J_m is the Bessel function of the first kind, $H_m^{(2)}$ is the Hankel function of the second kind, and

$$\tilde{q}_k^2 = q_k^2(p, \tilde{N}), \qquad q_k^2 \equiv -s_k^2 = q_k^2(p, N_a), \qquad q_k = \mathrm{i} e^{-\mathrm{i}\pi} s_k, \quad k = 1, 2. \tag{4.40}$$

The notation $q_k(p, N)$ means that this quantity is calculated in accordance with the formula (2.64), into which the tensor elements corresponding to the chosen density N are to be inserted.

Let $\tilde{\varepsilon}$, $\tilde{g}$, $\tilde{\eta}$ be components of the dielectric tensor for uniform plasma with the density $N = \tilde{N}$. Then, with the help of the particular solutions (4.29), and (4.35) to (4.38), we can construct the general solution for the duct core in the form:

$$E_\rho = -\sum_{k=1}^{2} B_k \left[\frac{\tilde{n}_k p + \tilde{g}}{\tilde{\varepsilon}} J_{m+1}(k_0 \tilde{q}_k \rho) + \tilde{\alpha}_k m \frac{J_m(k_0 \tilde{q}_k \rho)}{k_0 \tilde{q}_k \rho} \right], \tag{4.41}$$

$$E_\phi = \mathrm{i} \sum_{k=1}^{2} B_k \left[J_{m+1}(k_0 \tilde{q}_k \rho) + \tilde{\alpha}_k m \, \frac{J_m(k_0 \tilde{q}_k \rho)}{k_0 \tilde{q}_k \rho} \right], \qquad (4.42)$$

$$E_z = \frac{\mathrm{i}}{\tilde{\eta}} \sum_{k=1}^{2} B_k \, \tilde{n}_k \, \tilde{q}_k \, J_m(k_0 \tilde{q}_k \rho), \qquad (4.43)$$

$$\mathcal{H}_\rho = -\mathrm{i} \sum_{k=1}^{2} B_k \left[p J_{m+1}(k_0 \tilde{q}_k \rho) - \tilde{n}_k \, \tilde{\beta}_k m \, \frac{J_m(k_0 \tilde{q}_k \rho)}{k_0 \tilde{q}_k \rho} \right], \qquad (4.44)$$

$$\mathcal{H}_\phi = - \sum_{k=1}^{2} B_k \tilde{n}_k \left[J_{m+1}(k_0 \tilde{q}_k \rho) - \tilde{\beta}_k m \, \frac{J_m(k_0 \tilde{q}_k \rho)}{k_0 \tilde{q}_k \rho} \right], \qquad (4.45)$$

$$\mathcal{H}_z = - \sum_{k=1}^{2} B_k \tilde{q}_k \, J_m(k_0 \tilde{q}_k \rho), \qquad (4.46)$$

where

$$\tilde{\alpha}_{1,2} = \tilde{q}_{1,2}^2 \, \frac{1 - p \tilde{n}_{2,1}^{-1}}{p^2 - \tilde{\mathcal{P}}^2}, \qquad \tilde{\beta}_{1,2} = \tilde{q}_{1,2}^2 \, \frac{\tilde{n}_{2,1}^{-1} \tilde{\mathcal{P}}^2 - p}{\tilde{n}_{1,2}\,(p^2 - \tilde{\mathcal{P}}^2)},$$

$$\tilde{n}_{1,2} = - \frac{\tilde{\varepsilon}}{p \tilde{g}} \left(p^2 + \tilde{q}_{1,2}^2 + \frac{\tilde{g}^2}{\tilde{\varepsilon}} - \tilde{\varepsilon} \right), \qquad \tilde{\mathcal{P}}^2 = \tilde{\varepsilon} - \tilde{g}, \qquad (4.47)$$

and $B_{1,2}$ are constants. Note that in obtaining formulas (4.41) to (4.47) we made use of the fact that

$$\tilde{n}_1 \tilde{n}_2 = -\tilde{\eta}. \qquad (4.48)$$

This identity can be verified in a straightforward manner.

Before writing the field in the outer region, recall that we have adopted the notations ε_a, g_a, η_a for the tensor's elements in the uniform plasma that surrounds the core. Further, we find it more convenient to use the modified Bessel function of the second kind

$$K_m(k_0 s_k \rho) = \frac{\pi}{2} \mathrm{i}^{m+1} H_m^{(1)}(k_0 e^{\mathrm{i}\pi} q_k \rho) = -\frac{\pi}{2} \mathrm{i}^{m+1} e^{-\mathrm{i}m\pi} H_m^{(2)}(k_0 q_k \rho) \qquad (4.49)$$

rather than the Hankel functions $H_m^{(2)}(k_0 q_k \rho)$. Hence, the general solutions for the outer region become:

$$E_\rho = - \sum_{k=1}^{2} C_k \left[\frac{n_k p + g_a}{\varepsilon_a} K_{m+1}(k_0 s_k \rho) + \alpha_k m \, \frac{K_m(k_0 s_k \rho)}{k_0 s_k \rho} \right], \qquad (4.50)$$

$$E_\phi = \mathrm{i} \sum_{k=1}^{2} C_k \left[K_{m+1}(k_0 s_k \rho) + \alpha_k m \, \frac{K_m(k_0 s_k \rho)}{k_0 s_k \rho} \right], \qquad (4.51)$$

$$E_z = -\frac{i}{\eta_a} \sum_{k=1}^{2} C_k n_k s_k \, K_m(k_0 s_k \rho), \tag{4.52}$$

$$\mathcal{H}_\rho = -i \sum_{k=1}^{2} C_k \left[p K_{m+1}(k_0 s_k \rho) - n_k \beta_k m \, \frac{K_m(k_0 s_k \rho)}{k_0 s_k \rho} \right], \tag{4.53}$$

$$\mathcal{H}_\phi = -\sum_{k=1}^{2} C_k n_k \left[K_{m+1}(k_0 s_k \rho) - \beta_k m \, \frac{K_m(k_0 s_k \rho)}{k_0 s_k \rho} \right], \tag{4.54}$$

$$\mathcal{H}_z = \sum_{k=1}^{2} C_k s_k \, K_m(k_0 s_k \rho), \tag{4.55}$$

where

$$\alpha_{1,2} = -s_{1,2}^2 \, \frac{1 - p n_{2,1}^{-1}}{p^2 - \mathcal{P}^2} , \qquad \beta_{1,2} = -s_{1,2}^2 \, \frac{n_{2,1}^{-1} \mathcal{P}^2 - p}{n_{1,2} \left(p^2 - \mathcal{P}^2 \right)} ,$$

$$n_{1,2} = -\frac{\varepsilon_a}{p g_a} \left(p^2 - s_{1,2}^2 + \frac{g_a^2}{\varepsilon_a} - \varepsilon_a \right) , \qquad \mathcal{P}^2 = \varepsilon_a - g_a, \tag{4.56}$$

and $C_{1,2}$ are constants. In derivation of the above expressions we made use of the relation

$$n_1 \, n_2 = -\eta_a , \tag{4.57}$$

which is similar to (4.48).

In the case of axisymmetric modes, when $m = 0$, expressions (4.41) to (4.46), and (4.50) to (4.55) simplify to:

$\rho < a,$

$$E_\rho = -\sum_{k=1}^{2} B_k \, \frac{\tilde{n}_k p + \tilde{g}}{\tilde{\varepsilon}} \, J_1(k_0 \tilde{q}_k \rho),$$

$$E_\phi = i \sum_{k=1}^{2} B_k \, J_1(k_0 \tilde{q}_k \rho),$$

$$E_z = \frac{i}{\tilde{\eta}} \sum_{k=1}^{2} B_k \tilde{n}_k \tilde{q}_k \, J_0(k_0 \tilde{q}_k \rho),$$

$$\mathcal{H}_\rho = -p E_\phi = -ip \sum_{k=1}^{2} B_k \, J_1(k_0 \tilde{q}_k \rho), \tag{4.58}$$

$$\mathcal{H}_\phi = -\sum_{k=1}^{2} B_k \tilde{n}_k \, J_1(k_0 \tilde{q}_k \rho),$$

$$\mathcal{H}_z = -\sum_{k=1}^{2} B_k \tilde{q}_k \, J_0(k_0 \tilde{q}_k \rho);$$

$\rho > a$,

$$E_\rho = -\sum_{k=1}^{2} \dot{C}_k \, \frac{n_k p + g_a}{\varepsilon_a} \, K_1(k_0 s_k \rho),$$

$$E_\phi = \mathrm{i} \sum_{k=1}^{2} C_k \, K_1(k_0 s_k \rho),$$

$$E_z = -\frac{\mathrm{i}}{\eta_a} \sum_{k=1}^{2} C_k n_k s_k \, K_0(k_0 s_k \rho),$$

$$\mathcal{H}_\rho = -p E_\phi = -\mathrm{i} p \sum_{k=1}^{2} C_k \, K_1(k_0 s_k \rho), \qquad (4.59)$$

$$\mathcal{H}_\phi = -\sum_{k=1}^{2} C_k n_k \, K_1(k_0 s_k \rho),$$

$$\mathcal{H}_z = \sum_{k=1}^{2} C_k s_k \, K_0(k_0 s_k \rho).$$

It is seen from (4.58) and (4.59) that the axisymmetric modes of ducts are hybrid, unlike those of dielectric waveguides, and cannot be decomposed into separable TE and TM modes. We mentioned earlier that the gyrotropic properties of the magnetized plasma are responsible for the hybrid character of axisymmetric modes. The effect of these properties is built into the field equations (see, for example, Equations (4.14) and (4.15)) through the g term, which is the off-diagonal element of the dielectric tensor.

It is also useful to discuss the waveguide solutions of Equations (4.21) and (4.22) for the plane duct with the density profile $N(|x|)$. The modes of such a duct may be divided into even modes whose field components E_x, E_y, $\mathcal{H}_x$, $\mathcal{H}_y$ are symmetric about the plane $x = 0$, and odd modes whose field components E_x, E_y, $\mathcal{H}_x$, $\mathcal{H}_z$ are antisymmetric about this plane.

The solution here can again be represented in the analytic form for the uniform duct only. Let the density profile be

$$N(x) = \begin{cases} \tilde{N} & \text{if } |x| < a, \\ N_a & \text{if } |x| > a, \end{cases} \qquad (4.60)$$

where $\tilde{N}$ and N_a are constants. We keep the previous notations for the tensor elements, that is $\tilde{\varepsilon}$, $\tilde{g}$, $\tilde{\eta}$ when $N = \tilde{N}$, and ε_a, g_a, η_a when $N = N_a$. Also, the previous notations for $\tilde{n}_{1,2}$, $n_{1,2}$, $\tilde{q}_{1,2}$, and $s_{1,2}$ will be used. The waveguide solutions of Equations (4.21) and (4.22) for the plane duct are then as follows.

For the even modes:

$|x| < a$,

$$E_y = \mathrm{i} \sum_{k=1}^{2} B_k \cos(k_0 \tilde{q}_k x), \qquad E_z = -\frac{\mathrm{i}}{\tilde{\eta}} \sum_{k=1}^{2} B_k \tilde{n}_k \, \tilde{q}_k \sin(k_0 \tilde{q}_k x); \qquad (4.61)$$

$|x| > a$,

$$E_y = i \sum_{k=1}^{2} C_k \exp(-k_0 s_k |x|), \qquad E_z = -\frac{i}{\eta_a} \sum_{k=1}^{2} C_k n_k s_k \exp(-k_0 s_k |x|) \operatorname{sgn} x. \quad (4.62)$$

For the odd modes:

$|x| < a$,

$$E_y = i \sum_{k=1}^{2} B_k \sin(k_0 \tilde{q}_k x), \qquad E_z = \frac{i}{\tilde{\eta}} \sum_{k=1}^{2} B_k \tilde{n}_k \tilde{q}_k \cos(k_0 \tilde{q}_k x); \quad (4.63)$$

$|x| > a$,

$$E_y = i \sum_{k=1}^{2} C_k \exp(-k_0 s_k |x|) \operatorname{sgn} x, \qquad E_z = -\frac{i}{\eta_a} \sum_{k=1}^{2} C_k n_k s_k \exp(-k_0 s_k |x|). \quad (4.64)$$

As before, in the above expressions (4.61) to (4.64), B_k and C_k are constants. The remaining field components E_x, $\mathcal{H}_x$, $\mathcal{H}_y$, $\mathcal{H}_z$ can then be easily found with the help of expressions (4.23) and (4.24).

4.2.4. The dispersion equation for the modes of a uniform duct

The solutions of the source-free Maxwell equations must satisfy the boundary conditions at the duct interface, which reduce to continuity of the tangential components of the electric and magnetic fields at the surface $\rho = a$:

$$\begin{aligned}
E_\phi(a-0) &= E_\phi(a+0), \\
E_z(a-0) &= E_z(a+0), \\
\mathcal{H}_\phi(a-0) &= \mathcal{H}_\phi(a+0), \\
\mathcal{H}_z(a-0) &= \mathcal{H}_z(a+0).
\end{aligned} \quad (4.65)$$

If solutions (4.42), (4.43), (4.45), (4.46) for the duct core, and solutions (4.51), (4.52), (4.54), (4.55) for the outer region are substituted into (4.65), we obtain the following four simultaneous equations in the four quantities B_1, B_2, C_1, and C_2:

$$\left[J_{m+1}(\tilde{Q}_1) + \tilde{\alpha}_1 m \frac{J_m(\tilde{Q}_1)}{\tilde{Q}_1} \right] B_1 + \left[J_{m+1}(\tilde{Q}_2) + \tilde{\alpha}_2 m \frac{J_m(\tilde{Q}_2)}{\tilde{Q}_2} \right] B_2$$

$$- \left[K_{m+1}(S_1) + \alpha_1 m \frac{K_m(S_1)}{S_1} \right] C_1 - \left[K_{m+1}(S_2) + \alpha_2 m \frac{K_m(S_2)}{S_2} \right] C_2 = 0,$$

$$\tilde{n}_1 \tilde{Q}_1 J_m(\tilde{Q}_1) B_1 + \tilde{n}_2 \tilde{Q}_2 J_m(\tilde{Q}_2) B_2 + \frac{\tilde{\eta}}{\eta_a} n_1 S_1 K_m(S_1) C_1$$

$$+ \frac{\tilde{\eta}}{\eta_a} n_2 S_2 K_m(S_2) C_2 = 0, \quad (4.66)$$

$$\tilde{n}_1 \left[J_{m+1}(\tilde{Q}_1) - \tilde{\beta}_1 m \frac{J_m(\tilde{Q}_1)}{\tilde{Q}_1} \right] B_1 + \tilde{n}_2 \left[J_{m+1}(\tilde{Q}_2) - \tilde{\beta}_2 m \frac{J_m(\tilde{Q}_2)}{\tilde{Q}_2} \right] B_2$$

$$- n_1 \left[K_{m+1}(S_1) - \beta_1 m \frac{K_m(S_1)}{S_1} \right] C_1 - n_2 \left[K_{m+1}(S_2) - \beta_2 m \frac{K_m(S_2)}{S_2} \right] C_2 = 0,$$

$$\tilde{Q}_1 J_m(\tilde{Q}_1) B_1 + \tilde{Q}_2 J_m(\tilde{Q}_2) B_2 + S_1 K_m(S_1) C_1 + S_2 K_m(S_2) C_2 = 0,$$

where

$$\tilde{Q}_{1,2} = k_0 a \tilde{q}_{1,2}, \qquad S_{1,2} = k_0 a s_{1,2}.$$

If the quantities B_1, B_2, C_1, C_2 are eliminated, we obtain the determinantal equation that can be reduced after a considerable amount of algebraic manipulation to the following form:

$$\sum_{k=1}^{2} \left(D_m^{(k)} J_m^{(k)} + F_m^{(k)} K_m^{(k)} \right) + \frac{\tilde{\eta}}{\eta_a} J_m^{(1)} J_m^{(2)} + K_m^{(1)} K_m^{(2)} + G_m = 0, \qquad (4.67)$$

where

$$J_m^{(1)} = \frac{J_{m+1}(\tilde{Q}_1)}{\tilde{Q}_1 J_m(\tilde{Q}_1)} + m \frac{\tilde{\alpha}_1}{\tilde{Q}_1^2}, \qquad J_m^{(2)} = \frac{J_{m+1}(\tilde{Q}_2)}{\tilde{Q}_2 J_m(\tilde{Q}_2)} - m \frac{\tilde{\beta}_2}{\tilde{Q}_2^2},$$

$$K_m^{(1)} = \frac{K_{m+1}(S_1)}{S_1 K_m(S_1)} + m \frac{\alpha_1}{S_1^2}, \qquad K_m^{(2)} = \frac{K_{m+1}(S_2)}{S_2 K_m(S_2)} - m \frac{\beta_2}{S_2^2},$$

$$D_m^{(1)} = M_{12} L_{12} K_m^{(2)} - M_{11} L_{22} K_m^{(1)}$$
$$- mM_0 \left[n_1 L_{22}\gamma_1 - \tilde{n}_1 L_{12}\gamma_2 + \frac{\tilde{\eta}}{\eta_a} \tilde{n}_1 (n_2 - n_1) \tilde{\gamma}_2 \right],$$

$$D_m^{(2)} = M_{21} L_{21} K_m^{(1)} - M_{22} L_{11} K_m^{(2)}$$
$$+ mM_0 \left[n_1 L_{21}\gamma_1 - \tilde{n}_2 L_{11}\gamma_2 + \frac{\tilde{\eta}}{\eta_a} \tilde{n}_1 (n_2 - n_1) \tilde{\gamma}_1 \right],$$

$$F_m^{(1)} = -mM_0 \left[n_1 (\tilde{n}_2 - \tilde{n}_1)\gamma_2 - \tilde{n}_1 L_{22}\tilde{\gamma}_1 + n_1 L_{21}\tilde{\gamma}_2 \right], \qquad (4.68)$$

$$F_m^{(2)} = mM_0 \left[n_1 (\tilde{n}_2 - \tilde{n}_1)\gamma_1 - \tilde{n}_1 L_{12}\tilde{\gamma}_1 + n_2 L_{11}\tilde{\gamma}_2 \right],$$

$$G_m = m^2 M_0 \left[\frac{\tilde{\eta}}{\eta_a} \tilde{n}_1 (n_2 - n_1)\tilde{\gamma}_1 \tilde{\gamma}_2 + n_1 (\tilde{n}_2 - \tilde{n}_1)\gamma_1 \gamma_2 \right.$$
$$\left. - \tilde{n}_1 L_{12}\tilde{\gamma}_1 \gamma_2 + n_1 L_{21}\gamma_1 \tilde{\gamma}_2 \right].$$

In the preceding,

$$M_{ij} = M_0(\tilde{n}_i - n_j), \qquad L_{ij} = \frac{\tilde{\eta}}{\eta_a} n_i - \tilde{n}_j, \quad i, j = 1, 2,$$

$$M_0 = (n_1 - n_2)^{-1}(\tilde{n}_1 - \tilde{n}_2)^{-1} \qquad (4.69)$$

and

$$\tilde{\gamma}_k = (\tilde{\alpha}_k + \tilde{\beta}_k)\tilde{Q}_k^{-2}, \qquad \gamma_k = (\alpha_k + \beta_k)S_k^{-2}, \quad k = 1,2. \tag{4.70}$$

Other notations occurring in formulas (4.68) to (4.70) are as in (4.40), (4.47), and (4.56). Relation (4.67) is the dispersion equation for the guided modes and its roots are propagation constants p of the modes.

For the case of axisymmetric modes, when $m = 0$, Equation (4.67) becomes

$$\left[M_{12}L_{12}\frac{K_1(S_2)}{S_2\,K_0(S_2)} - M_{11}L_{22}\frac{K_1(S_1)}{S_1\,K_0(S_1)} \right] \frac{J_1(\tilde{Q}_1)}{\tilde{Q}_1\,J_0(\tilde{Q}_1)}$$
$$+ \left[M_{21}L_{21}\frac{K_1(S_1)}{S_1\,K_0(S_1)} - M_{22}L_{11}\frac{K_1(S_2)}{S_2\,K_0(S_2)} \right] \frac{J_1(\tilde{Q}_2)}{\tilde{Q}_2\,J_0(\tilde{Q}_2)}$$
$$+ \frac{\tilde{\eta}}{\eta_a}\frac{J_1(\tilde{Q}_1)}{\tilde{Q}_1\,J_0(\tilde{Q}_1)}\frac{J_1(\tilde{Q}_2)}{\tilde{Q}_2\,J_0(\tilde{Q}_2)} + \frac{K_1(S_1)}{S_1\,K_0(S_1)}\frac{K_1(S_2)}{S_2\,K_0(S_2)} = 0. \tag{4.71}$$

For the modes of a uniform plane duct with the density profile (4.60), the dispersion equation takes the following forms.

For the even modes:

$$\left(\frac{M_{12}L_{12}}{S_2} - \frac{M_{11}L_{22}}{S_1} \right) \frac{\cot \tilde{Q}_1}{\tilde{Q}_1} + \left(\frac{M_{21}L_{21}}{S_1} - \frac{M_{22}L_{11}}{S_2} \right) \frac{\cot \tilde{Q}_2}{\tilde{Q}_2}$$
$$- \frac{\tilde{\eta}}{\eta_a}\frac{\cot \tilde{Q}_1}{\tilde{Q}_1}\frac{\cot \tilde{Q}_2}{\tilde{Q}_2} - \frac{1}{S_1 S_2} = 0. \tag{4.72}$$

For the odd modes:

$$\left(\frac{M_{12}L_{12}}{S_2} - \frac{M_{11}L_{22}}{S_1} \right) \frac{\tan \tilde{Q}_1}{\tilde{Q}_1} + \left(\frac{M_{21}L_{21}}{S_1} - \frac{M_{22}L_{11}}{S_2} \right) \frac{\tan \tilde{Q}_2}{\tilde{Q}_2}$$
$$+ \frac{\tilde{\eta}}{\eta_a}\frac{\tan \tilde{Q}_1}{\tilde{Q}_1}\frac{\tan \tilde{Q}_2}{\tilde{Q}_2} + \frac{1}{S_1 S_2} = 0. \tag{4.73}$$

These two forms are derived by applying the conditions of continuity to the tangential fields E_y, E_z, $\mathcal{H}_y$, $\mathcal{H}_z$ at the surfaces $x = \pm a$. The proof is left as an exercise for the reader.

So far we have considered ducts for which the difference between the plasma in the core and that in the outer region is only in the value of density. However, it is easy to verify that all the general results of this section will be valid if the duct is made up due to variation in any plasma parameter involved in the dielectric tensor. Therefore, the general expressions given above could be extended, in principle, to the case of ducts formed by modifications of plasma conductivity, external magnetic-field strength, etc. It is also clear that the foregoing analysis is applicable for arbitrary values of the wave frequency ω.

4.3. Bound and leaky modes

In § 4.1 the idea of "bound" and "leaky" modes was introduced. These two kinds of modes are of great importance in electrodynamics of ducts and it is therefore useful to summarize their main features.

Bound (or completely trapped) modes are modes that propagate without attenuation along a loss-free duct. There is no energy flow transverse to the duct axis in these modes and hence no radiation of energy outwards the duct. Each of them is an inhomogeneous plane wave traveling with a purely real propagation constant p. The field components in the bound mode decay outside the guide as distance from its axis increases. This means that if, for example, the Hankel functions of the second kind $H_m^{(2)}(k_0 q_k \rho)$ are used to represent the field solutions in the outer region, it is necessary to require $\mathrm{Im}\,(q_k) < 0$ (i.e. $\mathrm{Re}\,(s_k) > 0$) to provide zero of the field components at infinity. Such solutions individually satisfying the source-free field equations plus appropriate boundary conditions and the above stated condition at infinity are also termed *eigen (proper) modes* of the duct. It will be shown later (Chapter 6) that each eigenmode carries a finite energy flux along the duct and can therefore be individually excited by a given source. All together, they constitute a discrete part of a complete spectrum of waves that can be excited by some source in the presence of a duct.

Leaky (or quasi-trapped) modes are modes that propagate along a nonabsorbing duct with some attenuation arising because energy is leaking out of the guide into the outer region. The propagation constant for the leaky mode is complex and has a negative imaginary part, so that $p = p' - ip''$. The amplitudes of the field components in the mode therefore decrease exponentially with increasing z. In the ambient medium they grow indefinitely as distance ρ from the duct axis increases.* This is because for a large value of ρ in the outer region, the energy comes from a remote part of the duct where the amplitudes of the field components are large. The situation is depicted schematically in Figure 4.3 for a typical case. This is analogous to the behavior of a leaky mode in an open isotropic guide (see for details Marcuvitz, 1956; Budden, 1961 b; Tamir and Oliner, 1963; Felsen and Marcuvitz, 1973). For each leaky mode in the case of a magnetoplasma, at least one of the quantities $q_{1,2}$ occurring in Hankel functions $H_n^{(2)}(k_0 q_k \rho)$ for the outer region must have a positive imaginary part. Thus, the leaky modes satisfy the source-free field equations and the boundary conditions at the duct interface, but they do not satisfy the requisite condition at the singular point, infinity, nor do they carry finite power. Therefore, these solutions do not represent the true eigenmodes and cannot

---

* The growth may become noticeable beginning with a rather large value of ρ if the leakage is weak enough (see § 4.5).

be individually excited by a given source. Although they are not generally members of a complete spectrum of excited waves, the leaky modes may nevertheless be separated formally from an integral that represents the radiation field. It is often true that the leaky modes may be employed to obtain the approximate representation of field solutions in certain regions close to a source. A full account of this topic will be given in Chapter 6. In particular, it will be shown that one can sometimes work with leaky modes as, in a sense, with the eigenmodes. This unifies the formulation and simplifies the analysis of the field pattern.

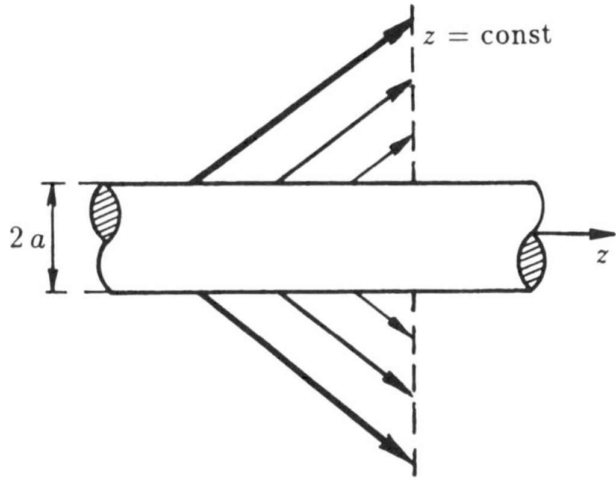

Figure 4.3. Qualitative representation of radiation from a duct excited by a remote source in its core. Arrows denote power flow in a leaky mode.

Before giving a detailed mathematical development for guided waves it is useful to have a general physical picture of what modes can be expected in a duct with prescribed parameters. For this, examine the relation between a dimensionless propagation constant p and a pair of transverse wavenumbers q_1 and q_2 for a uniform duct. As earlier, we shall confine our analysis to frequencies of the whistler band (2.69), where only the "extraordinary" wave is propagating.

We begin with the frequency interval (2.70) belonging to the whistler band. For this interval, the whistler-mode refractive index surface is closed, as shown in Figure 2.1. We now reproduce in Figure 4.4 two similar surfaces, one of which, I, corresponds to density N_a, while the other, II, corresponds to density $\tilde{N}$, the inequality $N_a < \tilde{N}$ being understood. Clearly the situation depicted is related to the case of a duct with enhanced density. As is seen from Figure 4.4, the quantity $\tilde{q}_1$ is real for $p < \tilde{\mathcal{P}}$, and imaginary for $p > \tilde{\mathcal{P}}$, whereas the quantity q_1 is real for $p < \mathcal{P}$, and imaginary for $p > \mathcal{P}$. The quantities $\tilde{q}_2$ and q_2 are imaginary here for all real values of p. This is directly associated with the fact that the "ordinary" wave is non-propagating in the frequency interval (2.70). Thus, it is evident that the completely trapped modes may exist for the frequency interval (2.70) if $p > \mathcal{P}$.[*] Later we shall show that there is an upper boundary for p, so that a propagation constant should lie in the range

$$\mathcal{P} < p < \tilde{\mathcal{P}} \tag{4.74}$$

for guided modes at these frequencies. From the refractive index surfaces given in Figure 4.4, one can also infer that the field of each eigenmode in the core would be synthesized of a volume wave with its real transverse wavenumber $\tilde{q}_1$ and a surface wave with its imaginary transverse wavenumber $\tilde{q}_2$, the contribution from the latter decaying inwards the duct. In the outer region, where both wavenumbers q_1 and q_2 are imaginary, the field would be of the surface character. For density troughs no propagating eigenmodes may be found at frequencies (2.70).

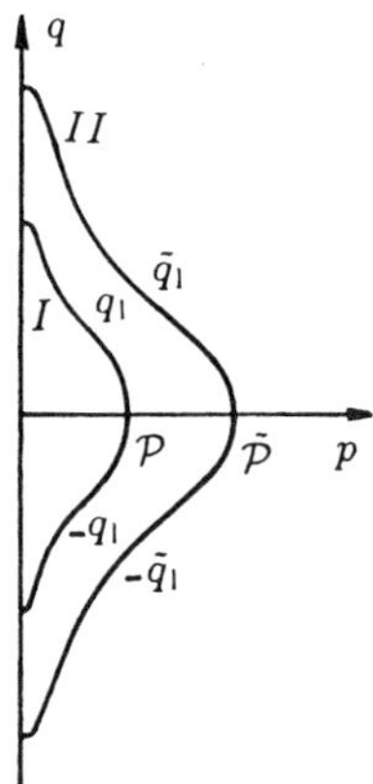

Figure 4.4. Refractive index surfaces for the two densities, N_a (I) and $\tilde{N}$ (II), when $N_a < \tilde{N}$ and $\Omega_\mathrm{H} \ll \omega < \omega_\mathrm{LH}$.

As explained in § 2.6, the whistler-mode refractive index surface undergoes a fundamental change in its form when the wave frequency ω passes over the lower-hybrid frequency ω_LH. At frequencies above the lower-hybrid frequency, the surface becomes open, as shown in Figure 2.2 for the frequency interval (2.71). Two such surfaces corresponding to two densities $\tilde{N}$ and N_a ($\tilde{N} > N_a$) are reproduced in Figure 4.5. Before analyzing what types of modes are possible, we introduce quantities $\mathcal{P}_c$ and $\tilde{\mathcal{P}}_c$ which are dimensionless propagation constants (along $\mathbf{B}_0$) of conical-refraction whistler-mode waves in plasmas with the densities N_a and $\tilde{N}$,

---

[*] Since the transverse wavenumbers $q_{1,2}$ and $\tilde{q}_2$ are here imaginary, it might be more convenient to utilize instead of them the real quantities $s_{1,2}$ and $\tilde{s}_2$ introduced by $s_{1,2}^2 = -q_{1,2}^2$ and $\tilde{s}_2^2 = -\tilde{q}_2^2$.

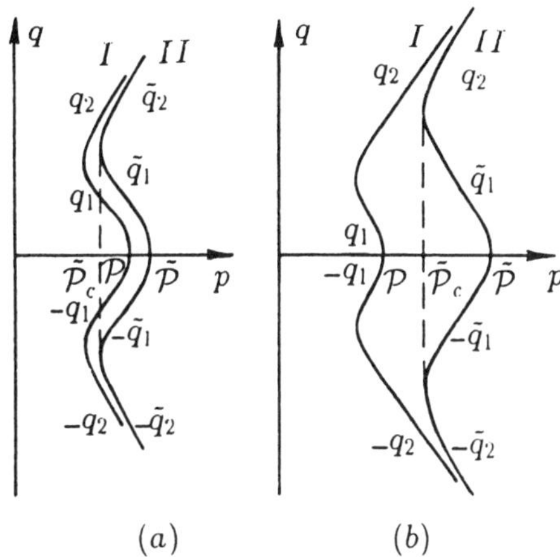

Figure 4.5. Refractive index surfaces for the two densities, N_a (I) and $\tilde{N}$ (II), when $N_a < \tilde{N}$ and $\omega_{\mathrm{LH}} < \omega < \omega_{\mathrm{H}}/2$; (a) $\tilde{\mathcal{P}}_c < \mathcal{P}$, (b) $\mathcal{P} < \tilde{\mathcal{P}}_c$.

respectively. There are two possible situations when the density is enhanced in the duct: (i) $\tilde{\mathcal{P}}_c < \mathcal{P}$ and (ii) $\mathcal{P} < \tilde{\mathcal{P}}_c$ (see respectively plots (a) and (b) in Figure 4.5). It is seen from Figure 4.5 that if the propagation constant p lies in the range $\mathcal{P} < p < \tilde{\mathcal{P}}$ for the former case, and in the range $\tilde{\mathcal{P}}_c < p < \tilde{\mathcal{P}}$ for the latter case, then the behavior of the branches $\tilde{q}_1$ and q_1 (or s_1) is nearly similar to that observed at frequencies (2.70) for a mode whose propagation constant lies slightly below the point $p = \tilde{\mathcal{P}}$. Hence contributions of these branches to the total field would be similar as well, and we may thus expect a guided mode in a density enhancement at frequencies above the lower-hybrid frequency. There are, however, markedly different contributions from branches $\tilde{q}_2$ and q_2 which now extend to infinity, remaining purely real if p is real. It is evident that the branch q_2 corresponds in the outer region to the escaping wave which carries energy away from the duct. Because of this the guided mode will be leaky. In the core its field is evidently of a volume character and consists of a "large-scale" part with its wavenumber $\tilde{q}_1$, and a "fine-scale" part with its wavenumber $\tilde{q}_2$ (note that $|\tilde{q}_2| > |\tilde{q}_1|$). In the outer region the field consists of a decaying part associated with q_1 (or, in other terms, s_1), and an escaping (leaking) part associated with q_2. It will be shown later (in § 4.4) that in many cases encountered in actual practice the leakage is weak enough, so that the attenuation constant $p'' = -\mathrm{Im}(p)$ for this type of mode is much less than the real propagation constant $p' = \mathrm{Re}(p)$. This, in turn, leads to the fact that the real parts of $\tilde{q}_{1,2}$, s_1 and q_2 will be much larger than their imaginary parts. Thus, the real propagation constants p' for leaky modes of this type should lie in the range

$$\max\{\mathcal{P}, \tilde{\mathcal{P}}_c\} < p' < \tilde{\mathcal{P}}. \qquad (4.75)$$

If p' shifts to the side of larger values ($p' > \tilde{\mathcal{P}}$), then the field in the core will consist of a surface part associated with $\tilde{q}_1$ and a volume part associated with $\tilde{q}_2$, whereas in the outer region it will not change. The "radiation" attenuation will then be appreciably higher than that in the range (4.75). Only for the step-like density profile with $\tilde{N} \gg N_a$ the attenuation remains rather small. Similarly, if p' shifts to the side of smaller values, becoming less than the lower boundary of the range (4.75), the attenuation rapidly increases. We stress that the transverse components of the wavenormal and the Poynting vector for the escaping wave (i.e. the branch q_2) have opposite signs. Therefore, for $\exp(i\omega t)$ time-dependence assumed, it is more convenient to employ the first kind instead of the second kind of Hankel functions for the escaping wave in the outer region. Actually, the use of $H_m^{(1)}(k_0 q_2 \rho)$, with $\mathrm{Re}\,(q_2) > 0$, will be in agreement with the radiation condition, which requires that the energy transport, not the phase progress of the wave, be outgoing. A close examination shows that for this branch, the inequality $\mathrm{Im}\,(q_2) < 0$ will then occur, and the amplitudes of the field components will increase as $\rho \to \infty$ that is typical of leaky modes.

Notice that when $p < \mathcal{P}_c$, a true bound mode with a real propagation constant may nevertheless be found at frequencies (2.71) for the duct within which the density is enhanced. This mode has complex transverse wavenumbers such that $\tilde{q}_1 = -\tilde{q}_2^*$ and $q_1 = -q_2^*$ (i.e. $s_1 = s_2^*$), and its field is localized in the vicinity of the duct interface. This mode, guided, in effect, by the interface, may exist only under some special conditions, and is essentially appropriate to the existence of a step-like discontinuity in the plasma density. Since such a mode seems to be mainly of academic interest it will not be considered here in detail.* For a discussion of its properties see Adachi (1966).

Now consider, for the same frequencies (2.71), the possible types of modes if the density within the duct is diminished. For this case, the reader should redraw Figure 4.5 keeping in mind that now $\tilde{N} < N_a$ and, hence, $\tilde{\mathcal{P}} < \mathcal{P}$ and $\tilde{\mathcal{P}}_c < \mathcal{P}_c$. The eigenmodes trapped in a density trough may exist for the range

$$\tilde{\mathcal{P}}_c < p < \mathcal{P}_c. \qquad (4.76)$$

The transverse wavenumbers of the modes in the outer region are now complex such that $q_1 = -q_2^*$ (i.e. $s_1 = s_2^*$). The modes are, however, of entirely volume character

---

*The authors do not know of any published work in which the experimental observation of such a mode is reported.

in the core if $\mathcal{P}_c < \tilde{\mathcal{P}}$, or if $\tilde{\mathcal{P}}_c < p < \tilde{\mathcal{P}} < \mathcal{P}_c$, in which cases $\tilde{q}_1$ and $\tilde{q}_2$ are real. If the propagation constant lies in the range $\tilde{\mathcal{P}} < p < \mathcal{P}_c$, that is possible in the case when $\tilde{\mathcal{P}} < \mathcal{P}_c$, the wavenumber $\tilde{q}_2$ remains real, while the other wavenumber $\tilde{q}_1$ becomes imaginary, and the field in the core would consist of a volume wave and a surface wave. Note that the eigenmode sustained by the duct interface may also exist here under some conditions, with p lying below $\tilde{\mathcal{P}}_c$ (Adachi, 1966).

In the remaining frequency interval (2.72) of the whistler band (2.69), the eigenmode guided by the duct interface exists in the domain $p < \min\{\mathcal{P}, \tilde{\mathcal{P}}\}$, regardless of whether the density is enhanced or diminished. For rather strong step-like enhancements, the leaky modes may be found when $p' > \mathcal{P}$ (see Figure 4.6 which is appropriate to the case of an enhanced density). If the density within the duct is diminished so that $\tilde{\mathcal{P}} < \mathcal{P}$, the eigenmodes trapped in the trough may exist, with real propagation constants lying in the domain

$$\tilde{\mathcal{P}} < p < \mathcal{P}. \tag{4.77}$$

We leave for the reader's exercise to verify what waves (volume, surface, or complex) are to be taken to synthesize the fields of modes possible at frequencies (2.72).

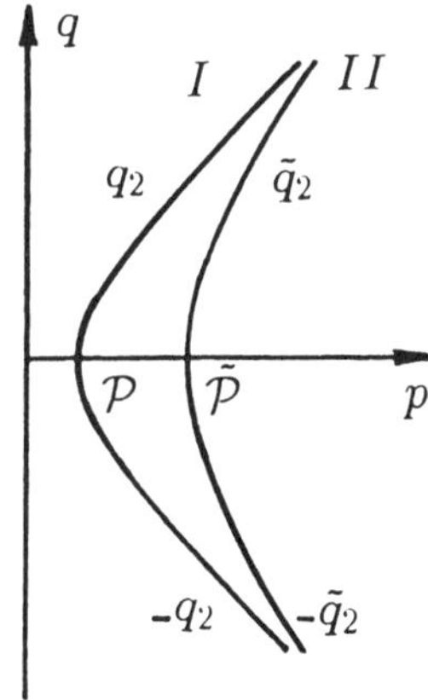

Figure 4.6. Refractive index surfaces for the two densities, N_a (I) and $\tilde{N}$ (II), when $N_a < \tilde{N}$ and $\omega_H/2 < \omega < \omega_H$.

A final point is that the inequalities (4.74) to (4.77), defining possible ranges of the propagation constants for the basic types of guided modes on sharply bounded uniform ducts with a step-like density profile, are applicable for nonuniform ducts with smooth monotonic variations of the density (Karpman and Kaufman, 1982 b). One should, however, remember that in the case of a nonuniform duct the quantities $\tilde{\mathcal{P}}$, $\tilde{\mathcal{P}}_c$ and $\tilde{N}$ must be related to the value of density at the duct axis, $\rho = 0$.

In what follows, some mathematical treatment will be given to confirm the physical picture delineated above. We shall discuss in detail the uniform duct with the simplest step-like density profile. Further, for more complicated profiles, attention will be given mainly to the new distinctive features of propagation which are caused by complication of the density distribution.

4.4. Uniform duct without collision damping — the dispersion properties of modes

In this section we shall discuss some important examples of the possible waveguide modes in a sharply bounded, uniform loss-free duct. This is only a crude model for a real duct with continuous variation of the density. Still, it permits a full analytic treatment of guiding systems filled by a magnetoplasma. The continuous variation of the density needs, in general, numerical computations of the modal characteristics. Only smooth, weakly inhomogeneous distributions of the density permit employing some asymptotic methods to obtain the field solutions in the analytic form. Some account of works on this topic will be given in the next sections.

The discussion, in §4.3, of the possible modes of density ducts in the whistler band suggests that at least three frequency intervals and two sorts of the density variation (enhancements and troughs) should be considered. Also, there are two important types of geometry of a duct that implies a separate consideration for a cylindrical duct and a plane duct. To give full results for each case would take up much space. Therefore we shall consider the following five topics:

1. The axisymmetric eigenmodes of a cylindrical enhancement
   (for the frequency interval (2.70), i.e. $\Omega_H \ll \omega < \omega_{LH}$, $\omega_H \ll \omega_p$).

2. The axisymmetric leaky modes of a cylindrical enhancement
   (for the frequency interval (2.71), i.e. $\omega_{LH} < \omega < \omega_H/2 \ll \omega_p$).

3. Some extension to the case of nonsymmetric modes
   (for a cylindrical enhancement and the frequency intervals (2.70) and (2.71)).

4. Modes of a plane duct with enhanced density.

5. Modes of a cylindrical trough (the axisymmetric eigenmodes for the frequency intervals (2.71) and (2.72), i.e. $\omega_{LH} < \omega < \omega_H \ll \omega_p$).

Each of these topics seems to be important for numerous interesting applications and will be useful in the later work.

4.4.1. The axisymmetric eigenmodes of a cylindrical enhancement

Consider the eigenmodes guided along a cylindrical enhancement of the plasma density at frequencies (2.70). The dimensionless propagation constants are roots of the dispersion equation (4.67), which, in general, must be investigated numerically. We begin with a study of the properties of axisymmetric eigenmodes, for which the dispersion relation takes the most simple form (4.71). In some special cases it may be solved analytically.

As mentioned in § 2.6, the general formula (2.64) giving transverse wavenumbers $q_{1,2}$ may be simplified under the condition (2.74). We assume this condition to be satisfied for the densities $\tilde{N}$ and N_a simultaneously, so that

$$|p^2 - \tilde{\varepsilon}| \gg \tilde{g}^2 / |\tilde{\eta}|, \qquad |p^2 - \varepsilon_a| \gg g_a^2 / |\eta_a|. \tag{4.78}$$

The branch q_1 is then approximated by (2.75), and q_2 is approximated by

$$q_2^2 \approx -\frac{\eta}{\varepsilon} (p^2 - \varepsilon) - \frac{g^2}{\varepsilon} \frac{p^2}{p^2 - \varepsilon}, \tag{4.79}$$

into which the relevant tensor elements are to be inserted for the core and the outer region. The elements of the dielectric tensor at the chosen frequencies are obtainable from general expressions (2.20) giving

$$\tilde{\varepsilon} = (1 - Y_{\mathrm{LH}}^2)\, \tilde{X}/Y^2, \quad \varepsilon_a = (1 - Y_{\mathrm{LH}}^2)\, X_a/Y^2,$$

$$\tilde{g} = -\tilde{X}/Y, \qquad\qquad g_a = -X_a/Y, \tag{4.80}$$

$$\tilde{\eta} = -\tilde{X}, \qquad\qquad \eta_a = -X_a,$$

where $Y_{\mathrm{LH}} = \omega_{\mathrm{LH}}/\omega$, and $\tilde{X}$ and X_a stand to denote the quantity X for plasmas with the densities $\tilde{N}$ and N_a, respectively. In addition to (4.78), assume that the simplifying inequality $p^2 \gg |\tilde{\varepsilon}|$ is satisfied. The quantities $\tilde{q}_{1,2}$ and $s_{1,2}$ then become

$$\tilde{q}_1^2 \approx \frac{\tilde{\mathcal{P}}^4}{p^2} - p^2, \quad \tilde{q}_2^2 \equiv -\tilde{s}_2^2 \approx -\frac{\tilde{\eta}}{\tilde{\varepsilon}} \left(p^2 + \frac{\tilde{g}^2}{\tilde{\eta}} - \tilde{\varepsilon} \right),$$

$$s_1^2 \approx p^2 - \frac{\mathcal{P}^4}{p^2}, \quad s_2^2 \equiv -q_2^2 \approx \frac{\eta_a}{\varepsilon_a} \left(p^2 + \frac{g_a^2}{\eta_a} - \varepsilon_a \right). \tag{4.81}$$

Substituting (2.75) and (4.79) into the expressions for $\tilde{n}_{1,2}$ and $n_{1,2}$ in (4.47) and (4.56) yields

$$\tilde{n}_1 \approx \frac{|\tilde{g}|}{p} \left(1 + \frac{\tilde{\varepsilon}}{p^2} \right), \quad \tilde{n}_2 \approx p\frac{\tilde{\eta}}{\tilde{g}} \left(1 - \frac{\tilde{\varepsilon}}{p^2} \right),$$

$$n_1 \approx \frac{|g_a|}{p} \left(1 + \frac{\varepsilon_a}{p^2} \right), \quad n_2 \approx p\frac{\eta_a}{g_a} \left(1 - \frac{\varepsilon_a}{p^2} \right). \tag{4.82}$$

In (4.81) and (4.82), we still retain small terms of the order ε/p^2 and $g^2/p^2\eta$ to represent the dispersion equation (4.71) in the form suitable for examining the accuracy of the approximations utilized. If relations (4.80) to (4.82) are used in Equation (4.71) it takes the form

$$
\left(\frac{J_1(\tilde{Q}_1)}{J_0(\tilde{Q}_1)} + \frac{\tilde{Q}_1}{S_1}\frac{K_1(S_1)}{K_0(S_1)}\right)\left(\frac{J_1(\tilde{Q}_2)}{J_0(\tilde{Q}_2)} + \frac{\eta_a\tilde{Q}_2}{\tilde{\eta}S_2}\frac{K_1(S_2)}{K_0(S_2)}\right)
$$
$$
= \text{æ}(p)\,\frac{J_1(\tilde{Q}_1)}{J_0(\tilde{Q}_1)}\frac{K_1(S_1)}{K_0(S_1)} + \sigma(p)\,\frac{J_1(\tilde{Q}_2)}{J_0(\tilde{Q}_2)}\frac{K_1(S_2)}{K_0(S_2)}, \qquad (4.83)
$$

where

$$
\text{æ}(p) = \frac{X_a\tilde{Q}_2}{\tilde{X}S_1}\frac{X_a}{p^2Y^2}\left(\frac{\tilde{X}}{X_a}-1\right)^2,
$$

$$
\sigma(p) = (Y_{\text{LH}}^2 - 1)^2\frac{\tilde{Q}_1}{S_2}\left(\frac{X_a}{p^2Y^2}\right)^3\left(\frac{\tilde{X}}{X_a}-1\right)^2. \qquad (4.84)
$$

Now show how solutions to Equation (4.83) can be found in the domain (4.74). As mentioned in §4.3, in that domain $\tilde{Q}_1$ and $S_{1,2}$ are real, whereas $\tilde{Q}_2$ is pure imaginary, and the inequalities $\tilde{q}_1 \ll |\tilde{q}_2|$ and $s_1 \ll s_2$ are valid. It is therefore useful to employ a quantity $\tilde{S}_2 = k_0 a\tilde{s}_2$ introduced by $\tilde{q}_2 = i\tilde{s}_2$, and this gives $\text{æ}(p) = i\,|\text{æ}(p)|$ and

$$
\frac{J_1(\tilde{Q}_2)}{J_0(\tilde{Q}_2)} = i\,\frac{I_1(\tilde{S}_2)}{I_0(\tilde{S}_2)},
$$

where $I_m(\zeta)$ is the mth-order modified Bessel function of the first kind.

For a rather *small drop in the plasma density* at the boundary $\rho = a$, when

$$
(\tilde{X} - X_a)/X_a \ll 1, \qquad (4.85)
$$

that is often encountered in practice, a solution to Equation (4.83) is conveniently sought in the form $p = \mathcal{P} + \Delta p$ where Δp is a small additive term, so that $|\Delta p| \ll \mathcal{P}$. When condition (4.85) holds, the quantities $\tilde{q}_{1,2}$ and $s_{1,2}$ reduce to

$$
\tilde{q}_1^2 \approx q_1^2(\mathcal{P},\tilde{N}) - \frac{1}{2}\left(\frac{\tilde{X}^2}{X_a^2}+1\right)s_1^2, \qquad \tilde{q}_2^2 = -\tilde{s}_2^2 \approx -(\mathcal{P}Y)^2(Y_{\text{LH}}^2 - 1)^{-1},
$$

$$
s_1^2 \approx 4\,\Delta p\,\mathcal{P}, \qquad s_2^2 \approx \tilde{s}_2^2, \qquad (4.86)
$$

and the quantity $\sigma(p)$ is a small parameter such that $|\sigma(p)| \ll 1$. In addition, suppose that the inequality

$$
|\text{æ}(p)| \simeq \frac{1}{2}\left(\frac{\mathcal{P}}{\Delta p}\right)^{1/2}|Y_{\text{LH}}^2 - 1|^{-1/2}\left(\frac{\tilde{X}}{X_a}-1\right)^2 \ll 1 \qquad (4.87)
$$

is satisfied. This would occur for the case (4.85) when the propagation constant is more close to $\tilde{\mathcal{P}}$ than to $\mathcal{P}$.* Equation (4.83) can then be solved by the use of the method of successive approximations. In a zero-order approximation, when the terms involving small parameter æ and σ on the right of Equation (4.83) are neglected, we obtain from (4.83) the following two equations:

$$\frac{J_1(\tilde{Q}_1)}{J_0(\tilde{Q}_1)} = -\frac{\tilde{Q}_1}{S_1}\frac{K_1(S_1)}{K_0(S_1)}\,, \tag{4.88}$$

$$\frac{I_1(\tilde{S}_2)}{I_0(\tilde{S}_2)} = -\frac{\eta_a \tilde{S}_2}{\tilde{\eta} S_2}\frac{K_1(S_2)}{K_0(S_2)}\,. \tag{4.89}$$

It is easy to show that only the equation (4.88) offers solutions corresponding to the eigenmodes of our guide.[†] For a sufficiently wide guide, when

$$\left(k_0 a \tilde{\mathcal{P}}\right)^2 \gg \mu_n^{(1)^2}\left(\frac{\tilde{X}}{X_a} - 1\right)^{-1}\,, \quad n = 1, 2, \ldots$$

($\mu_n^{(k)}$ is the nth zero of the Bessel function J_k, $\mu_n^{(k)} \neq 0$), the right side of Equation (4.88) would be small, and the solution to it can be found by an application once again of the method of successive approximations. The result is then

$$p_n = \mathcal{P} + \Delta p_n = \tilde{\mathcal{P}} - \frac{\mu_n^{(1)^2}}{4\left(k_0 a\right)^2 \tilde{\mathcal{P}}}\,. \tag{4.90}$$

For a more narrow guide, when $k_0 a \tilde{\mathcal{P}} \sim 1$, the solution takes the form which is too complicated to be given analytically.

In general, as the radius of the duct increases the propagation constant of a given eigenmode shifts to the value $\tilde{\mathcal{P}}$, while as the duct shrinks it shifts to the value $\mathcal{P}$. For a fixed radius a, with decreasing the wave frequency ω, each curve $p_n(\omega)$ meets the curve $\mathcal{P}(\omega)$ at some frequency $\omega_{c,n}$ which is called the *cutoff frequency* for the nth eigenmode. Each propagating eigenmode can exist only at the frequencies above its cutoff frequency. Note that near the cutoff frequency the term Δp vanishes and the condition (4.87) is no longer valid, so that the approximation (4.88) for the dispersion equation needs to be refined.

For a *large drop in the plasma density*, when $\tilde{N} \gg N_a$, we have $s_1 \approx p$ and the parameter æ becomes

$$æ(p) \approx \mathrm{i}\,\frac{\tilde{X}}{p^2 Y}\frac{1}{\sqrt{Y_{\mathrm{LH}}^2 - 1}} \approx \mathrm{i}\,\frac{\tilde{\mathcal{P}}^2}{p^2 \sqrt{Y_{\mathrm{LH}}^2 - 1}}\,.$$

---

* It will be clear from what follows that such a situation is possible for a sufficiently wide guide.

† In this case, the quantities $\tilde{\eta}$ and η_a have the same sign. Hence $\tilde{\eta}/\eta_a > 0$, and no solutions with real propagation constants can be found from Equation (4.89). If the ratio $\tilde{\eta}/\eta_a$ were negative, Equation (4.89) would give the solution corresponding to the surface eigenmode (see, for example, Kondratenko, 1976, § 18).

It is now clear that the inequality $|\ae| \ll 1$ would, in general, cease to be valid. The parameter σ, however, remains small. Here the quantities $\tilde{S}_2$ and S_2 are normally large enough, whence $I_1(\tilde{S}_2)/I_0(\tilde{S}_2) \approx 1$ and $K_1(S_2)/K_0(S_2) \approx 1$. Then Equation (4.83) becomes

$$\frac{J_1(\tilde{Q}_1)}{J_0(\tilde{Q}_1)} = -\left(1 - \frac{\tilde{P}^2}{p^2\sqrt{Y_{\text{LH}}^2 - 1}}\frac{K_1(S_1)}{K_0(S_1)}\right)^{-1}\frac{\tilde{Q}_1}{S_1}\frac{K_1(S_1)}{K_0(S_1)}. \qquad (4.91)$$

The qualitative behavior of solutions to this equation, when the guide is wide enough, is similar to that as for the case (4.85). The investigation to be made here may be left as an exercise for the reader. For the general case, the solution of the dispersion equation must be found numerically.

Before continuing the discussion of the propagation of modes for other intervals of the whistler band, one important topic must be dealt with: numbering the modes. It seems natural to number the eigenmodes of a plasma cylinder with enhanced density in the order of increase of their transverse wavenumbers $\tilde{q}_1$, starting with the minimum value of $\tilde{q}_1$ and using numbers $n = 1, 2, \ldots$. Note that this method was utilized in the expression (4.90) appropriate to axisymmetric modes. When considering nonsymmetric solutions, the azimuthal index m must be added to order the modes. For the fixed index m, the mode with the least magnitude of $\tilde{q}_1$, that is with the minimum number n, will be called the *lower order mode of index m.* This numbering system is nearly similar to that usually used for open dielectric waveguides (Marcuse, 1972; Snyder and Love, 1983). We shall, however, see later that it cannot be applied without making some corrections to leaky modes guided by strong enhancements of plasma density in the frequency interval (2.71).

4.4.2. The axisymmetric leaky modes of a cylindrical enhancement

The same method as used above for the eigenmodes can be employed to investigate the dispersion properties of the leaky modes propagating along a cylindrical density enhancement in the frequency interval (2.71). For this interval, $Y_{\text{LH}} < 1$, so that $\tilde{\varepsilon}$ and ε_a are positive. We shall henceforth assume that the wave frequency ω is well above the lower hybrid frequency ω_{LH}, so that $Y_{\text{LH}} \ll 1$. This allows the use of the formula (2.79) for q_k that simplifies the algebra. Thence $\mathcal{P} = [X_a/(Y-1)]^{1/2}$, $\tilde{\mathcal{P}} = \left[\tilde{X}/(Y-1)\right]^{1/2}$, $\mathcal{P}_c = 2X_a^{1/2}/Y$, and $\tilde{\mathcal{P}}_c = 2\tilde{X}^{1/2}/Y$. Consideration can be formulated without these simplifications but is then much longer. Now we shall assume that the following inequality is satisfied:

$$|p|^2 \gg \tilde{\mathcal{P}}_c^2. \qquad (4.92)$$

This inequality has an important physical meaning. When it holds, the wavenumbers of a "trapped" part of the leaky mode and those of its escaping part differ markedly from each other, so that $|\tilde{q}_1| \ll |\tilde{q}_2|$ and $|s_1| \ll |s_2|$. As we shall see from the later analysis, this fact ensures a small leakage of the mode out of a guide with enhanced density. It is important that in the case (4.92) the simple approximations (2.80) and (2.81) are applicable to represent $q_{1,2}(p, N)$. In addition, one can easily verify that the formulas (2.80) and (2.81) give the same results as do the expressions (4.81). Hence for studying the properties of the axisymmetric leaky modes the previous formulas (4.82) to (4.84) may be utilized, in which one should neglect the terms of order Y_{LH} and take, instead of functions $K_m(S_2)$, more convenient Hankel functions $H_n^{(1)}(Q_2)$ with $Q_2 = iS_2 = k_0 a q_2$, where $\mathrm{Re}\,(q_2) > 0$ (see §4.3). If this is used in Equation (4.83) it gives

$$
\left(\frac{J_1(\tilde{Q}_1)}{J_0(\tilde{Q}_1)} + \frac{\tilde{Q}_1}{S_1} \frac{K_1(S_1)}{K_0(S_1)} \right) \left(\frac{J_1(\tilde{Q}_2)}{J_0(\tilde{Q}_2)} - \frac{\eta_a \tilde{Q}_2}{\tilde{\eta} Q_2} \frac{H_1^{(1)}(Q_2)}{H_0^{(1)}(Q_2)} \right)
$$
$$
= æ(p) \frac{J_1(\tilde{Q}_1)}{J_0(\tilde{Q}_1)} \frac{K_1(S_1)}{K_0(S_1)} + i\sigma(p) \frac{J_1(\tilde{Q}_2)}{J_0(\tilde{Q}_2)} \frac{H_1^{(1)}(Q_2)}{H_0^{(1)}(Q_2)}, \qquad (4.93)
$$

where $æ(p)$ and $\sigma(p)$ are obtainable from (4.84) by putting $S_2 = -iQ_2$ and $\left(Y_{\mathrm{LH}}^2 - 1 \right)^2 \approx 1$.

Show that when p' lies in a range (4.75), Equation (4.93) has roots $p = p' - ip''$ with $p'' \ll p'$, which correspond to slightly-leaky modes. To distinguish them from bound modes, we shall use the subscript ν ($\nu = 1, 2, \ldots$), instead of n, for the mode numbering.

Let a drop in the plasma density at the interface $\rho = a$ be rather *small*, so that inequality (4.85) holds. Then the condition (4.92) of validity of Equation (4.93) is satisfied if

$$
4\tilde{X}/X_a \ll Y. \qquad (4.94)
$$

In this case $\max\{\mathcal{P}, \tilde{\mathcal{P}}_c\} = \mathcal{P}$, and $æ$ and σ are small parameters such that $|\sigma| \ll |æ| \ll 1$ (it is assumed that $(\mathcal{P}/|\Delta p|)^{1/2}(\tilde{X}/X_a - 1)^2/2 \ll 1$). Hence the dispersion equation (4.93) can be solved approximately as was made for the frequency interval (2.70). In a zero-order approximation (viz. $æ \to 0$, $\sigma \to 0$), Equation (4.93) becomes

$$
\left(\frac{J_1(\tilde{Q}_1)}{J_0(\tilde{Q}_1)} + \frac{\tilde{Q}_1}{S_1} \frac{K_1(S_1)}{K_0(S_1)} \right) \left(\frac{J_1(\tilde{Q}_2)}{J_0(\tilde{Q}_2)} - \frac{X_a \tilde{Q}_2}{\tilde{X} Q_2} \frac{H_1^{(1)}(Q_2)}{H_0^{(1)}(Q_2)} \right) = 0. \qquad (4.95)
$$

The approximate dispersion equation (4.95) has a multiplicative form, and its roots are divided into two families. One family is constituted of the roots of Equation (4.88) and gives the "whistler resonances" of the plasma cylinder. These roots,

$p = p'_\nu$, correspond to the leaky modes whose fields involve appreciable contributions from the components appropriate to quasi-longitudinal whistlers. The zero-order solutions $p_\nu^{(0)} = p'_\nu$ are now inserted in the right side of Equation (4.93) and in the second multiplier on its left side, and the dispersion equation is solved to give a first-order approximation. This process enables us to determine the attenuation constants p''_ν for the leaky modes. Before proceeding in this way, note that the zero-order equation (4.88) permits solutions when $\tilde{Q}_1(\mathcal{P}) > \mu_1^{(0)} = 2.405\ldots$ (recall that $\mu_l^{(k)}$ stands for the lth zero of the Bessel function J_k). Since $|Q_2| \simeq |\tilde{Q}_2| \gg |\tilde{Q}_1|$, then $|Q_2| \gg 1$ whence $H_1^{(1)}(Q_2)/H_0^{(1)}(Q_2) \approx -\mathrm{i}$. Making use of this fact, we rewrite Equation (4.93) in the following form:

$$\left(\frac{J_1(\tilde{Q}_1)}{J_0(\tilde{Q}_1)} + \frac{\tilde{Q}_1}{S_1} \frac{K_1(S_1)}{K_0(S_1)} - \sigma(p) \right) \frac{J_1(\tilde{Q}_2)}{J_0(\tilde{Q}_2)}$$

$$= -\mathrm{i}\, \frac{X_a \tilde{Q}_2}{\tilde{X} Q_2} \left(\frac{J_1(\tilde{Q}_1)}{J_0(\tilde{Q}_1)} + \frac{\tilde{Q}_1}{S_1} \frac{K_1(S_1)}{K_0(S_1)} \right) + \ae(p) \frac{J_1(\tilde{Q}_1)}{J_0(\tilde{Q}_1)} \frac{K_1(S_1)}{K_0(S_1)}. \qquad (4.96)$$

A close examination of (4.96) shows that attenuation constant p''_ν of each mode varies with the frequency ω between the lower $p''_{\nu,\min}(\omega)$ and upper $p''_{\nu,\max}(\omega)$ boundaries, that is $p''_{\nu,\min}(\omega) \le p''_\nu(\omega) \le p''_{\nu,\max}(\omega)$. It turns out that the minima and maxima of the functions $p''_\nu(\omega)$ for axisymmetric modes correspond to the conditions $\tilde{Q}_2(p'_\nu) = \mu_l^{(0)}$ and $\tilde{Q}_2(p'_\nu) = \mu_l^{(1)}$, respectively. Here we restrict ourselves to calculation of the quantity $p''_{\nu,\max}(\omega)$ only. To do this, we make a series expansion of the left and right sides of Equation (4.96) in powers $\Delta p_\nu^{(1)} = p - p_\nu^{(0)}$ at the point $p = p_\nu^{(0)}$ such that $\tilde{Q}_2(p_\nu^{(0)}) = \mu_l^{(1)}$, assuming that $p'' \ll (k_0 a Y)^{-1}$ (this assumption will be confirmed by the result to be found). Retaining terms proportional to the first power of $\Delta p^{(1)}$ and taking into account the inequality $|\sigma| \ll |\ae|$, we find that $\Delta p_\nu^{(1)} = -\mathrm{i}p''_{\nu,\max}$, where

$$p''_{\nu,\max}(\omega) = \frac{1}{2k_0 a\, \beta(p'_\nu)} \frac{\mathcal{P}^2}{p'^2_\nu} \left(\frac{\tilde{X}}{X_a} - 1 \right)^2. \qquad (4.97)$$

In the preceding,

$$\beta(p) = \left(1 + \frac{2}{\tilde{Q}_1^2\, K(S_1)} + \frac{1}{\tilde{Q}_1^2\, K^2(S_1)} \right) \left(1 + \frac{\tilde{Q}_1^2}{2(k_0 a p)^2} \right)$$

$$+ \frac{1}{2} \left(\frac{1}{S_1^2\, K^2(S_1)} + \frac{2}{S_1^2\, K(S_1)} - 1 \right) \left(1 + \frac{\mathcal{P}^4}{p^4} \right), \qquad (4.98)$$

where $K(S_1) = K_1(S_1)/S_1\, K_0(S_1)$. Formula (4.97), together with (4.98), is applicable when $\left(\dfrac{\tilde{X}}{X_a} - 1 \right)^2 \ll 2\beta(p'_\nu)/Y$, and it implies the inequality $p''_{\max} \ll p'$.

In addition, it can be shown that $p''_{\max} \gg p''_{\min}$ (see, for details, Zaboronkova *et al.*, 1993 b).

Thus, under the condition (4.94) in the range (4.75), the leakage of energy from a duct with increased density at frequencies (2.71) turns out to be very slight (even in the case of a rather narrow guide, i.e., $a \sim \lambda_{\rm w} = 2\pi/p'$), while the dependence of the attenuation constant $p''(\omega)$ on frequency has a pronounced resonant character. Such a character of the function $p''(\omega)$ may be associated with resonances of fine-scale escaping waves. Note that in the case considered it is satisfactory to use for leaky modes the same numbering system as was suggested for the eigenmodes at frequencies below $\omega_{\rm LH}$.

Apart from the slightly leaky modes discussed above, there exist solutions of the second family. They may be found as zeros of the second multiplier in the approximate dispersion equation (4.95) and correspond to some modes with a comparatively strong leakage. These are of no practical interest and, therefore, will not be discussed here.

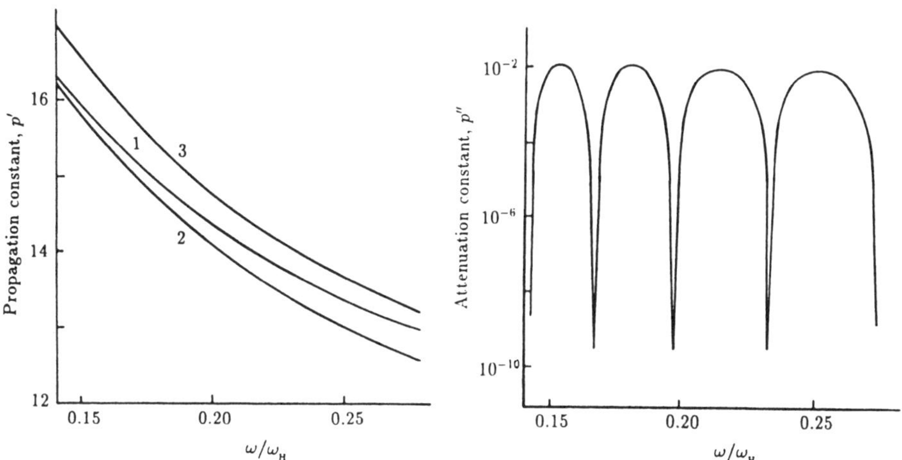

Figure 4.7. Plots of p' (curve 1) and p'' versus $\omega/\omega_{\rm H}$ for the lower-order axisymmetric leaky mode of a cylindrical duct with enhanced density, assuming $\tilde{X}/X_a = 1.1$, $X_a^{1/2}/Y = 5.6$, and $k_0 a Y = 2.5$. Curves 2 and 3 are for $\mathcal{P}(\omega)$ and $\tilde{\mathcal{P}}(\omega)$, respectively.

The rigorous dispersion equation (4.71) was solved numerically, and the results are in good agreement with those obtained analytically. As an example, Figure 4.7 shows the results of numerical solution of equation (4.71) for the case $\tilde{X}/X_a = 1.1$, $X_a^{1/2}/Y = 5.6$, $k_0 a Y = 2.5$. We also show the curves $\mathcal{P}(\omega)$ and $\tilde{\mathcal{P}}(\omega)$ in Figure 4.7 (a). Here, only one lower-order leaky mode exists, with its propagation constant p_1' lying in the range (4.75). Observe that $p_1'' \ll p_1'$ for this mode, and the minima and maxima of p_1'' correspond indeed with a high accuracy to the relations $\tilde{Q}_2(p_1') = \mu_l^{(0)}$ and $\tilde{Q}_2(p_1') = \mu_l^{(1)}$, respectively.

Note that as ω decreases, each dispersion curve moves to the lower boundary $\max\{\mathcal{P}, \tilde{\mathcal{P}}_c\} = \mathcal{P}$ of the range (4.75) and eventually meets it at the frequency called the lower cutoff frequency. As ω increases, each dispersion curve leaves the domain where it lies between $\max\{\mathcal{P}, \tilde{\mathcal{P}}_c\} = \mathcal{P}$ and $\tilde{\mathcal{P}}$, and appears in the domain where it lies between $\max\{\mathcal{P}, \tilde{\mathcal{P}}_c\} = \tilde{\mathcal{P}}_c$ and $\tilde{\mathcal{P}}$; with further increasing ω, it eventually crosses the upper boundary $\tilde{\mathcal{P}}$ at the frequency called the upper cutoff frequency, which is situated somewhat below half the electron gyrofrequency. Although outside the range (4.75) the leaky modes may exist, their attenuation becomes so large there that they are of no practical interest.

Now consider the case of a *strong enhancement of the density* within the duct. We shall again seek solutions of the dispersion equation in the range (4.75). There are two special cases, namely

$$\tilde{X} \gg X_a, \qquad Y \gg 4\tilde{X}/X_a \tag{4.99}$$

and

$$4\tilde{X}/X_a > Y \gg 1, \tag{4.100}$$

when some analytic treatment is possible. For the latter case, the condition (4.92) is also assumed. For the former case, that condition is satisfied automatically. For these two cases the approximate expressions (4.81) and (4.82) preserve and hence the approximate form (4.93) of the dispersion equation can again be employed. Now Equation (4.93) contains small parameters $|\sigma|$ and $X_a/\tilde{X}$. In a zero-order approximation (viz. $\sigma \to 0$, $X_a/\tilde{X} \to 0$) it reduces to

$$\left(\frac{J_1(\tilde{Q}_1)}{J_0(\tilde{Q}_1)} + \frac{\tilde{Q}_1}{S_1} \frac{K_1(S_1)}{K_0(S_1)} \right) J_1(\tilde{Q}_2) = \text{æ}(p) \frac{J_1(\tilde{Q}_1)}{J_0(\tilde{Q}_1)} \frac{K_1(S_1)}{K_0(S_1)} J_0(\tilde{Q}_2). \tag{4.101}$$

Note that the parameter $\text{æ}(p) \simeq \tilde{\mathcal{P}}^2/(ps_1)$ is no longer small, and because of this fact, the zero-order equation does not take the multiplicative form as was observed for a small drop in the plasma density. In other words, a coupling between fine-scale quasi-electrostatic (QE) waves and large-scale whistlers described by the quasi-longitudinal (QL) approximation cannot be neglected for rather large value of ratio $\tilde{X}/X_a$, in contrast to (4.85), where this coupling does not appear in the zero-order approximation. The coupling between QL- and QE-waves may become apparent even for moderate ratios $\tilde{X}/X_a$ (about $2-\text{to}-3$) that is confirmed by the results of numerical solution of Equation (4.71). The pertinent results are given in Figure 4.8 where the dispersion characteristics $p_\nu'(\omega)$ and $p_\nu''(\omega)$ obtained numerically for the leaky modes of numbers $\nu = 1, 2, \ldots, 7$ (curves $1-\text{to}-7$) are plotted for the following parameters: $\tilde{X}/X_a = 3.2$, $X_a^{1/2}/Y = 5.6$, and $k_0 aY = 1$.* Dot-and-dash lines 8 and 9 in Figure 4.8 (a) show the behavior of the lower and upper boundaries of range (4.75) with the frequency ω. In addition, dashed curves 10 and 11 plotted in Figure 4.8 (a) represent the behavior of solutions of zero-order Equation (4.88) neglecting a coupling between the waves with two distinct transverse wavenumbers. At the points where the dispersion curves $p'_\nu(\omega)$ cross the dashed lines, the relation $\tilde{Q}_2(p'_\nu) = \mu_l^{(0)}$ occurs, and the attenuation constants $p''_\nu(\omega)$ reach their minima $p''_{\nu,\min}$. With receding p'_ν from these crossing points, the quantities p''_ν increase and reach their maxima $p''_{\nu,\max}$ at the points where $\tilde{Q}_2(p'_\nu) = \mu_l^{(1)}$. Following the steps similar to those made for a weak enhancement, it is possible to deduce that

$$p''_{\nu,\max}(\omega) = \frac{X_a}{\tilde{X}} \frac{1}{k_0 aY\, \gamma(p'_\nu)} , \tag{4.102}$$

where

$$\gamma(p) = 1 + \ae(p) \frac{S_1}{k_0 apY} \left[1 + 2\left(\frac{k_0 ap}{\tilde{Q}_1}\right)^2 \right] \tag{4.103}$$

(Zaboronkova *et al.*, 1993 b). It is evident from Figure 4.8 as well as from (4.102) and (4.103) that even for a rather narrow duct such that $k_0 ap' \sim 1$, the inequalities $p''_{\min} \ll p''_{\max} \ll p'$ are satisfied. For clarity, the parts of curves 4 to 7 in Figure 4.8 (a) which are closest to solutions of Equation (4.88) are darkened. As seen from Figure 4.8 (b), the attenuation constants p''_ν corresponding to these darkened parts are well below $p''_{\nu,\max}$. Notice that modes whose curves p'_ν lie in the vicinity of line 10 in Figure 4.8 (a), have one large-scale variation of the field components over the radius; the curves lying in the vicinity of line 11 are appropriate to modes with two large-scale variations.

Thus it is seen that leaky modes guided in the frequency interval (2.71) by comparatively strong enhancements of density can be distinguished from each other by both the number of large-scale variations over the radius and the number of fine-scale oscillations. Therefore, the customary numbering system similar to that used for open dielectric waveguides is not suitable here. To provide a unique numbering system, it is satisfactory to number the modes in order of increasing the magnitude of their propagation constants, starting from the lower boundary of range (4.75) and using the numbers $\nu = 1, 2, \ldots$, as in Figure 4.8. This numbering system allows unique ordering, but it does not yet seem to be in general use.

Among leaky modes having one large-scale variation over the radius, we distinguish especially a mode which is the least attenuated. We shall call it the *least*

---

* The parameters are chosen to be relevant to the conditions of the laboratory experiment of Vdovichenko *et al.* (1986).

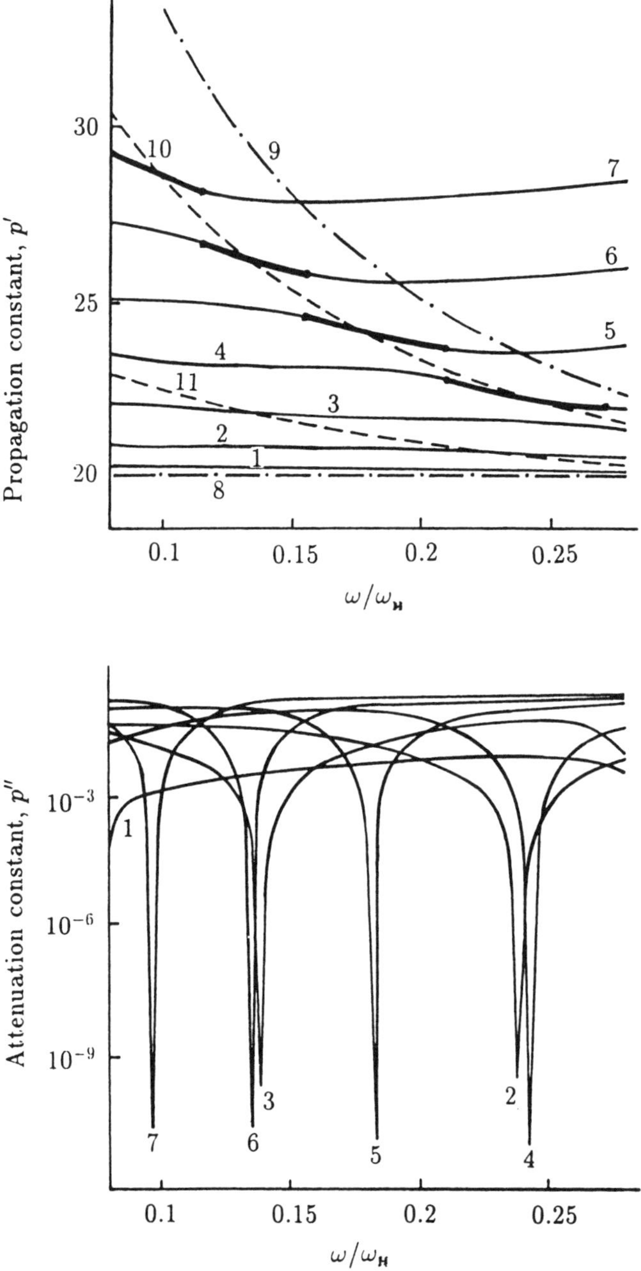

Figure 4.8. Plots of p' and p'' versus ω/ω_H for the axisymmetric leaky modes of a cylindrical duct with enhanced density, assuming $\tilde{X}/X_a = 3.2$, $X_a^{1/2}/Y = 5.6$, and $k_0 a Y = 1$ (curves 1 to 7). Curves 8 and 9 are for the lower and upper boundaries of range (4.75). Curves 10 and 11 are for the solutions of Equation (4.88).

attenuated lower-order leaky mode. Figure 4.8 shows that the number ν of the least attenuated lower-order mode depends on ω. Thus, for $\omega = 0.10\,\omega_{\mathrm{H}}$ its number is $\nu = 7$, for $\omega = 0.13\,\omega_{\mathrm{H}}$ the number is $\nu = 6$, etc.

We notice that as the wave frequency ω decreases, each curve p'_ν crosses the line $p = \mathcal{P}(\omega)$, as in the case of a smaller enhancement of density, while with increasing ω the dispersion curves leave the range (4.75), getting into the region

$$p' > \tilde{\mathcal{P}}(\omega). \tag{4.104}$$

The results presented in Figure 4.8 show that though the leakage is much higher in domain (4.104) than in (4.75), it still remains small whenever the ratio $\tilde{X}/X_a$ is large enough. This may be demonstrated analytically when $\tilde{X} \gg X_a$, $Y \gg 1$, $k_0 a\,|p|\,Y \gg 1$, and $p' \gg \mathcal{P}$, in which case the dispersion equation (4.96) can be transformed into the approximate form

$$\frac{J_1(\tilde{Q}_2)}{J_0(\tilde{Q}_2)} = -\mathrm{i}\,\frac{X_a}{\tilde{X}}\,. \tag{4.105}$$

Here we retain the small parameter $X_a/\tilde{X}$ since it is responsible for producing the imaginary part of the complex propagation constant $p = p' - \mathrm{i}p''$. We notice that under the above conditions, the parameter æ becomes small again, and Equation (4.105) could therefore be derived directly from Equation (4.95). Also note that $\tilde{Q}_2$ now obeys the relation $\tilde{Q}_2 \approx k_0 a p Y$. Solving Equation (4.105) using the well-known large-argument approximation for the Bessel functions then yields

$$p_\nu \approx \frac{\pi}{k_0 a Y}\left(\nu + \frac{1}{4}\right) - \mathrm{i}\,\frac{X_a}{\tilde{X}}\,\frac{1}{k_0 a Y}\,, \tag{4.106}$$

where ν is an integer such that $\nu \gg 1$ (Zaboronkova *et al.*, 1993 b). We stress that leaky modes whose propagation constants lie in the domain (4.104) may be slightly attenuated only for the duct with a strong enhancement of density and a rather sharp boundary. If we choose a smooth density profile, the leakage would remain small only for the modes with propagation constants lying in the range (4.75).

4.4.3. Some extension to the case of nonsymmetric modes

In the preceding work the dispersion properties only of axisymmetric modes on cylindrical density enhancements have been considered. We now discuss some features of nonsymmetric modes. A convenient method of studying their properties is similar to that described above.

First consider the duct of a large radius such that the inequalities

$$a \gg \frac{|m|}{k_0}\,\max\left(\left|\frac{\tilde{\alpha}_k}{\tilde{q}_k}\right|,\left|\frac{\tilde{\beta}_k}{\tilde{q}_k}\right|\right)\left|\frac{J_m(\tilde{Q}_k)}{J_{m+1}(\tilde{Q}_k)}\right|,$$

$$a \gg \frac{|m|}{k_0}\,\max\left(\left|\frac{\alpha_k}{s_k}\right|,\left|\frac{\beta_k}{s_k}\right|\right)\left|\frac{K_m(S_k)}{K_{m+1}(S_k)}\right| \tag{4.107}$$

are simultaneously true ($k = 1, 2$), in which case the quantities $J_m^{(1,2)}$ and $K_m^{(1,2)}$ in the general dispersion equation (4.67) reduce approximately to

$$J_m^{(1,2)} \approx \frac{J_{m+1}(\tilde{Q}_{1,2})}{\tilde{Q}_{1,2}\, J_m(\tilde{Q}_{1,2})}, \qquad K_m^{(1,2)} \approx \frac{K_{m+1}(S_{1,2})}{S_{1,2}\, K_m(S_{1,2})}.$$

Equation (4.67) then simplifies to the following form:

$$\left(M_{12} L_{12} \frac{K_{m+1}(S_2)}{S_2\, K_m(S_2)} - M_{11} L_{22} \frac{K_{m+1}(S_1)}{S_1\, K_m(S_1)} \right) \frac{J_{m+1}(\tilde{Q}_1)}{\tilde{Q}_1\, J_m(\tilde{Q}_1)}$$

$$+ \left(M_{21} L_{21} \frac{K_{m+1}(S_1)}{S_1\, K_m(S_1)} - M_{22} L_{11} \frac{K_{m+1}(S_2)}{S_2\, K_m(S_2)} \right) \frac{J_{m+1}(\tilde{Q}_2)}{\tilde{Q}_2\, J_m(\tilde{Q}_2)}$$

$$+ \frac{\tilde{\eta}}{\eta_a} \frac{J_{m+1}(\tilde{Q}_1)}{\tilde{Q}_1\, J_m(\tilde{Q}_1)} \frac{J_{m+1}(\tilde{Q}_2)}{\tilde{Q}_2\, J_m(\tilde{Q}_2)} + \frac{K_{m+1}(S_1)}{S_1\, K_m(S_1)} \frac{K_{m+1}(S_2)}{S_2\, K_m(S_2)} = 0. \qquad (4.108)$$

It is worth noting that Equation (4.108) is obtainable from Equation (4.71) for axisymmetric modes by making the formal replacements

$$\frac{J_1(\tilde{Q}_k)}{J_0(\tilde{Q}_k)} \rightarrow \frac{J_{m+1}(\tilde{Q}_k)}{J_m(\tilde{Q}_k)}, \qquad \frac{K_1(S_k)}{K_0(S_k)} \rightarrow \frac{K_{m+1}(S_k)}{K_m(S_k)}. \qquad (4.109)$$

This implies that useful approximate equations, which could be derived after some lengthy algebra from Equations (4.108) for the same special cases as analyzed earlier, may be obtained in a simpler way, namely by making the replacements (4.109) in the corresponding equations of sections 4.4.1 and 4.4.2. Solutions to the equations which are to be derived in such a manner can then be deduced in a similar way to that of axisymmetric modes, and the concrete expressions for propagation constants need not be given here.

If the duct is of a moderate radius such that inequalities (4.107) are no longer satisfied, but the quantities $\tilde{Q}_1$ and S_1 are not still much less than unity, a more accurate approximation must be used to study the dispersion equation. For a *small drop in the plasma density* at $\rho = a$, when inequality (4.85) holds, Equation (4.67) in a zero-order approximation takes the multiplicative form

$$\left(J_m^{(1)} + K_m^{(1)} \right) \left(J_m^{(2)} + \frac{\eta_a}{\tilde{\eta}}\, K_m^{(2)} \right) = 0 \qquad (4.110)$$

that leads to the equations

$$\frac{J_{m+1}(\tilde{Q}_1)}{J_m(\tilde{Q}_1)} + \frac{m\tilde{\alpha}_1}{\tilde{Q}_1} = -\frac{\tilde{Q}_1}{S_1} \left(\frac{K_{m+1}(S_1)}{K_m(S_1)} + \frac{m\alpha_1}{S_1} \right) \qquad (4.111)$$

and

$$\frac{J_{m+1}(\tilde{Q}_2)}{J_m(\tilde{Q}_2)} - \frac{m\tilde{\beta}_2}{\tilde{Q}_2} = -\frac{\eta_a \tilde{Q}_2}{\tilde{\eta} S_2} \left(\frac{K_{m+1}(S_2)}{K_m(S_2)} - \frac{m\beta_2}{S_2} \right). \qquad (4.112)$$

It is fairly clear that the last two approximate equations represent a generalization of Equations (4.88) and (4.89) (see also Equation (4.95)) to the case of nonsymmetric modes. An accurate derivation of zero-order equations (4.111) and (4.112) from the rigorous equation (4.67) needs a considerable amount of algebraic manipulation. There is, however, a simple instructive method to obtain them without directly applying to Equation (4.67). By inspecting the behavior of the terms involved in the field expressions (4.41) to (4.46), and (4.50) to (4.55), it can be shown that the part associated with branches $\tilde{q}_1$ and s_1 represents a quasi-TE wave in the full field, while the part associated with branches $\tilde{q}_2$ and s_2 represents a 'quasi-electrostatic' wave, in which $\tilde{q}_2^2 \approx -\tilde{\eta}p^2/\tilde{\varepsilon}$ and $s_2^2 \approx \eta_a p^2/\varepsilon_a$. Recall that a coupling between these parts is practically insignificant when a drop in the density is small enough, provided that the wave frequency is not very close to the cutoff frequency. Hence, to obtain Equation (4.111), it is satisfactory to put $B_2 = C_2 = 0$ everywhere in the field expressions and then provide continuity of the E_ϕ- and $\mathcal{H}_z$-components at the surface $\rho = a$. By analogy, Equation (4.112) is deduced by putting $B_1 = C_1 = 0$ in the field expressions and then satisfying the boundary conditions for the E_z- and D_ρ-components. Such a procedure is suitable whenever the coupling may be neglected, at least as a zero-order approximation, in frequency intervals (2.70) and (2.71). At frequencies (2.71), the coupling is also negligible when the leaky-mode propagation constants are well above $\tilde{\mathcal{P}}$, even if $\tilde{N} \gg N_a$. The dispersion properties for such modes are defined by Equation (4.112) where $\tilde{\beta}_2 \approx \beta_2 \approx 1$. By using the quasi-static theory method, we may obtain the same result more simply. We first express the electric field via $\mathbf{E} = -\nabla\psi$ (see § 2.6) and then satisfy the requisite boundary conditions for the E_z- and D_ρ-components.

There are situations, however, when the coupling appears even within a zero-order approximation. When $\tilde{N} \gg N_a$, this is true in the range (4.74) for the eigenmodes at frequencies (2.70), and in the range (4.75) for the leaky modes at frequencies (2.71). Now, to construct the zero-order dispersion equations, we notice the absence of an appreciable contribution from the branch s_2 (or, in other terms, q_2) to the full field in the *outer region*. This may be verified by inspection of the field solutions and will be fairly clear from the plots of the field components presented in the next section. From the formal viewpoint, this results in the absence of any contribution from that branch to the zero-order equations (4.91) and (4.101). Therefore, the method of a simple derivation of those equations would consist in putting $C_2 = 0$ in the field expressions throughout and then satisfying the boundary conditions at $\rho = a$ for the E_ϕ-, $\mathcal{H}_\phi$-, and $\mathcal{H}_z$-components. This immediately gives Equations (4.91) and (4.101) for $m = 0$. The appropriate equations for the general case $m \neq 0$ can be derived in a similar manner but are too complicated to give the useful result immediately. The derivation may be left for the reader's exercise.

In the procedure described above the condition of continuity for the E_z-component has been omitted. Some small imprecision thereby appearing in the E_z-component here is unimportant as it just lies within an accuracy of a zero-order approximation for strong density enhancements.

Because of gyrotropic properties of a magnetoplasma, the modes with positive and negative indices m are different. For a weak density enhancement, when the dispersion properties are defined by Equation (4.111), it can be shown that $\tilde{\alpha}_1 \approx \alpha_1 \approx -2$. If this and the recurrence relations for cylindrical functions are used, Equation (4.111) becomes

$$\frac{J_{m-1}(\tilde{Q}_1)}{J_m(\tilde{Q}_1)} = \frac{\tilde{Q}_1}{S_1} \frac{K_{m-1}(S_1)}{K_m(S_1)} \, .$$

When the duct is wide enough, $|\tilde{Q}_1| \ll |S_1|$ and the right side of the above equation is small. Thus we may set $J_{m-1}(\tilde{Q}_1) \approx 0$. Hence a considerable similarity would be observed for propagation constants of modes with $m = 1 \pm M$ where $M = 1, 2, \ldots$. Near symmetry about $m = 1$ in the properties of modes in wide cylindrical ducts was noticed by Laird and Nunn (1975).

Thus, no significant differences between the properties of axisymmetric and non-symmetric modes have been observed. The only noticeable feature of nonsymmetric solutions to be pointed out here is that the lower-order mode with the azimuthal index $m = 1$ still propagates and shows no cutoff when the duct becomes so narrow or weak that it disappears. We shall hereinafter call this mode the *fundamental mode*. It may be unambiguously identified, regardless of the numbering system used, as it is the only mode which propagates for all values of a and $\tilde{N}$. Such a behavior is nearly similar to that of the fundamental ("dipolar") eigenmode of a circular dielectric fiber (Snyder and Love, 1983). At frequencies (2.70) the fundamental mode is bound, at frequencies (2.71) it is leaky. The leakage, however, remains small if $Y \gg 1$, even when the duct radius tends to zero.

The propagation constant of the fundamental mode may be found explicitly for the narrow duct. Consider, as an example, the dispersion equation (4.67) with $m = 1$ for frequencies (2.71). As earlier, in the dispersion equation, we put $S_2 = -iQ_2$, with $\mathrm{Re}\,(Q_2) > 0$, and utilize the identities (4.49). When Y is sufficiently large such that $Y > 2/[1 - (1 - X_a/\tilde{X})^{1/2}]$, the propagation constant p of the fundamental mode, for small a, would be slightly above the quantity $\mathcal{P}$. We shall therefore seek the propagation constant in the form $p = \mathcal{P} + \Delta p$, assuming that $|\Delta p| \ll \mathcal{P}$. The transverse wavenumbers would then be given approximately by

$$\tilde{q}_{1,2} \approx q_{1,2}(\mathcal{P}, \tilde{N}), \quad s_1 \approx 2(\Delta p\, \mathcal{P})^{1/2}(1 + \eta_a^{-1} \mathcal{P}^2)^{-1/2}, \quad q_2 \approx q_2(\mathcal{P}, N_a).$$

Suppose that the duct radius a is so small that the quantities $\tilde{Q}_{1,2}$, S_1 and Q_2 satisfy the following inequalities:

$$|\tilde{Q}_{1,2}| \ll 1, \qquad |S_1| \ll 1, \qquad |Q_2| \ll 1.$$

The cylindrical functions $J_{1,2}(\tilde{Q}_{1,2})$, $K_{1,2}(S_1)$ and $H_{1,2}^{(1)}(Q_2)$ in Equation (4.67) for the case $m = 1$ may then be replaced by their small-argument approximations:

$$J_1(\tilde{Q}_{1,2}) \approx \frac{\tilde{Q}_{1,2}}{2}, \qquad J_2(\tilde{Q}_{1,2}) \approx \frac{\tilde{Q}_{1,2}^2}{8},$$

$$K_1(S_1) \approx \frac{1}{S_1}, \qquad K_2(S_1) \approx \frac{2}{S_1^2} + \ln \frac{2}{S_1} - \gamma,$$

$$H_1^{(1)}(Q_2) \approx -\mathrm{i}\,\frac{2}{\pi Q_2}, \qquad H_2^{(1)}(Q_2) \approx -\mathrm{i}\,\frac{4}{\pi Q_2^2},$$

where $\gamma = 0.5772\ldots$ is Euler's constant. If these are inserted in the dispersion equation and a pass to the limit $a \to 0$ is made in it, all the terms arising from functions $K_{1,2}(S_1)$ and containing powers of S_1 will cancel, and only $\ln(2/S_1)$ will remain in the resulting form of the dispersion equation. Having expressed Δp via S_1, it is possible to come to the following result for the real propagation constant:

$$p' = \mathcal{P} + \frac{1 + \eta_a^{-1}\mathcal{P}^2}{(k_0 a\, \mathcal{C})^2 \mathcal{P}} \exp\left[-\frac{2}{(k_0 a)^2}\left(\frac{1 + \eta_a^{-1}\tilde{\mathcal{P}}^2}{\tilde{\mathcal{P}}^2 - \mathcal{P}^2} - \frac{\mathcal{P}^2}{2\varepsilon_a g_a}\frac{\left[1 - \eta_a^{-1}(\varepsilon_a + g_a)\right]^2}{1 + \eta_a^{-1}\mathcal{P}^2}\right)\right], \quad (4.113)$$

where $\mathcal{C} = e^\gamma = 1.7811\ldots$. Although the formula (4.113) is derived in a straightforward fashion, the proof is too tedious to be presented here. Note that we have found only the real part p' of the complex propagation constant $p = p' - \mathrm{i}p''$. To obtain the attenuation constant p'', it is necessary to employ a better approximation for the cylindrical functions. If this is used, it gives the expression for p'', which shows that indeed $p'' \ll p'$ when $Y \gg 1$. As the duct becomes narrower or weaker p tends to $\mathcal{P}$, and the large-scale part of the fundamental-mode field becomes a plane wave, traveling along the ambient magnetic field with the propagation constant $p = \mathcal{P}$; the fine-scale part becomes a conical wave with the same p and the transverse wavenumber $q_2(\mathcal{P}, N_a)$.

4.4.4. Modes of a plane duct with enhanced density

By analogy, it is possible to consider the modes in a plane duct (slab). To do this, one should repeat the preceding analysis, employing everywhere the pertinent trigonometric functions instead of the cylindrical functions. For example, when conditions (4.85) and (4.87) hold well, the zero-order approximate equations defining the propagation constants of the even and odd modes in a uniform, sharply bounded plane duct with enhanced density are written thus:

(i) for the even modes,

$$\cot \tilde{Q}_1 = \tilde{Q}_1/S_1; \qquad (4.114)$$

(ii) for the odd modes,

$$\tan \tilde{Q}_1 = -\tilde{Q}_1/S_1. \qquad (4.115)$$

Now the fundamental mode is the lower-order even mode. For a detailed discussion of the properties of modes in a uniform slab of ionization see Adachi (1965, 1966) and Zaboronkova *et al.* (1992 a).

4.4.5. Modes of a cylindrical trough

Finally, we give some results for a cylindrical *trough*. Recall that propagation constants of the eigenmodes trapped in a trough lie in the range (4.76) when $Y > 2$ (i. e. $\omega < \omega_H/2$). If ω is well above the lower hybrid frequency, we may employ the approximations (2.78) for elements of the dielectric tensor, in which case the boundaries $p = \tilde{\mathcal{P}}_c$ and $p = \mathcal{P}_c$ of the domain (4.76) are independent of the wave frequency ω, as seen from (2.79). At $Y = 2$, we have that $\tilde{\mathcal{P}}_c = \tilde{\mathcal{P}} \equiv [\tilde{X}/(Y - 1)]^{1/2}$ and $\mathcal{P}_c = \mathcal{P} \equiv [X_a/(Y - 1)]^{1/2}$. For smaller values of Y, when $1 < Y < 2$ (i. e. $\omega_H/2 < \omega < \omega_H$), the boundaries for propagation constants are given by (4.77). This implies that with increasing ω dispersion curves would leave (at $\omega = \omega_H/2$) the domain (4.76) to appear in the domain (4.77). This is confirmed by a numerical investigation of the dispersion equation. Figure 4.9 shows the dispersion curves $p_n(\omega)$ calculated numerically by using Equation (4.71) for the case of the axisymmetric eigenmodes, for the following parameters: $X_a/\tilde{X} = 1.5$, $X_a^{1/2}/Y = 29.3$, and $k_0 a Y = 0.42$. It is also useful to give the values of the lower and upper boundaries of range (4.76): $\tilde{\mathcal{P}}_c = 47.8$ and $\mathcal{P}_c = 58.6$. Note that the parameters taken are typical of the conditions of the laboratory experiment reported by Zaboronkova *et al.* (1992 a) and mentioned in § 1.3. The curve labelled 1 in Figure 4.9 represents the behavior of the propagation constant of the lower order mode, that is, the mode with the minimum number of variations of the field along the transverse coordinate. In contrast to the case of density enhancements, the number of variations may be greater than unity for the lower order mode at frequencies below half the electron gyrofrequency. This is explained by the fact that the quantities $\tilde{Q}_{1,2}$, which are now approximated by $\tilde{Q}_1 \simeq \tilde{Q}_2 \simeq k_0 a \tilde{q}_c$, with $\tilde{q}_c = q_{1,2}(\tilde{\mathcal{P}}_c, \tilde{N})$, may appreciably exceed unity if the guide radius is large enough. Cutoff for each mode of this type appears when the dispersion curve crosses the upper boundary $p = \mathcal{P}_c$ of domain (4.76).

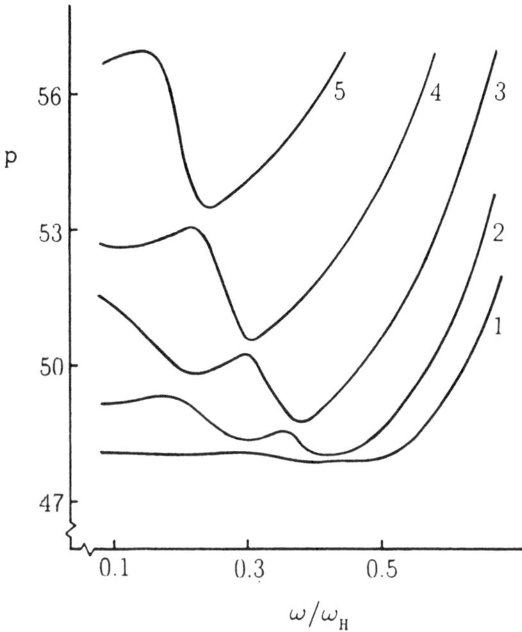

Figure 4.9. Plots of p versus ω/ω_H for the axisymmetric eigenmodes of a cylindrical duct with depressed density, assuming $X_a/\tilde{X} = 1.5$, $X_a^{1/2}/Y = 29.3$, and $k_0 a Y = 0.42$.

4.5. Uniform duct without collision damping — field distribution

Examine how the field components in modes vary across the duct. In this examination, we shall follow the same order as used in the previous sections.

We start by presenting the field distribution in modes existing in a cylindrical enhancement at frequencies (2.70). In Figure 4.10 the field components for the lower order axisymmetric eigenmode are shown. In this example $\tilde{X}/X_a = 2$, $X_a^{1/2}/Y = 4.5$, $Y = 350$, $Y_{\mathrm{LH}} = 1.25$, and $k_0 a = 4.2 \times 10^{-2}$.[*] The propagation constant of the mode is equal to $p = 110.85$. In these conditions, the behavior of the field is largely determined by the part associated with the wavenumbers $\tilde{q}_1$ and s_1. The presence of the other part associated with $\tilde{q}_2 = i\tilde{s}_2$ and s_2 results in sharp peaks of the electric components E_ρ and E_z and in a rapid variation of the magnetic components $\mathcal{H}_\phi$ and $\mathcal{H}_z$ in the near vicinity of the duct interface. Note that the quantity iE_ρ,

---

[*] These parameters are close to those that can be observed in large plasma devices such as described in the work of Golubyatnikov *et al.* (1995) reviewed in §1.3.

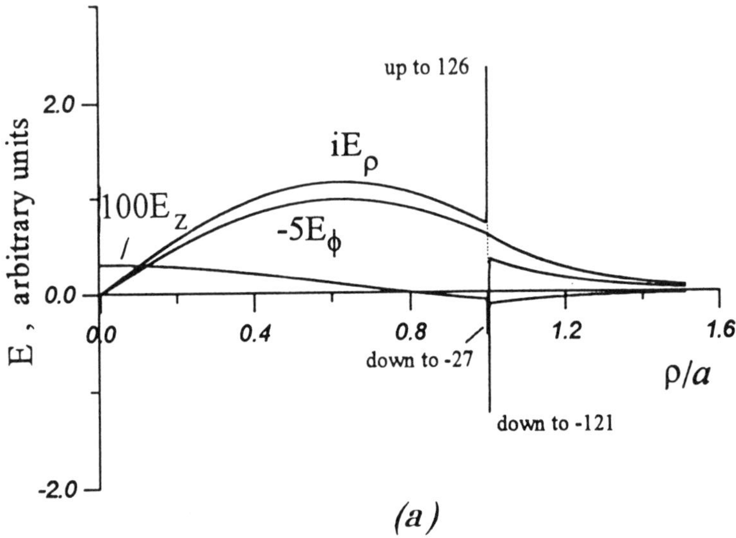

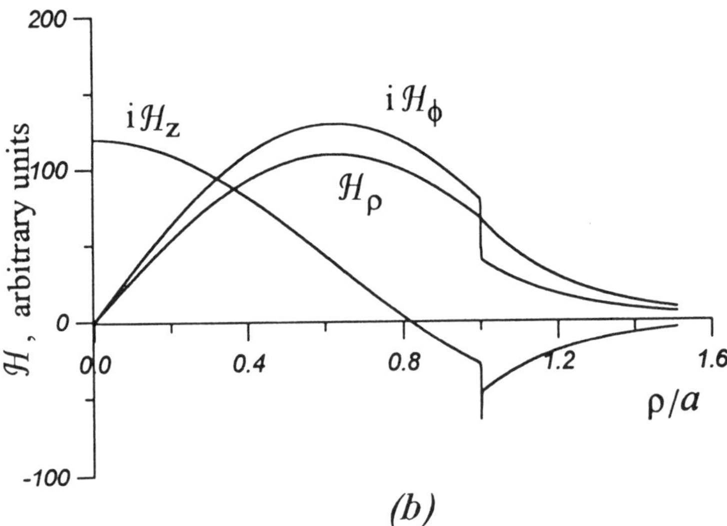

Figure 4.10. Field components in the lower-order axisymmetric eigen-mode of a cylindrical density enhancement. $\tilde{X}/X_a = 2$, $X_a^{1/2}/Y = 4.5$, $k_0 aY = 14.7$, $Y_{\mathrm{LH}} = 1.25$, $Y = 350$.

reaching the value $+126$ at $\rho = a - 0$ and the value -121 at $\rho = a + 0$, undergoes a discontinuity at the surface $\rho = a$. This is because a model of a sharply bounded duct is used here (because of this, a similar discontinuity will appear on further plots computed for such a model). The field component E_z decreases sharply down to -0.27 at the surface $\rho = a$ and then its magnitude tends to zero as ρ increases. Note that the curves showing the behavior of E_ρ and E_z are broken in the vicinity near to the duct boundary $\rho = a$ because their peaks cannot be clearly plotted in the same scale as used for the other components in Figure 4.10. We emphasize that the part associated with the wavenumbers $\tilde{s}_2$ and s_2 is important only in an extremely narrow layer adjacent to the duct interface. This becomes evident by comparing the values of the transverse wavenumbers in the eigenmode discussed:

$$\tilde{Q}_1 = 2.968, \quad \tilde{S}_2 = 2.168 \times 10^3,$$

$$S_1 = 3.741, \quad S_2 = 2.169 \times 10^3.$$

Figure 4.10 indicates a comparatively small integral contribution from the neighborhood of the duct interface, where E_ρ is discontinuous, to the total energy flux along the duct axis. In addition, component E_z is much less than the other components, so that the mode is of the quasi-TE type. It is of importance that the major features for this type of mode would not change significantly when considering a smooth profile of the density.

Now some aspects of the behavior of the field distribution in leaky modes at frequencies (2.71) will be discussed for the density enhancement. We first note, however, that since the modes are leaky, their propagation constants are complex, and hence each field component would have both real and imaginary parts. In reality, though, there is no need to present the distributions of both parts. The field in each modes is defined to an arbitrary (complex) constant. When examining the results of our computations, we noticed that if this constant is chosen to make a maximum of the ratio $|\mathrm{Re}(E_\phi)/\mathrm{Im}(E_\phi)|$, the field quantities $\mathrm{Re}(E_\rho)$, $\mathrm{Im}(E_z)$, $\mathrm{Im}(\mathcal{H}_\rho)$, and $\mathrm{Re}(\mathcal{H}_{\phi,z})$ become appreciably smaller in absolute value than the corresponding quantities $\mathrm{Im}(E_\rho)$, $\mathrm{Re}(E_z)$, $\mathrm{Re}(\mathcal{H}_\rho)$, and $\mathrm{Im}(\mathcal{H}_{\phi,z})$. This was used in plotting the field distributions in leaky modes, given in Figures 4.11 to 4.13.

Figure 4.11 shows the field components of the least attenuated lower-order axisymmetric mode for the following values of the parameters: $\tilde{X}/X_a = 3.2$, $X_a^{1/2}/Y = 5.6$, $k_0 a Y = 1$, and $Y = 10$. This mode exactly corresponds to point $\omega/\omega_\mathrm{H} = 0.1$ at curves 7 in Figure 4.8 and its propagation constant is $p_\nu = 28.56 - \mathrm{i}\,3.95 \times 10^{-4}$ ($\nu = 7$). One can easily verify that p_ν' belongs to the range (4.75). The large-scale part of the field and the short period oscillations showing the presence of the fine-scale escaping wave are clearly seen on the plots of the field components. Notwithstanding that the amount of energy leakage is small, some field components (especially $\mathrm{Im}\,(E_\rho)$ and $\mathrm{Re}\,(E_z)$) are heavily affected by the escaping wave.[*]

Figure 4.12 shows the field components for the mode of the same number $\nu = 7$ when $Y = 4$, while the parameters of the duct are like those in Figure 4.11. For this example $p = 28.18 - \mathrm{i}2.02 \times 10^{-1}$, so that p' lies in the domain (4.104). From the plots, we see that now the fine-scale (quasi-electrostatic) part predominates in the total field everywhere. In light of this, the mode may be termed 'quasi-electrostatic'.

The field distribution in the least attenuated lower-order nonsymmetric mode with the azimuthal index $m = 1$ is shown in Figure 4.13. The conditions are the same as for Figure 4.11. The propagation constant of the mode was found to be $p = 30.76 - \mathrm{i}6.91 \times 10^{-2}$. The large-scale part of each transverse component in the mode reaches its maximum absolute value at the duct axis and the polarization in this part of the field is not significantly different from circular. This is probably not seen well from Figure 4.12 because the fine-scale oscillations with comparatively large amplitudes are added to the large-scale variations (for the transverse magnetic components, the two parts with different characteristic scales are added at $\rho = 0$ with opposite signs but close amplitudes, so that $\mathrm{Im}\,(\mathcal{H}_\phi)$ and $\mathrm{Re}\,(\mathcal{H}_\rho)$ turn out to be very small near the duct axis). Unlike the case of axisymmetric modes, for which the transverse fields vanish at $\rho = 0$, the longitudinal fields here are equal to zero at the duct axis, in accordance with (4.43) and (4.46).

Finally, we examine how the field components vary across the density trough. We take the values of the parameters $X_a/\tilde{X} = 1.5$, $X_a^{1/2}/Y = 29.3$, and $k_0 a Y = 0.42$, which are just the same as those used in Figure 4.9, and give the distributions of the field components in the lower-order axisymmetric eigenmode for $Y = 10$. Figure 4.14 shows the plots for this mode, with its propagation constant $p = 48.08$, corresponding to point $\omega/\omega_{\mathrm{H}} = 10$ at curve 1 in Figure 4.9. Since here $\tilde{Q}_1 \simeq \tilde{Q}_2 \simeq k_0 a\,\tilde{q}_c$, the only characteristic scale is seen in the distribution of each field component along the transverse coordinate.

---

[*] Many authors, who computed the fields for a smooth duct (for example, Laird and Nunn, 1975), presented only the plots of the components which did not contain an appreciable contribution from the fine-scale escaping wave. However, if all the components were presented, a more noticeable contribution from the escaping wave could become apparent for some of them.

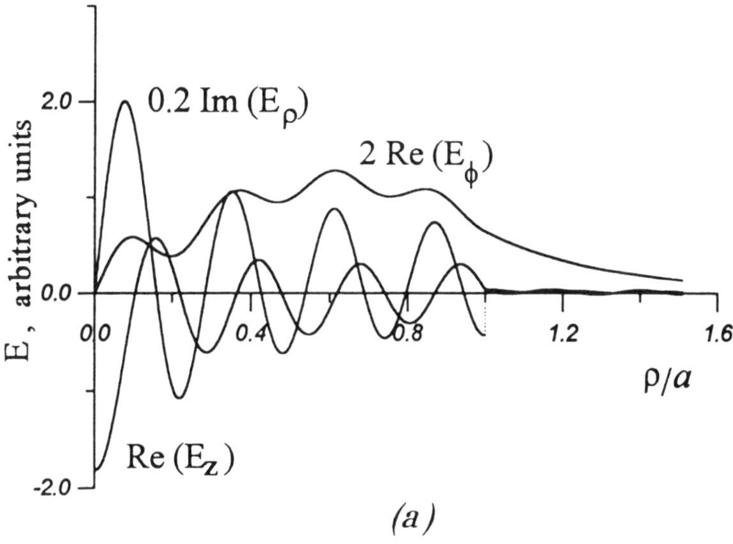

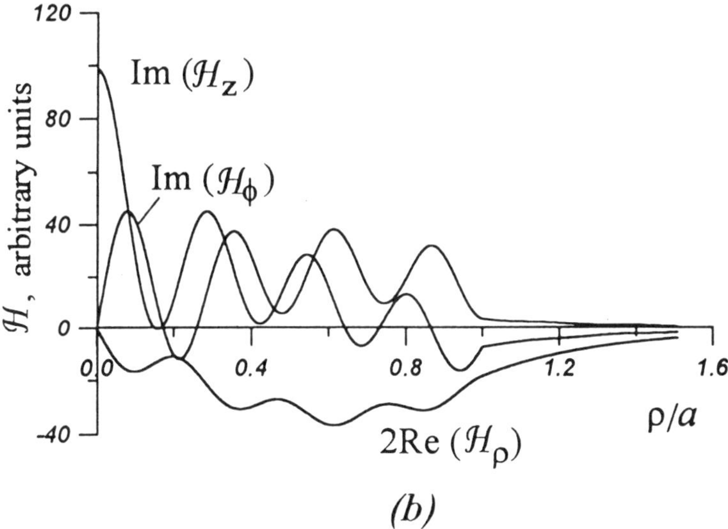

Figure 4.11. Field components in the least attenuated lower-order axisymmetric leaky mode of a cylindrical density enhancement. $\tilde{X}/X_a = 3.2$, $X_a^{1/2}/Y = 5.6$, $k_0 aY = 1$, $Y = 10$.

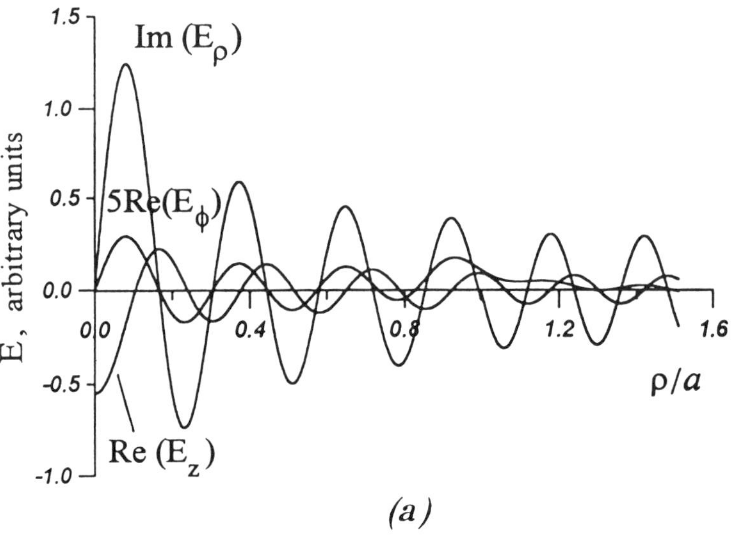

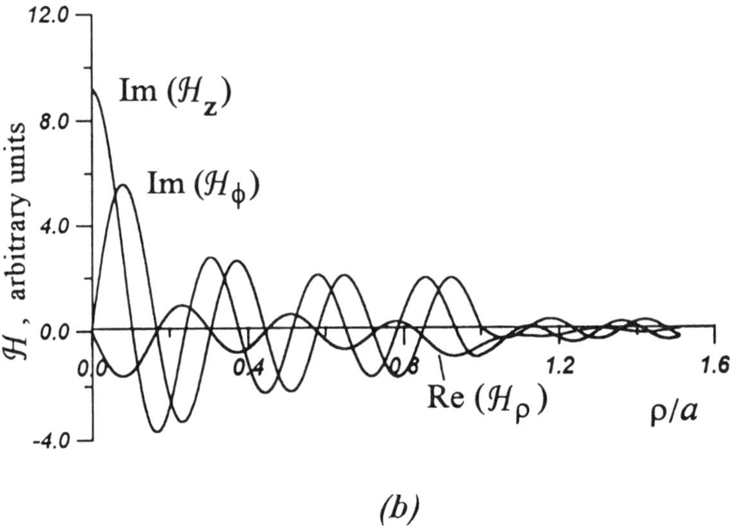

Figure 4.12. Field components in the axisymmetric leaky mode of number $\nu = 7$. The parameters $\tilde{X}/X_a$, $X_a^{1/2}/Y$ and k_0aY are the same as for Figure 4.11; $Y = 4$. See text for discussion.

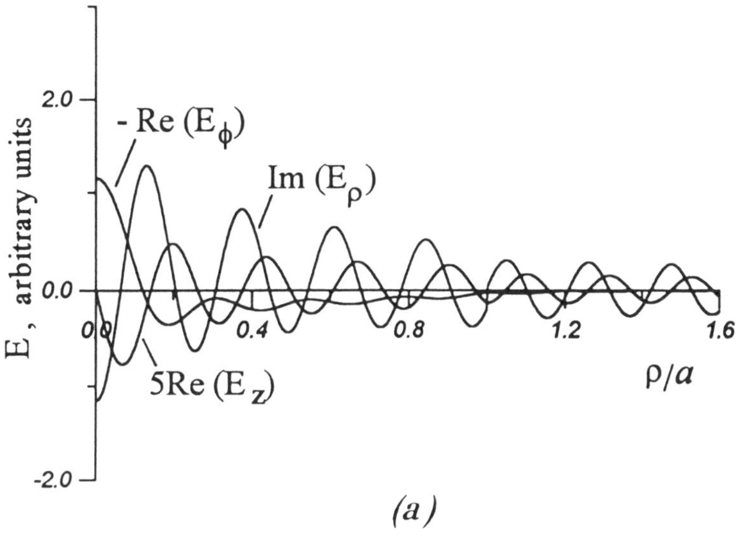

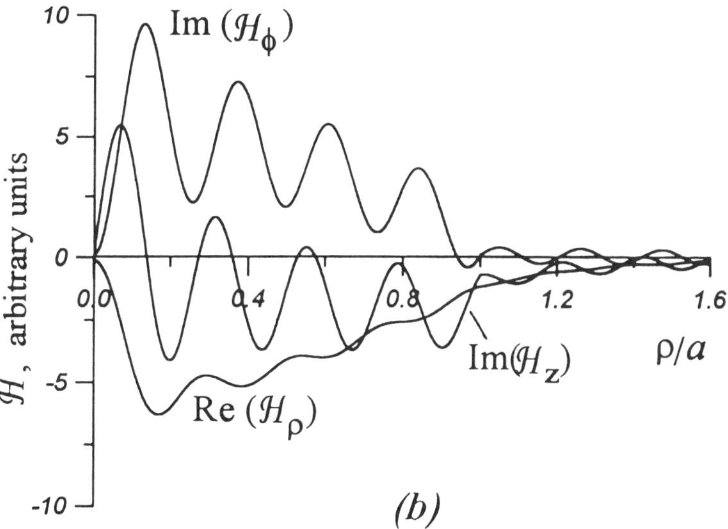

Figure 4.13. Field components in the least attenuated lower-order non-symmetric leaky mode with $m = 1$. $\tilde{X}/X_a = 3.2$, $X_a^{1/2}/Y = 5.6$, $k_0 a Y = 1$, $Y = 10$.

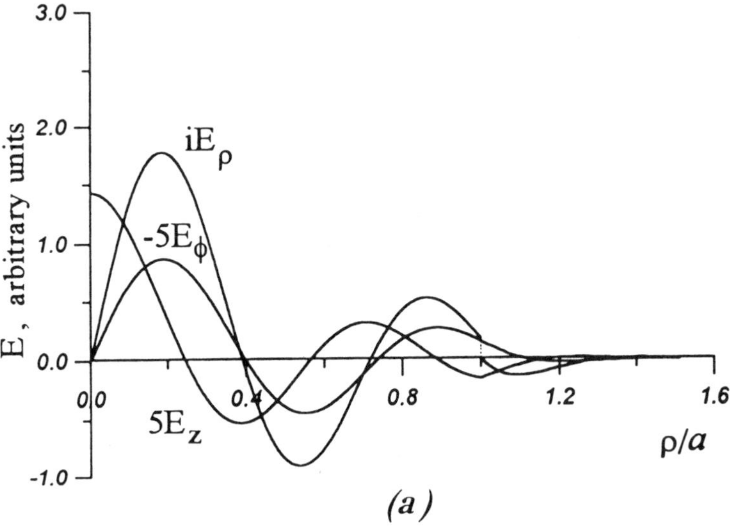

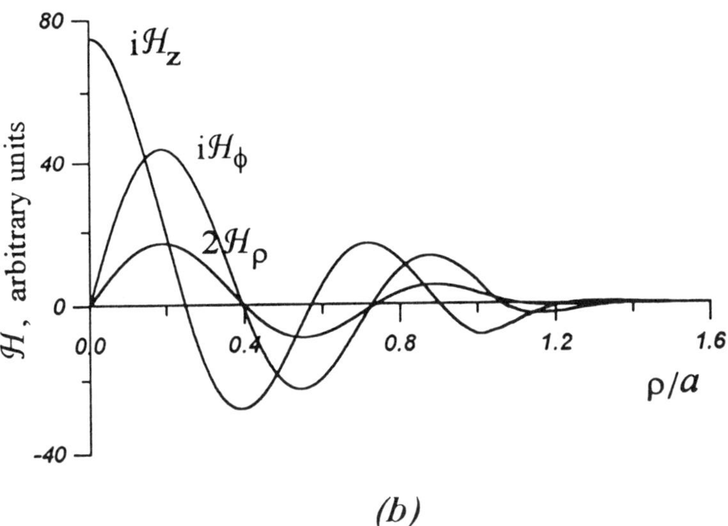

Figure 4.14. Field components in the lower-order axisymmetric eigen-mode of a density trough. $X_a/\tilde{X} = 1.5$, $X_a^{1/2}/Y = 29.3$, $k_0 a Y = 0.42$, $Y = 10$.

4.6. Mode synthesizing using Brillouin's concept

As is known from the electromagnetic theory, any waveguide mode may be synthesized by crossing characteristic waves of the medium that fills the guide. An interpretation of the mode propagation with the help of such waves multiply reflected from boundaries of the guide is commonly called the *concept of Brillouin*. It is discussed here for leaky modes guided by a cylindrical enhancement of density in the frequency interval (2.71). Although this type of mode is taken only as an example illustrating the basic ideas, it will be useful for some important topics to be dealt with in the later work. When considering a plane geometry, the crossing plane waves are needed to synthesize a waveguide mode. For a cylindrical geometry, the conical waves are to be used.

Let a conical wave travel outwards with its axial wavenumber p, transverse wavenumber $\tilde{q}_1$, and azimuthal index m. Then the field components E_z and $\mathcal{H}_z$ of the wave are

$$E_z = iE_0^{(1)} \frac{\tilde{n}_1 \, \tilde{q}_1}{\tilde{\eta}} \, H_m^{(2)}(k_0\tilde{q}_1\rho) \exp(-im\phi - ik_0 pz),$$

$$(4.116)$$

$$\mathcal{H}_z = -E_0^{(1)}\tilde{q}_1 \, H_m^{(2)}(k_0\tilde{q}_1\rho) \exp(-im\phi - ik_0 pz),$$

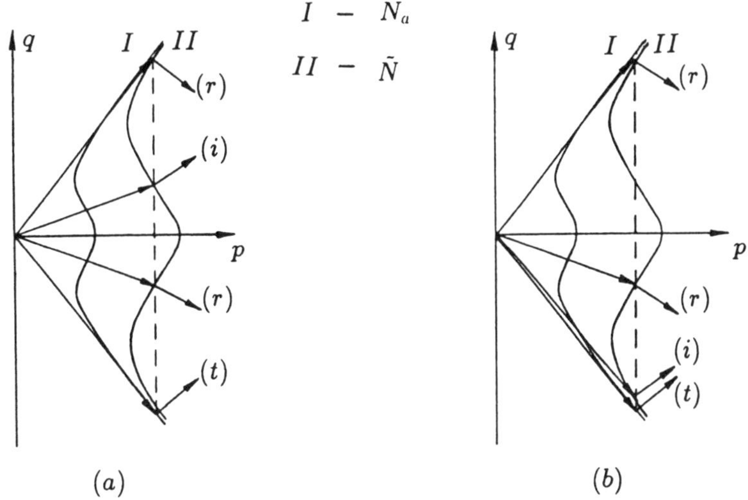

Figure 4.15. Possible configurations of the wavenormals and rays in incident ('i'), reflected ('r') and transmitted ('t') fields. (a) incidence of a 'large-scale' wave. (b) incidence of a 'fine-scale' wave.

with $E_0^{(1)}$ being the wave amplitude. These are obtainable from the general field expressions summarized in § 4.2 if we set $N = \tilde{N}$. Now suppose that there is a sharp boundary at $\rho = a$ between the plasma with density $\tilde{N}$, for $\rho < a$ (the duct core), and the ambient uniform plasma with its density N_a, for $\rho > a$ (the outer region). When the incident wave meets the boundary at $\rho = a$ it gives two reflected waves with two different transverse wavenumbers $\tilde{q}_1$ and $\tilde{q}_2$. The situation is depicted schematically in Figure 4.15 (a) where, for definiteness, the condition (4.75) is assumed to be satisfied. In Figure 4.15 we made use of the fact that because of Snell's law, all field quantities contain a factor $\exp(-im\phi - ik_0 p z)$. This will hereinafter be omitted, in the same way as the time factor $\exp(i\omega t)$ is omitted. Then the components of the reflected field are

$$E_z = iE_0^{(1)} \left[R_{11} \frac{\tilde{n}_1 \tilde{q}_1}{\tilde{\eta}} H_m^{(1)}(k_0 \tilde{q}_1 \rho) + R_{21} \frac{\tilde{n}_2 \tilde{q}_2}{\tilde{\eta}} H_m^{(2)}(k_0 \tilde{q}_2 \rho) \right],$$

$$(4.117)$$

$$\mathcal{H}_z = -E_0^{(1)} \left[R_{11}\tilde{q}_1 H_m^{(1)}(k_0 \tilde{q}_1 \rho) + R_{21}\tilde{q}_2 H_m^{(2)}(k_0 \tilde{q}_2 \rho) \right].$$

Use of the second instead of the first kind of Hankel function for the wave with the transverse wavenumber $\tilde{q}_2$ is due to the fact that it is a backward wave in ρ direction, as is seen from Figure 4.15. Therefore, the inward directed energy flow corresponds to the outgoing phase progress of this wave. To find the components appropriate to the total field in the core one must add (4.116) and (4.117). Clearly there are also transmitted waves in the outer region (see Figure 4.15), and the E_z and $\mathcal{H}_z$ components of the total transmitted field are

$$\begin{aligned} E_z &= -iE_0^{(1)} \left[T_{11} \frac{n_1 s_1}{\eta_a} K_m(k_0 s_1 \rho) - T_{21}^{(H)} \frac{n_2 q_2}{\eta_a} H_m^{(1)}(k_0 q_2 \rho) \right] \\ &= -iE_0^{(1)} \left[T_{11} \frac{n_1 s_1}{\eta_a} K_m(k_0 s_1 \rho) + T_{21} \frac{n_2 s_2}{\eta_a} K_m(k_0 s_2 \rho) \right], \\ \\ \mathcal{H}_z &= E_0^{(1)} \left[T_{11} s_1 K_m(k_0 s_1 \rho) - T_{21}^{(H)} q_2 H_m^{(1)}(k_0 q_2 \rho) \right] \\ &= E_0^{(1)} \left[T_{11} s_1 K_m(k_0 s_1 \rho) + T_{21} s_2 K_m(k_0 s_2 \rho) \right]. \end{aligned}$$

$$(4.118)$$

In the above expressions, $R_{11,21}$ and $T_{11,21}$ are pairs of respectively reflection and transmission coefficients appropriate to an incidence of a wave with its transverse wavenumber $\tilde{q}_1$; $T_{21}^{(H)} = -\frac{\pi}{2} i^m T_{21}$; the other notations are like those in §4.2. To unify the formulation, we shall further employ the modified Bessel function of the second kind $K_m(k_0 s_2 \rho)$ throughout, instead of the Hankel function $H_m^{(1)}(k_0 q_2 \rho)$ though the latter seems to be more suitable for the escaping wave in the outer region.

The tangential components of the field in the core, given by adding the incident wave and two reflected waves, and those of the field in the outer region, represented by two transmitted waves, must satisfy boundary conditions such as (4.65). These give four simultaneous linear equations in the quantities R_{11}, R_{21}, T_{11}, and T_{21}. Solving these equations yields the reflection and transmission coefficients. The general expressions for them are rather complicated, and we therefore write down the resulting formulas for R_{11} and R_{21} only, for the case $m = 0$:

$$
\begin{aligned}
R_{11} &= -\left[\tilde{D}_1 H^{(2)}(\tilde{Q}_1) + \tilde{D}_2 H^{(2)}(\tilde{Q}_2) + \frac{\tilde{\eta}}{\eta_a}\, H^{(2)}(\tilde{Q}_1)\, H^{(2)}(\tilde{Q}_2) + K(S_1)\, K(S_2)\right] \\
&\quad \times \frac{H_0^{(2)}(\tilde{Q}_1)}{H_0^{(1)}(\tilde{Q}_1)}\, \tilde{\Delta}^{-1},
\end{aligned}
$$

$$
R_{21} = \left[H^{(2)}(\tilde{Q}_1) - H^{(1)}(\tilde{Q}_1)\right] \tilde{F}_1\, \frac{\tilde{Q}_1 H_0^{(2)}(\tilde{Q}_1)}{\tilde{Q}_2 H_0^{(2)}(\tilde{Q}_2)}\, \tilde{\Delta}^{-1},
$$

(4.119)

where

$$
\tilde{\Delta} = \tilde{D}_1 H^{(1)}(\tilde{Q}_1) + \tilde{D}_2 H^{(2)}(\tilde{Q}_2) + \frac{\tilde{\eta}}{\eta_a}\, H^{(1)}(\tilde{Q}_1)\, H^{(2)}(\tilde{Q}_2) + K(S_1)\, K(S_2),
$$

$$
\tilde{D}_1 = M_{12} L_{12}\, K(S_2) - M_{11} L_{22}\, K(S_1), \quad \tilde{D}_2 = M_{21} L_{21}\, K(S_1) - M_{22} L_{11}\, K(S_2),
$$

$$
\tilde{F}_1 = M_{12} L_{11}\, K(S_2) - M_{11} L_{21}\, K(S_1),
$$

$$
K(\zeta) = \frac{K_1(\zeta)}{\zeta\, K_0(\zeta)}, \quad H^{(k)}(\zeta) = \frac{H_1^{(k)}(\zeta)}{\zeta\, H_0^{(k)}(\zeta)}, \quad k = 1, 2;
\tag{4.120}
$$

the other notations are the same as in (4.69).

A close examination of expressions (4.119) shows that when the wave frequency is well below half the electron gyrofrequency, $|R_{11}|$ is very close to unity, whereas $|R_{21}|$ is much smaller than unity, if, of course, the propagation constants lie in the range (4.75) and satisfy the condition (4.92). Thus, the nearly total internal reflection would be observed for the whistler wave appropriate to branch $\tilde{q}_1$. Small magnitude of R_{21} implies a negligible transformation of it into the fine-scale wave appropriate to branch $\tilde{q}_2$. Therefore, a very small amplitude of the fine-scale oscillations is observed for the E_ϕ- and $\mathcal{H}_\rho$-components in leaky modes. The other field components would be more heavily affected by the fine-scale oscillations. This is explained by the fact that though $|R_{21}| \ll |R_{11}|$ in (4.117), but $|\tilde{q}_2| \gg |\tilde{q}_1|$ and $|\tilde{n}_2| \gg |\tilde{n}_1|$. Thence the contribution arising from the fine-scale branch of q becomes apparent in $\mathcal{H}_\phi$ and $\mathcal{H}_z$, and, moreover, it predominates in E_ρ and E_z (see Figure 4.11).

Now, in the plasma with density $\tilde{N}$, let a conical wave travel with its axial wavenumber p, transverse wavenumber $\tilde{q}_2$, and the azimuthal index m. Let the energy flow in the wave be outgoing. Then the field components E_z and $\mathcal{H}_z$, for $\rho < a$, are

$$
E_z = \mathrm{i} E_0^{(2)}\, \frac{\tilde{n}_2 \tilde{q}_2}{\tilde{\eta}}\, H_m^{(1)}(k_0 \tilde{q}_2 \rho),
$$

$$
\mathcal{H}_z = -E_0^{(2)} \tilde{q}_2\, H_m^{(1)}(k_0 \tilde{q}_2 \rho),
$$

(4.121)

with $E_0^{(2)}$ being the wave amplitude. The situation is depicted in Figure 4.15(b). The corresponding components of the reflected field are

$$E_z = \mathrm{i}E_0^{(2)} \left[R_{12} \frac{\tilde{n}_1 \tilde{q}_1}{\tilde{\eta}} H_m^{(1)}(k_0 \tilde{q}_1 \rho) + R_{22} \frac{\tilde{n}_2 \tilde{q}_2}{\tilde{\eta}} H_m^{(2)}(k_0 \tilde{q}_2 \rho) \right],$$

$$\mathcal{H}_z = -E_0^{(2)} \left[R_{12} \tilde{q}_1 H_m^{(1)}(k_0 \tilde{q}_1 \rho) + R_{22} \tilde{q}_2 H_m^{(2)}(k_0 \tilde{q}_2 \rho) \right].$$

$$(4.122)$$

In the outer region, the components of the transmitted field are

$$E_z = -\mathrm{i}E_0^{(2)} \left[T_{12} \frac{n_1 s_1}{\eta_a} K_m(k_0 s_1 \rho) + T_{22} \frac{n_2 s_2}{\eta_a} K_m(k_0 s_2 \rho) \right],$$

$$\mathcal{H}_z = E_0^{(2)} \left[T_{12} s_1 K_m(k_0 s_1 \rho) + T_{22} s_2 K_m(k_0 s_2 \rho) \right].$$

$$(4.123)$$

The complex coefficients R_{12}, R_{22}, T_{12}, and T_{22} may be found by satisfying the boundary conditions at $\rho = a$, by analogy with the preceding discussion for the incident wave with the transverse wavenumber $\tilde{q}_1$. We give the expressions for R_{12} and R_{22}, when $m = 0$:

$$R_{12} = \left[H^{(1)}(\tilde{Q}_2) - H^{(2)}(\tilde{Q}_2) \right] \tilde{F}_2 \frac{\tilde{Q}_2 H_0^{(1)}(\tilde{Q}_2)}{\tilde{Q}_1 H_0^{(1)}(\tilde{Q}_1)} \tilde{\Delta}^{-1},$$

$$(4.124)$$

$$R_{22} = -\left[\tilde{D}_1 H^{(1)}(\tilde{Q}_1) + \tilde{D}_2 H^{(1)}(\tilde{Q}_2) + \frac{\tilde{\eta}}{\eta_a} H^{(1)}(\tilde{Q}_1) H^{(1)}(\tilde{Q}_2) + K(S_1) K(S_2) \right]$$
$$\times \frac{H_0^{(1)}(\tilde{Q}_2)}{H_0^{(2)}(\tilde{Q}_2)} \tilde{\Delta}^{-1},$$

where $\tilde{F}_2 = M_{21} L_{22} K(S_1) - M_{22} L_{12} K(S_2)$.

Here the same notations are used as in (4.69) and (4.120). By inspection of (4.124) it is possible to establish that $|R_{12}|$ would be appreciably less than unity. Magnitude of R_{22}, would be of the order unity for a sharp boundary in the case $\tilde{N} \gg N_a$. In this situation, the fine-scale wave reflects well from the boundary. Under the condition (4.104), this process leads to the appearance of the leaky modes as illustrated by Figure 4.12. As $\tilde{N}$ tends to N_a, or as the density profile becomes smoothed-out, $|R_{22}|$ would fall rapidly and modes of this sort cease to exist.

When the ratio $(\tilde{N} - N_a)/N_a$ is not small, two sorts of leaky modes discussed above correspond in reality to different parts of the same dispersion curves, as shown in Figure 4.8. Clearly these two sorts can be delimited conditionally only when the propagation constant is well away from $p = \tilde{P}$. In the following, we shall establish an accurate formalism for construction of modes, regardless of what sort they belong to. For doing this, we employ a reflection coefficient matrix formed by the four complex coefficients R_{11}, R_{12}, R_{21}, and R_{22}:

$$\hat{\mathbf{R}} = \begin{pmatrix} R_{11} & R_{12} \\ R_{21} & R_{22} \end{pmatrix}.$$

$$(4.125)$$

First construct the general form for the total incident field, by adding (4.116) and (4.121). Its components E_z and $\mathcal{H}_z$ are

$$E_z = \mathrm{i} \left[E_0^{(1)} \frac{\tilde{n}_1 \tilde{q}_1}{\tilde{\eta}} H_m^{(2)}(k_0 \tilde{q}_1 \rho) + E_0^{(2)} \frac{\tilde{n}_2 \tilde{q}_2}{\tilde{\eta}} H_m^{(1)}(k_0 \tilde{q}_2 \rho) \right],$$

$$\mathcal{H}_z = - \left[E_0^{(1)} \tilde{q}_1 H_m^{(2)}(k_0 \tilde{q}_1 \rho) + E_0^{(2)} \tilde{q}_2 H_m^{(1)}(k_0 \tilde{q}_2 \rho) \right].$$

(4.126)

The components of the total reflected field given by adding (4.117) and (4.122) are then

$$E_z = \mathrm{i} \left[\mathcal{E}_0^{(1)} \frac{\tilde{n}_1 \tilde{q}_1}{\tilde{\eta}} H_m^{(1)}(k_0 \tilde{q}_1 \rho) + \mathcal{E}_0^{(2)} \frac{\tilde{n}_2 \tilde{q}_2}{\tilde{\eta}} H_m^{(2)}(k_0 \tilde{q}_2 \rho) \right],$$

$$\mathcal{H}_z = - \left[\mathcal{E}_0^{(1)} \tilde{q}_1 H_m^{(1)}(k_0 \tilde{q}_1 \rho) + \mathcal{E}_0^{(2)} \tilde{q}_2 H_m^{(2)}(k_0 \tilde{q}_2 \rho) \right],$$

(4.127)

where $\mathcal{E}_0^{(1)}$ and $\mathcal{E}_0^{(2)}$ are given by

$$\begin{pmatrix} \mathcal{E}_0^{(1)} \\ \mathcal{E}_0^{(2)} \end{pmatrix} = \begin{pmatrix} R_{11} & R_{12} \\ R_{21} & R_{22} \end{pmatrix} \begin{pmatrix} E_0^{(1)} \\ E_0^{(2)} \end{pmatrix}.$$

(4.128)

Now the two fields (4.126) and (4.127) are to form a self-consistent waveguide mode. Clearly its field must be finite everywhere in the core. This requires that $\mathcal{E}_0^{(1)} = E_0^{(1)}$ and $\mathcal{E}_0^{(2)} = E_0^{(2)}$, which leads to

$$\det \| R_{ij} - \delta_{ij} \| = 0,$$

whence

$$(R_{11} - 1)(R_{22} - 1) = R_{12} R_{21} \tag{4.129}$$

(see also Laird and Nunn, 1975). In terms of the waveguide mode theory of wave propagation (Budden, 1961 b), this is the condition for a self-consistent mode. Since the elements of the reflection matrix (4.125) may be regarded as functions of p, Equation (4.129) is, in effect, the dispersion equation in p, written in terms of R_{ij}. It may be rewritten to give the more customary form. For example, substituting for R_{ij} from (4.119) and (4.124) yields after some lengthy algebra the dispersion equation appropriate to axisymmetric modes in its original form (4.71).

Equation (4.129) applies for any "boundaries" whose reflection coefficients are known, and for arbitrary frequencies. It would apply, for example, to a smooth duct which has no real sharp boundary. In many cases the density profile for the smooth duct may be specified as follows: the density is a constant, $\tilde{N}$, in the inner core $\rho < a$, a constant, N_a, in the outer region $\rho > a + d$, and varies smoothly from $\tilde{N}$ to N_a when ρ ranges from $\rho = a$ to $\rho = a + d$. In this case we could choose the surface $\rho = a$ as the "boundary", and determine the matrix $\hat{\mathbf{R}}$ at this surface. The reflection coefficients R_{ij} may now be found with the help a numerical solution of the second-order coupled equations (4.14) and (4.15) in the range $a < \rho < a + d$. Methods of doing this have been described for cylindrical ducts by Laird and Nunn (1975). These methods are similar to those used for finding the reflection coefficients from a plane stratified slab. It should be noted that methods of a full-wave numerical solution of the field equations in a stratified magnetoplasma received much careful study, and there are many good accounts of them. Discussion of these topics is beyond the scope of this book, but the interested reader should consult works of Budden (1961 a), Johler and Harper (1962), Pitteway (1965), Inoue and Horowitz (1966), Pitteway and Jespersen (1966), Altman and Cory (1969), Scarabucci (1969), Smith and Pitteway (1974), and Nagano *et al.* (1975).

Once the matrix $\hat{\mathbf{R}}$ is known, the problem becomes one of solving Equation (4.129). The procedure of finding zeros of this equation as well as subsequent constructing the total field in the modes are discussed, for example, in the above-mentioned work of Laird and Nunn (1975).

As follows from the results of the relevant works (see, for example, Scarabucci and Smith, 1971; Laird and Nunn, 1975; Laird, 1992), the general picture of the waveguide propagation in smooth ducts for the whistler band is in a good qualitative agreement with that drawn in § 4.4 for sharply bounded ducts. The main difference to notice here is that the radiation attenuation of the leaky modes corresponding to the range (4.75) is usually much less for a smooth-walled duct. So it can convey whistler modes over longer distances. Conversely, the leaky modes corresponding to the range (4.104) become heavily attenuated in a smooth duct and cease to exist as waveguide-like solutions.

4.7. The effect of collisions on the characteristics of modes

Given that the medium must always absorb energy from the electromagnetic field, this can be accounted for by collisions. For whistler waves, it is satisfactory to consider only the collisions of electrons. If electron collisions are allowed for, some collision damping appears for bound modes, and their propagation constants become complex. For leaky modes, collisions lead to the appearance of additional damping. But if the electron collision-frequency ν_e is rather small compared with the angular frequency ω, the effect of collisions on the characteristics of modes is believed to be negligible. There may be situations, however, for which this is not true, and even slight collisions cause significant changes in the field distributions for some modes. To show this, consider the case when

$$\omega_{\mathrm{LH}} \ll |\omega - i\nu_e| \ll \omega_{\mathrm{H}} \ll \omega_{\mathrm{p}}. \qquad (4.130)$$

Now the elements of the dielectric tensor may be written, from (2.28):

$$\varepsilon = X(1 - iZ)/Y^2, \quad g = -X/Y, \quad \eta = -X/(1 - iZ), \qquad (4.131)$$

where $Z = \nu_e/\omega$, ion collisions being neglected. For vanishing ν_e, the frequencies obeying (4.130) belong to the interval (2.71), for which slightly leaky modes, with their propagation constants lying in the range (4.75), exist in density enhancements. The aim here is to elucidate what happens with these modes when collisions are included.

Let the density profile be given, as before, by (4.39), with $\tilde{N} > N_a$, and let the collision frequency ν_e be a certain constant such that the pertinent inequalities in (4.130) are satisfied. Then the imaginary parts of the quantities $\tilde{q}_2$ and q_2, under the condition $|p|^2 \gg |\tilde{\varepsilon}|$, which is also assumed, would be, from (4.81):

$$\mathrm{Im}\,(\tilde{Q}_2) \approx \mathrm{Im}\,(Q_2) \approx \mathrm{Im}\left(k_0 a p \sqrt{-\frac{\eta_a}{\varepsilon_a}} \right) = k_0 a p' \,\frac{Y}{1 + Z^2}\, \left(Z - \frac{p''}{p'} \right). \qquad (4.132)$$

In the absence of collisions, the ratio p''/p' was found to be very small (see § 4.4), so that the inequalities $|\mathrm{Im}\,(\tilde{Q}_2)| \ll 1$ and $|\mathrm{Im}\,(Q_2)| \ll 1$ were valid. When collisions are included, the complex propagation constants $p = p' - ip''$ of modes can be found through the same argument as was utilized in § 4.4 for non-absorbing ducts. If the drop in the plasma density is small such that the inequality $(\tilde{N}/N_a - 1)^2 \ll \min\{1, k_0 a p' Z\}$ holds, the attenuation turns out to be determined mainly by collision damping rather than radiation losses, and the attenuation constant is readily found to be

$$p'' \simeq \frac{\nu_e}{2\omega_\mathrm{H}}\, p'. \qquad (4.133)$$

This coincides with the well-known result for the attenuation of a whistler plane wave traveling along the ambient static magnetic-field in a uniform collisional magnetoplasma. If (4.133) is substituted in (4.132), and if the inequality $\omega \ll \omega_\mathrm{H}$ is used, then

$$\mathrm{Im}\,(\tilde{Q}_2) \approx \mathrm{Im}\,(Q_2) \simeq k_0 a p' Y \,\frac{Z}{1 + Z^2}\,.$$

It was shown in § 4.4 for the propagating axisymmetric modes that when $Y \gg 1$, the parameter $k_0 a p' Y$ is much greater than unity. If it is large enough, the inequality

$$k_0 a p' Y \,\frac{Z}{1 + Z^2} \gg 1 \qquad (4.134)$$

may hold even for small Z, thereby leading to

$$\mathrm{Im}\,(\tilde{Q}_2) \gg 1, \quad \mathrm{Im}\,(Q_2) \gg 1. \qquad (4.135)$$

The last two inequalities imply a damping rate sufficient to make the fine-scale part of the mode field, but not its large-scale part, to be localized in the vicinity of duct walls. Thus, the slight electron collisions may appreciably affect the field distribution in mode, and hence the conventional inequality $\nu_e \ll \omega$ cannot, in general, be considered as a sufficient condition for neglecting collisions in a duct.

Figure 4.16 shows the distributions of the field components in the lower-order axisymmetric mode for the parameters $\tilde{X}/X_a = 1.5$, $X_a^{1/2}/Y = 6.4$, $Y = 8.8$, $Z = 0.25$ and $k_0 a = 0.12$. In this example $p = 21.18 - i\,0.79$ and

$$\tilde{Q}_1 = 2.69 + i\,5.92 \times 10^{-2}, \quad \tilde{Q}_2 = 17.23 + i\,4.79,$$

$$S_1 = 1.12 - i\,2.83 \times 10^{-1}, \quad Q_2 = 18.57 + i\,4.39.$$

Note that the attenuation for the mode is somewhat higher than that given by (4.133). This is because p' is not well away from the branch point $p = \tilde{\mathcal{P}}_c$, at which $\tilde{q}_1^2 = \tilde{q}_2^2$. For fixed $\tilde{X}/X_a$, as radius a increases, the propagation constant for this mode shifts away from $p = \tilde{\mathcal{P}}_c$ to $p = \tilde{\mathcal{P}}$, and the attenuation constant becomes more close to (4.133).

For comparison, Figure 4.17 plots the field for the same mode when the parameters $\tilde{X}/X_a$, $X_a^{1/2}/Y$, Y, and $k_0 a$ are as in the preceding case, but no collisions are included ($Z = 0$). Now $p = 21.10 - i\,1.34 \times 10^{-2}$ and

$$\tilde{Q}_1 = 2.72 + i\,6.33 \times 10^{-3}, \quad \tilde{Q}_2 = 18.40 - i\,1.76 \times 10^{-2},$$

$$S_1 = 1.06 - i\,7.84 \times 10^{-3}, \quad Q_2 = 19.82 - i\,1.59 \times 10^{-2}.$$

The difference between imaginary parts of $\tilde{Q}_2$ and Q_2 in the former and latter cases is clearly seen in the different behavior of a fine-scale part of the field for both examples.

There is a simple physical explanation for the above results. It has been mentioned in § 3.3, that under conditions like (4.130) collisional damping for the branch q_2 is much higher than that for the branch q_1. As follows from Brillouin's concept (§ 4.6), the waves attributed to branch 2 appear as a result of the multiple reflections of the waves of branch 1 from the duct interface. In the absence of collisions, the excited fine-scale waves (branch 2) travel away the duct interface, thereby contributing to the total field in the bulk of the plasma within the duct. When collisions are included, these waves undergo rapid damping and, therefore, decay in the vicinity of duct walls (see Figure 4.16).

For the example illustrated by Figures 4.16 and 4.17, a moderate drop of the density was used. It is not difficult to verify that similar behavior is observed for modes in ducts with more appreciable variations of the plasma density. However, the characteristics of modes in the corresponding ducts are affected by collisions in much more complicated manner than for the above simple example, and this topic calls for a separate consideration which is beyond the scope of the present book.

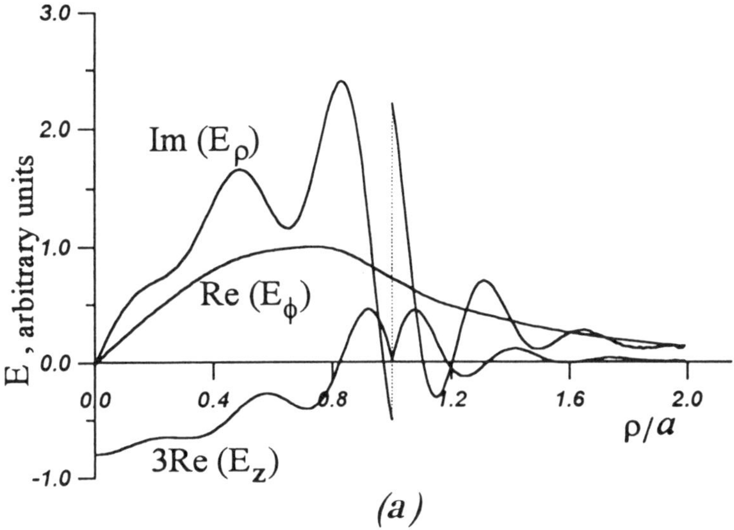

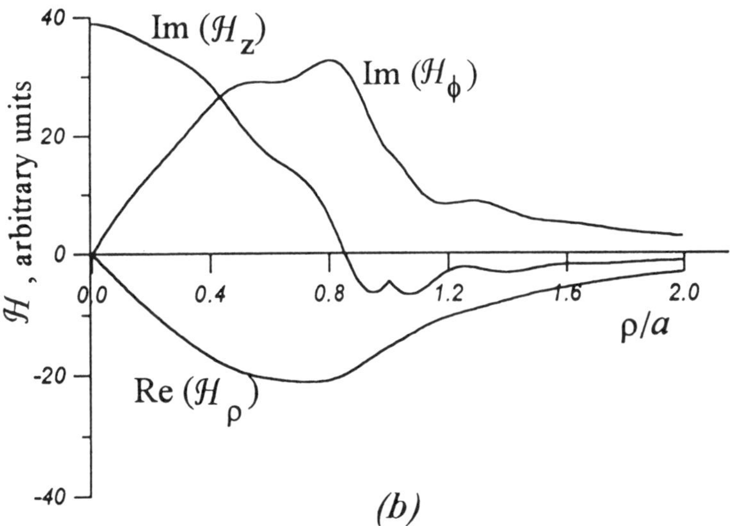

Figure 4.16. Field components in the lower-order axisymmetric mode of a cylindrical enhancement, for collisional plasma with $Z = 0.25$. $\tilde{X}/X_a = 1.5$, $X_a^{1/2}/Y = 6.4$, $Y = 8.8$, $k_0 a = 0.12$.

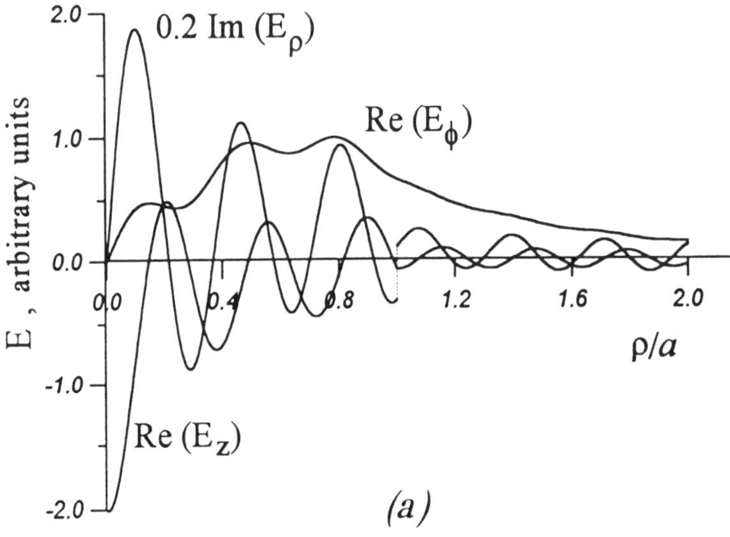

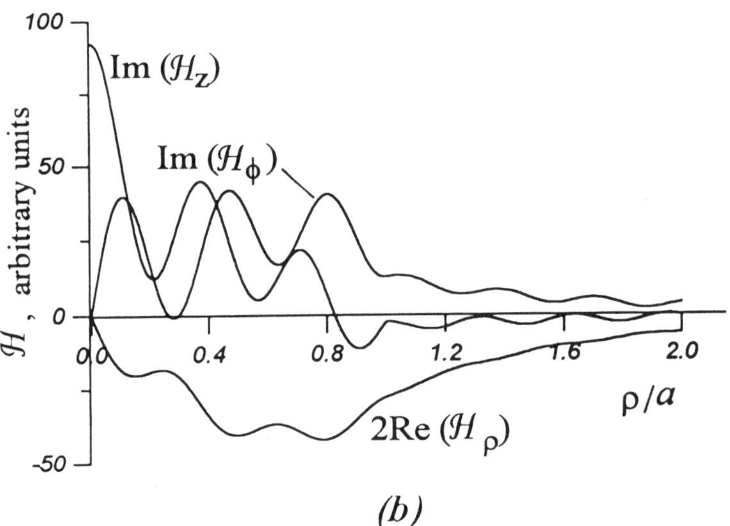

Figure 4.17. Field components in the lower-order axisymmetric mode of a cylindrical enhancement, for collisionless plasma ($Z = 0$). The other parameters are the same as for Figure 4.16.

4.8. Radially nonuniform duct with a monotonic density profile

In §§ 4.2–4.4 we showed how modes of uniform, sharply bounded ducts are constructed from solutions of the field equations in a homogeneous medium. It was also mentioned that for a general nonuniform duct, numerical methods must be resorted to find the propagation constants and modal fields (see § 4.6). Here we pay particular attention to some useful analytical methods allowing approximate solution of the pertinent field equations. Consideration will be made for axisymmetric modes of a circularly symmetric duct, in which the plasma density is a monotonic smooth function, $N(\rho)$, of the radial coordinate only. As ρ increases, the density is assumed to tend to the background value N_a. The density at the duct axis $\rho = 0$ will be denoted by $\tilde{N}$. The previous notations $\mathcal{P}$, $\mathcal{P}_c$ and $\tilde{\mathcal{P}}$, $\tilde{\mathcal{P}}_c$ are implied for the densities N_a and $\tilde{N}$, respectively. We shall be discussing the properties of axisymmetric modes in density enhancements for frequencies well below half the electron gyrofrequency (i. e., when $Y \gg 1$), in order to better appreciate the nature of limitations introduced earlier for this important case.

Now let the plasma density $N(\rho)$ diminish monotonically with ρ, with its characteristic scale a. Define quantities

$$\hat{q}_1^2 = \frac{g^2}{p^2 - \varepsilon} - p^2 + \varepsilon, \quad \hat{q}_2^2 = -\frac{\eta}{\varepsilon}\left(p^2 - \varepsilon\right),$$

$$\chi = \left(k_0 a p Y\right)^{-1}, \quad \xi = \rho/a. \tag{4.136}$$

By using these notations Equations (4.18) and (4.19) may be rewritten to give

$$\frac{\mathrm{d}^2 E_\phi}{\mathrm{d}\xi^2} + \frac{1}{\xi}\frac{\mathrm{d}E_\phi}{\mathrm{d}\xi} - \frac{E_\phi}{\xi^2} + \left(k_0 a \hat{q}_1\right)^2 E_\phi = k_0 a p \, \frac{g}{p^2 - \varepsilon} \frac{\mathrm{d}E_z}{\mathrm{d}\xi}, \tag{4.137}$$

$$\chi^2(1 - Y_{\mathrm{LH}}^2)\left(\frac{\mathrm{d}^2 E_z}{\mathrm{d}\xi^2} + \frac{1}{\xi}\frac{\mathrm{d}E_z}{\mathrm{d}\xi} + \frac{p^2}{(p^2 - \varepsilon)\varepsilon}\frac{\mathrm{d}\varepsilon}{\mathrm{d}\xi}\frac{\mathrm{d}E_z}{\mathrm{d}\xi}\right) - \frac{\varepsilon}{p^2 \eta}\hat{q}_2^2 E_z$$

$$= -\chi \frac{p^2 - \varepsilon}{g}\frac{1}{\xi}\frac{\mathrm{d}}{\mathrm{d}\xi}\left(\frac{\xi g E_\phi}{p^2 - \varepsilon}\right). \tag{4.138}$$

Here we made use of the expressions (4.80) for elements of the dielectric tensor.

It was shown, in § 4.4, for the axisymmetric modes propagating along a uniform duct that the parameter $k_0 a p$ is above or of the order unity. Since in the case considered $Y \gg 1$, we now have $\chi \ll 1$ and (in view of the evident inequality $Y_{\mathrm{LH}} \ll Y$) $\chi Y_{\mathrm{LH}} \ll 1$. This enables the use of a perturbation technique to solve Equations (4.137) and (4.138).

Now consider whistler modes at frequencies (2.70) for the range (4.74), assuming that $p^2 \gg |\varepsilon|$. Neglecting small terms of the order ε/p^2, in zero-order perturbation theory (viz. $\chi \approx 0$ and $\chi^2 Y_{\mathrm{LH}}^2 \approx 0$) we obtain, from (4.137) and (4.138):

$$\frac{\mathrm{d}^2 E_\phi}{\mathrm{d}\rho^2} + \frac{1}{\rho}\frac{\mathrm{d}E_\phi}{\mathrm{d}\rho} - \frac{E_\phi}{\rho^2} + k_0^2\left(\frac{g^2}{p^2} - p^2\right)E_\phi = 0 \qquad (4.139)$$

and

$$E_z = 0.$$

To get a non-trivial result for E_z, the next order of the perturbation theory (viz. $\chi \neq 0$ but $\chi^2(1 - Y_{\mathrm{LH}}^2) \approx 0$) is used in (4.138). This gives

$$E_z = -\frac{1}{k_0 p \eta}\frac{1}{\rho}\frac{\mathrm{d}}{\mathrm{d}\rho}(\rho g E_\phi). \qquad (4.140)$$

The remaining field components may be written in the zero-order approximation, from (4.16) and (4.17):

$$E_\rho = -\mathrm{i}g p^{-2} E_\phi,$$

$$\mathcal{H}_\rho = -p E_\phi, \quad \mathcal{H}_\phi = -\mathrm{i}g p^{-1} E_\phi, \quad \mathcal{H}_z = \frac{\mathrm{i}}{k_0\,\rho}\frac{\mathrm{d}}{\mathrm{d}\rho}(\rho E_\phi). \qquad (4.141)$$

A simple physical interpretation for the derivation of Equation (4.139) and the corresponding relations (4.140) and (4.141) exists. When examining the field behavior for a sharply bounded duct, we noticed that the modes at frequencies under consideration were nearly TE waves (see Figure 4.10). So, it is natural to employ the purely TE approximation, namely $E_z \to 0$, and then, if necessary, to find perturbations to it. This was done in obtaining (4.139) and (4.140). In fact, when passing to Equation (4.139), we have discarded the contribution from branch q_2. This impression in description of the field is, however, unimportant because a more accurate examination shows that a contribution from that branch is apparent only in a very thin annular layer, somewhere near the surface at which $\hat{q}_1 = 0$.

Equation (4.139) is much simpler than the previous coupled equations (4.137) and (4.138), and there are few density profiles which lead to closed-form solutions to it. In the case of a smooth nonuniform duct the parabolic approximation is usually satisfactory to represent $N(\rho)$ in the near vicinity of the duct axis. That is

$$N(\rho) = \tilde{N}\left(1 - \frac{\rho^2}{2\tilde{a}^2}\right). \qquad (4.142)$$

This implies that we take $g^2 \approx \tilde{g}^2(1 - \rho^2/\tilde{a}^2)$ to substitute for g^2 into Equation (4.139), a notation $\tilde{g} = g(0)$ being understood. The solution $E_\phi(\rho)$ must be finite everywhere and vanish at infinity so as to be an absolutely square integrable function. It is conveniently sought in the form

$$E_\phi = \zeta^{1/2} f(\zeta) \exp\left(-\frac{\zeta}{2}\right),$$

where $\zeta = \rho^2 k_0 |\tilde{g}|/(p\tilde{a})$, that leads to the following equation in f:

$$\zeta f''(\zeta) + (2 - \zeta) f'(\zeta) + (n - 1) f(\zeta) = 0,$$

with $n = (\tilde{g}^2 - p^4) k_0 \tilde{a}/(4 |\tilde{g}| p)$. The solutions satisfying all the prescribed conditions exist only when n is a positive integer. This gives

$$p^4 + \frac{4n |\tilde{g}|}{k_0 \tilde{a}} p - \tilde{g}^2 = 0, \quad n = 1, 2, \ldots. \tag{4.143}$$

The solutions can be found to be expressible via the confluent hypergeometric function $\Phi(a, c; x)$ by $f(\zeta) = \Phi(1 - n, 2; \zeta)$ (see Bateman Manuscript Project, 1953, Vol. 1, Chapter 6). The functions $\Phi(-k, 1 + \alpha; x)$ where k is a nonnegative integer reduce to generalized Laguerre polynomials $L_k^\alpha(x)$, which are defined as

$$L_k^\alpha(x) = \frac{1}{k!} e^x x^{-\alpha} \frac{\mathrm{d}^k}{\mathrm{d}x^k} (e^{-x} x^{\alpha+k})$$

(Bateman Manuscript Project, 1953, Vol. 2, p. 188). Thus, the E_ϕ component for the nth mode may be written in the form

$$E_\phi = E_0 \frac{\rho}{\varrho_n} L_{n-1}^1 \left(\frac{\rho^2}{\varrho_n^2}\right) \exp\left(-\frac{\rho^2}{2\varrho_n^2}\right), \tag{4.144}$$

where E_0 is an arbitrary constant, $\varrho_n = (p_n \tilde{a}/k_0 |\tilde{g}|)^{1/2}$ is the characteristic transverse scale of the field distribution in the nth mode, and p_n is its propagation constant which is found from Equation (4.143). For the lower order mode, i.e. $n = 1$, (4.144) becomes

$$E_\phi = E_0 \frac{\rho}{\varrho_1} \exp\left(-\frac{\rho^2}{2\varrho_1^2}\right). \tag{4.145}$$

It is evident that the parabolic approximation of the density is accurate while the inequality $\varrho_n \ll \tilde{a}$ holds. In addition, the propagation constant cannot be less than the lower boundary of range (4.74). These two conditions are satisfied when $k_0 a \tilde{\mathcal{P}} \gg 4n$. For narrower ducts (i.e. $k_0 a \tilde{\mathcal{P}} \sim 1$), the approximation (4.142) fails its accuracy, and Equation (4.139) must be solved for the particular density profile by using numerical methods.

Provided that the duct is weakly nonuniform, by setting $g \approx \tilde{g}$ in (4.141), it is easy to obtain a useful approximate expression for the power P_n transmitted by the nth mode along the duct. That is

$$
\begin{aligned}
P_n &= \frac{1}{2} Z_0^{-1} \operatorname{Re} \int_0^{2\pi} \mathrm{d}\phi \int_0^{\infty} (\mathbf{E}_n \times \boldsymbol{\mathcal{H}}_n^*) \cdot \hat{z}_0 \rho \, \mathrm{d}\rho \\
&= Z_0^{-1} |E_0|^2 \frac{\pi}{2} p_n \varrho_n^2 \left(1 + \frac{\tilde{g}^2}{p_n^4}\right) \int_0^{\infty} e^{-\xi} \xi \left(L_{n-1}^1(\xi)\right)^2 \mathrm{d}\xi. \quad (4.146)
\end{aligned}
$$

The integral on the right of (4.146) can be found in *Tables* of Gradshteyn and Ryzhik (1965):

$$
\int_0^{\infty} e^{-\xi} \xi \left(L_k^1(\xi)\right)^2 \mathrm{d}\xi = k + 1.
$$

Finally, we arrive at

$$
P_n = Z_0^{-1} |E_0|^2 \frac{\pi n}{2} p_n \varrho_n^2 \left(1 + \frac{\tilde{g}^2}{p_n^4}\right). \quad (4.147)
$$

In the above considerations, a cylindrical geometry has been assumed. In applying to a plane model of duct, a similar problem for whistler modes of the same frequencies (2.70) has been discussed by Shvartsburg and Stenflo (1990).

Now examine whether the foregoing method could be extended to the frequency interval (2.71). As in § 4.4, we assume that the propagation constants for the modes of interest lie in the range (4.75), being well away from the branch point $p = \tilde{\mathcal{P}}_c$. Also, to simplify matters, we take the case when $\omega \gg \omega_{\mathrm{LH}}$, so that we may neglect Y_{LH} in comparison with unity. It is clear that Equation (4.139) and the pertinent expressions (4.140) and (4.141) would be applicable to collisional ducts such as discussed in § 4.6. For such ducts, the branch 2 represents the rapidly damping wave and may therefore be discarded in a first approximation (see Figure 4.16).

In the absence of collisions, when the second branch represents an escaping wave, Equation (4.139) gives a good approximation for the E_ϕ component (see Figure 4.11) but the expressions (4.140) and (4.141) fail to represent some other components which are more heavily affected by that wave. However, if the duct is *weakly nonuniform* there is a simple way to estimate the contribution from the escaping wave to the full field and then to find the radiation attenuation of mode. To do this, we take Equation (4.138) in the form

$$
\frac{\mathrm{d}^2 E_z}{\mathrm{d}\rho^2} + \frac{1}{\rho} \frac{\mathrm{d}E_z}{\mathrm{d}\rho} + \frac{p^2}{(p^2 - \varepsilon)\varepsilon} \frac{\mathrm{d}\varepsilon}{\mathrm{d}\rho} \frac{\mathrm{d}E_z}{\mathrm{d}\rho} + k_0^2 \hat{q}_2^2(\rho) E_z = F(\rho), \quad (4.148)
$$

where

$$
F(\rho) = k_0 \frac{p}{\varepsilon} (p^2 - \varepsilon) \frac{1}{\rho} \frac{\mathrm{d}}{\mathrm{d}\rho} \left(\frac{\rho g E_\phi}{p^2 - \varepsilon}\right), \quad (4.149)
$$

and substitute the solution of Equation (4.139) for E_ϕ into (4.149). The right side of Equation (4.148) is then known, and the method of variation of parameters may be applied to give the E_z component in the escaping wave. Let $\mathcal{E}_z^{(+)}$ and $\mathcal{E}_z^{(-)}$ be two independent solutions of Equation (4.148) with its zero right side. Here a superscript $(+)$ is used to denote the independent solution for which the power flow is outgoing, and the superscript $(-)$ is used to denote the independent solution for which the power flow is ingoing. The general solution to the inhomogeneous equation (4.148) is then

$$E_z(\rho) = \mathcal{E}_z^{(+)}(\rho) \left[-\int_0^\rho F(\rho')\mathcal{E}_z^{(-)}(\rho')W^{-1}(\rho')\,d\rho' + C^{(+)} \right]$$
$$+ \ \mathcal{E}_z^{(-)}(\rho) \left[\int_0^\rho F(\rho')\mathcal{E}_z^{(+)}(\rho')W^{-1}(\rho')\,d\rho' + C^{(-)} \right], \qquad (4.150)$$

where $C^{(+)}$ and $C^{(-)}$ are undetermined constants and W is the Wronskian of $\mathcal{E}_z^{(+)}$, $\mathcal{E}_z^{(-)}$:

$$W(\rho) = \mathcal{E}_z^{(+)}(\rho)\frac{d}{d\rho}\mathcal{E}_z^{(-)}(\rho) - \mathcal{E}_z^{(-)}\frac{d}{d\rho}\mathcal{E}_z^{(+)}(\rho). \qquad (4.151)$$

If the density in a weakly nonuniform duct does not vary appreciably within a scale $(k_0|\hat{q}_2|)^{-1}$, the solutions $\mathcal{E}_z^{(\pm)}$ may be found using the WKB method. Making routine steps well described in standard texts (Heading, 1962; Fröman and Fröman, 1965; Mathews and Walker, 1965) yields

$$\mathcal{E}_z^{(\pm)}(\rho) = \left(\frac{p^2 - \varepsilon(\rho)}{\varepsilon(\rho)} \right)^{1/2} \frac{1}{(k_0\hat{q}_2(\rho)\rho)^{1/2}} \exp\left(\pm ik_0 \int_0^\rho \hat{q}_2(\rho')\,d\rho' \right). \qquad (4.152)$$

The WKB approximation is applicable in regions where $k_0|\hat{q}_2|\rho \gg 1$. As ρ decreases, it gradually fails to accurately represent the solution. On the other hand, for small values of ρ, the solution must have the form $E_z(\rho) = \text{const} \times J_0(k_0\hat{q}_2\rho)$. Assume that the plasma nonuniformity in the duct is so weak that it is still negligible up to large values of ρ where $k_0|\hat{q}_2|\rho \gg 1$, for which the Bessel function $J_0(k_0\hat{q}_2\rho)$ allows its large-argument approximation

$$J_0(k_0\hat{q}_2\rho) \approx \left(\frac{1}{2\pi k_0\hat{q}_2\rho} \right)^{1/2} \left(e^{ik_0\hat{q}_2\rho - i\pi/4} + e^{-ik_0\hat{q}_2\rho + i\pi/4} \right). \qquad (4.153)$$

We can link the solution in (4.150) with (4.153) using the connection formulas

$$C^{(+)} = Ce^{-i\pi/4}, \quad C^{(-)} = Ce^{i\pi/4}. \qquad (4.154)$$

The constant C is obtainable by making use of the fact that for $\rho \to \infty$ only the solution $\mathcal{E}^{(+)}$ must occur, whence

$$C = C^{(-)}e^{-i\pi/4} = -e^{-i\pi/4}\int_0^\infty F(\rho')\mathcal{E}_z^{(+)}(\rho')W^{-1}(\rho')\,d\rho'. \qquad (4.155)$$

Combining the expressions (4.150) to (4.152), (4.154), and (4.155), it follows without difficulty that, for rather large values of ρ,

$$E_z(\rho) = -iE_0 \left(\frac{2}{\pi k_0\,\hat{q}_2(\rho)\,\rho} \right)^{1/2} A\exp\left(ik_0\int_0^\rho \hat{q}_2(\rho')\,d\rho' - i\frac{\pi}{4} \right), \qquad (4.156)$$

where

$$
\begin{aligned}
A \;=\; & \left(\frac{\pi \left(p^2 - \varepsilon_a \right)}{2\varepsilon_a} \right)^{1/2} \int_0^\infty \rho' \left(\frac{\varepsilon(\rho')}{p^2 - \varepsilon(\rho')} \right)^{1/2} \frac{F(\rho')}{E_0} \frac{1}{\left(k_0 \hat{q}_2(\rho')\,\rho' \right)^{1/2}} \\
& \times \; \cos \left(k_0 \int_0^{\rho'} \hat{q}_2(\varrho)\,\mathrm{d}\varrho - \frac{\pi}{4} \right) \mathrm{d}\rho'.
\end{aligned}
\tag{4.157}
$$

In deriving these formulas, use was made of the fact that to within an accuracy of the WKB method, the Wronskian (4.151) can be taken as $W = -2\mathrm{i}(p^2 - \varepsilon)/\varepsilon\rho$.

Now the attenuation constant p'' can readily be estimated. For this, we employ the so-called "power" formula for the attenuation rate

$$
p'' = \Sigma/2k_0\,P,
\tag{4.158}
$$

which gives the relationship between P, the power transmitted by a mode along the guide, and Σ, the power loss per unit length of the guide (Jackson, 1962, Chapter 8, §5). The quantity P for modes of various orders is obtainable from (4.147). The quantity Σ is given here by

$$
\Sigma \approx -\frac{1}{2} Z_0^{-1} \lim_{\rho \to \infty} \left[2\pi\rho\,\mathrm{Re}\left(E_z \mathcal{H}_\phi^* \right) \right].
\tag{4.159}
$$

Note that the term $\mathrm{Re}\left(E_\phi \mathcal{H}_z^* \right)$ also contributing to Σ can be shown to be negligible for the escaping wave in which $|\hat{q}_2|^2 \gg |\eta|$. A close examination of the field expression in the outer region (see, for example, (4.52) and (4.54), with allowance for $m = 0$ and $s_2 \approx -\mathrm{i}\hat{q}_2 \approx -\mathrm{i}pY$) gives the relation $\mathcal{H}_\phi \approx E_z \eta_a/\hat{q}_2$ in the escaping wave, provided that $k_0\,|\hat{q}_2|\,\rho \gg 1$.[*] Then one obtains

$$
\Sigma \approx 2Z_0^{-1} |E_0|^2 \frac{X_a}{k_0\,(pY)^2} |A|^2,
\tag{4.160}
$$

whence

$$
p'' \approx \frac{X_a}{\tilde{X}} \frac{1}{k_0 \tilde{a} Y} \frac{2}{\pi n \left(1 + p^4/\tilde{g}^2 \right)} |A|^2 \Bigg|_{p = p_n'}.
\tag{4.161}
$$

To determine A, the solution of Equation (4.139) is inserted in (4.149) to give the appropriate integrand of (4.157); the propagation constants $p = p_n'$ are found, as before, from (4.143). A straightforward evaluation of A shows that $|A| \ll 1$ when $Y \gg 1$ and $|p|^2 \gg 4\varepsilon$, that leads to the inequality $p'' \ll p'$.

In the above, a weakly nonuniform duct has been assumed. A similar but much more complicated treatment for leaky modes in a smooth duct with a strong density enhancement, of radius a, gives the inequality $p''_{\max} < (X_a/\tilde{X})(k_0 a Y)^{-1} \ll p'$ which is accurate at frequencies $\omega \ll \omega_{\mathrm{H}}$, provided that p and $\tilde{\mathcal{P}}_c$ are not too close together. The last restriction implies that branches q_1 and q_2 are well separated (i.e. $|\hat{q}_2| \gg |\hat{q}_1|$). In view of this, it becomes clear that only a small amount of the power flow associated with the "trapped" branch q_1 can convert into the leaking power flow appropriate to the escaping branch q_2.

---

[*] Bearing this in mind, it is possible to show that the factors outside the exponential in (4.152) account for power conservation.

It is to be noted that if a duct is so wide that the condition $k_0 a |\hat{q}_1| \gg 1$ is satisfied (because of the inequality $|\hat{q}_2| \gg |\hat{q}_1|$, then $k_0 a |\hat{q}_2| \gg 1$) the whistler waveguide propagation along it may be treated by using such well-known asymptotic methods as ray theory (Helliwell, 1965), the parabolic equation method (Karpman and Kaufman, 1982 a; Washimi, 1976), the WKB and phase integral methods (e. g., Walker, 1972; Karpman and Kaufman, 1982 b, 1984), etc. Discussion of them is beyond the scope of this book, but the interested reader should consult the works mentioned.

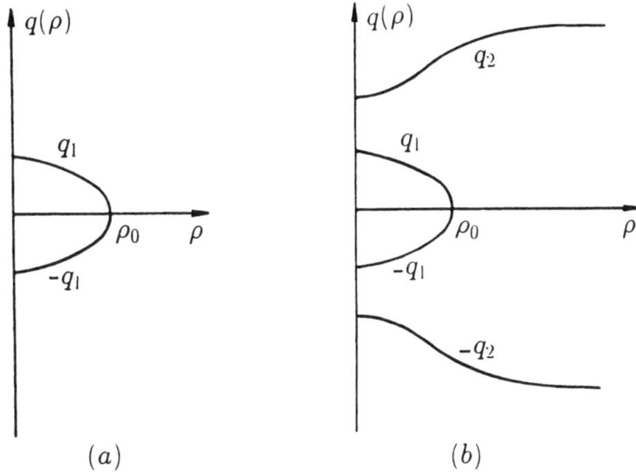

Figure 4.18. Plots of the transverse wavenumbers q_k as functions of the cylindrical coordinate ρ for a smooth density enhancement when (a) $\Omega_H \ll \omega < \omega_{LH}$, $\mathcal{P} < p < \tilde{\mathcal{P}}$, and (b) $\omega_{LH} < \omega < \omega_H/2$, $\max\{\mathcal{P}, \tilde{\mathcal{P}}_c\} < \mathrm{Re}\,(p) < \tilde{\mathcal{P}}$.

When the density is a slowly varying function of position, the fields, according to ray theory, behave locally as a bundle of rays, with each ray transporting energy along the geometrical optics trajectory (Kelso, 1964, Chapter 6). This has an especially important bearing on multimode ducts such as may exist in the earth's magnetosphere. The waveguide propagation along them can be described using the ray-tracing methods of classical geometrical optics. The geometrical optics approximation enables us to determine the variation of transverse wavenumbers q_k with ρ in such ducts by employing the local relation $q_k(\rho) = q_k(p, N(\rho))$. Some typical plots of $q_{1,2}$ versus ρ are shown in Figures 4.18 and 4.19 for the case of an enhancement and a trough, respectively (see Karpman and Kaufman, 1982 b). These allow us to gain a better qualitative insight into the topic concerning the general conditions for the existence of various sorts of modes (see also §4.3). Part (a) of Figure 4.17 shows the behavior of q_1 for density enhancement, at frequencies (2.70), when p belongs to the range (4.74). The purely imaginary branch q_2 is not given here. The curve $q_1(\rho)$ for this example is closed, thereby representing the case of a trapped wave with its turning point at $\rho = \rho_0$. Part (b) of Figure 4.18 relates to the frequency interval (2.71), the condition (4.75) for the propagation constant being understood. Although $q_1(\rho)$ does not change qualitatively in comparison with the preceding case, the untrapped propagating branch q_2 appears here. The two branches coalesce nowhere for purely real ρ because the branch point, ρ_1, at which $q_1(\rho_1) = q_2(\rho_1)$ is evidently complex. Hence no coupling between the branches can be established within the pure geometrical optics approximation. In reality, the coupling exists but its accurate description requires a more exact theory. Also, this is necessary in the neighborhood of the turning point $\rho = \rho_0$ where the geometrical optics approximation becomes inapplicable.

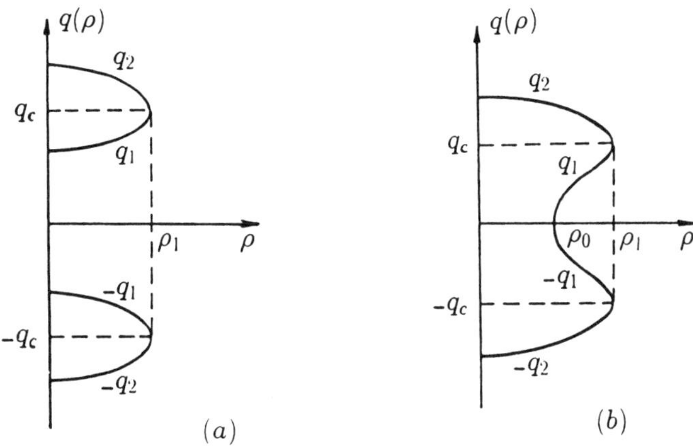

Figure 4.19. Plots of the transverse wavenumbers q_k as functions of the cylindrical coordinate ρ for a smooth density trough, at frequencies $\omega_{\mathrm{LH}} < \omega < \omega_{\mathrm{H}}/2$, when (a) $\tilde{\mathcal{P}}_c < p < \min\{\mathcal{P}_c, \tilde{\mathcal{P}}\}$, and (b) $\tilde{\mathcal{P}} < p < \mathcal{P}_c$.

The behavior of curves $q_k(\rho)$ for a trough, at frequencies (2.71), is illustrated by Figure 4.19. It is assumed that p lies in the range (4.76). Under these circumstances two distinct situations are possible, namely $\tilde{\mathcal{P}}_c < p < \min\{\mathcal{P}_c, \tilde{\mathcal{P}}\}$ and $\tilde{\mathcal{P}} < p < \mathcal{P}_c$, which correspond to parts (a) and (b), respectively, of Figure 4.19. Although the geometrical optics approximation fails in the vicinity of the branch point ρ_1, which is now real, the modes are completely trapped, as seen from Figure 4.19.

Finally, it is to be noted that all the examples indicated in Figures 4.18 and 4.19 are in full agreement with the previous discussion made in § 4.3.

4.9. Radially nonuniform duct with a nonmonotonic density profile

It has been mentioned in § 1.3 that when thermal nonlinear effects become apparent in the vicinity of an electromagnetic source, a cylindrical duct can arise in which the density variation transverse to an ambient magnetic field turns out essentially nonmonotonic, as shown, for example, in Figure 1.3. In this section, we discuss possible types of the whistler waveguide propagation along such ducts. We confine ourselves to the frequency interval (2.71) and the density profile as in Figure 1.3. In other words, it is assumed that the density distribution is characterized by the presence of an inner core with decreased density on the duct axis and a surrounding annular layer with an enhanced density, in comparison with an ambient plasma density.

As before, let N_a be an ambient plasma density and $\tilde{N}$ be a density on the duct axis. Also, let N_m be a maximum density within the annular layer. We shall use the notations $\mathcal{P}$, $\tilde{\mathcal{P}}$, and $\mathcal{P}^{(m)}$ for propagation constants of the waves traveling exactly along the static magnetic field in the uniform plasmas with densities N_a, $\tilde{N}$, and N_m, respectively. Similarly, the notations $\mathcal{P}_c$, $\tilde{\mathcal{P}}_c$, and $\mathcal{P}_c^{(m)}$ will stand for longitudinal propagation constants of the conical-refraction whistler-mode waves in such plasmas. In this case, that is, when $\tilde{N} < N_a < N_m$, four possible situations may occur which are specified by the following sets of inequalities:

$$
\begin{array}{lll}
\text{I.} & \mathcal{P}_c^{(m)} < \tilde{\mathcal{P}}; & \\[1.5ex]
\text{II.} & \tilde{\mathcal{P}} < \mathcal{P}_c, & \mathcal{P}_c^{(m)} < \mathcal{P}; \\[1.5ex]
\text{III.} & \mathcal{P}_c < \tilde{\mathcal{P}}, & \mathcal{P} < \mathcal{P}_c^{(m)}; \\[1.5ex]
\text{IV.} & \tilde{\mathcal{P}} < \mathcal{P}_c, & \mathcal{P} < \mathcal{P}_c^{(m)}.
\end{array}
\tag{4.162}
$$

To gain a qualitative insight into the general behavior of guided waves under these circumstances, consider the curves q_k versus ρ, plotted within the geometrical optics approximation.

A simple analysis similar to that made in § 4.3 shows that the eigenmodes can now exist only under the condition $\tilde{\mathcal{P}}_c < p < \mathcal{P}_c$ which coincides with (4.76). Two distinct cases appear here, namely

$$
\tilde{\mathcal{P}}_c < p < \min\{\mathcal{P}_c, \tilde{\mathcal{P}}\}
\tag{4.163}
$$

and

$$
\tilde{\mathcal{P}} < p < \mathcal{P}_c.
\tag{4.164}
$$

The range (4.163) exists for all the situations listed in (4.162), and the corresponding plot of $q_{1,2}(\rho)$ is given in part (a) of Figure 4.20. The range (4.164) may exist only under the circumstances specified by the inequalities II and IV in (4.162). The plot for that range is shown in part (b) of Figure 4.20. It is evident that the above two plots, (a) and (b), are quite similar to those in Figure 4.19 for a trough with a monotonic density variation. Hence one may expect that the eigenmodes whose propagation constants lie in the ranges (4.163) and (4.164), look like the eigenmodes of a 'single' trough. If the annular layer is so narrow that its characteristic width, $\Delta\rho$, satisfies the condition $\Delta\rho \ll (k_0 \,|\, \mathrm{Im}\, q_{1,2}(p, N_m)\,|)^{-1}$, it may be neglected in the waveguide mode analysis. Conversely, for a wide layer, when $\Delta\rho \gg (k_0 \,|\, \mathrm{Im}\, q_{1,2}(p, N_m)\,|)^{-1}$, the effect of the ambient medium on the eigenmode characteristics would be negligible.

Apart from the eigenmodes completely trapped in the inner core with decreased density, there may exist some leaky modes partially trapped in it. These are possible in either of the following ranges:

$$\mathcal{P}_c < \mathrm{Re}\,(p) < \min\{\mathcal{P}_c^{(m)}, \tilde{\mathcal{P}}\}, \tag{4.165}$$

$$\max\{\mathcal{P}_c, \tilde{\mathcal{P}}\} < \mathrm{Re}\,(p) < \min\{\mathcal{P}_c^{(m)}, \mathcal{P}\}, \tag{4.166}$$

$$\mathcal{P} < \mathrm{Re}\,(p) < \mathcal{P}_c^{(m)}. \tag{4.167}$$

The range (4.165) may be found for the cases I and III in (4.162), the range (4.166) for II, III, IV, and the range (4.167) for III and IV. Parts (c), (d), and (e) in Figure 4.20 show the curves $q_k(\rho)$ for the ranges (4.165), (4.166), and (4.167), respectively. As is seen from these plots, the modal fields in the region $\rho_1 < \rho < \rho_2$ would look like the evanescent ones. The leakage is now caused by transformation of core waves into propagating waves which exist for $\rho > \rho_2$. The physical mechanism whereby core power is transmitted across the evanescent region is known as *tunneling*. In general, the larger the width of the evanescent region, the smaller is the leakage caused by tunneling.

Apart from the leaky modes mentioned above, the leaky modes partially trapped in the layer with enhanced density can exist in the range

$$\max\{\mathcal{P}_c^{(m)}, \mathcal{P}\} < \mathrm{Re}\,(p) < \mathcal{P}^{(m)}, \tag{4.168}$$

which case may occur for each of the conditions listed in (4.162). These are leaky modes of the same type as discussed in §§ 4.3 and 4.4, where the similar range (4.75) was indicated. The curves $q_{1,2}(\rho)$ are shown for the range (4.168) in part (f) of Figure 4.20. If the radius of the layer with a fixed width increases, the properties of these modes become similar to those corresponding to a plane duct.

Thus, in the above discussion, we have indicated how modal solutions for the duct with a nonmonotonic density profile are related to the types of solutions considered in the previous sections. The discussion was qualitative, being restricted to the framework of the geometrical optics approximation. A quantitative solution for a general case, involving the situation when that approximation fails to represent the field behavior, may be found numerically, as described in § 4.6. Sometimes various simplifications are possible. For example, Zaboronkova *et al.* (1992 a) made use of a piecewise constant approximation to the nonmonotonic density distribution in the duct observed in the laboratory experiments on thermal heating of the plasma electrons (see § 1.3). In particular, Zaboronkova *et al.* (1992 a) found that the experimental values for the wavelengths for guided propagation in the central decreased-density part of the weakly nonuniform duct and in the enhanced-density annular layer that surrounds this central part corresponded to the propagation constants which satisfied the conditions (4.163) and (4.168), respectively.

Sometimes, one has to deal with more complicated nonmonotonic density variations than the simple case for which the plots of Figure 4.20 were prepared. One relevant example is the ducting of whistler-mode waves within multi-duct structures in the earth's magnetosphere. Fortunately, very often the parameters of the magnetospheric multi-duct structures vary slowly with position, so that ray tracing may be used to investigate the

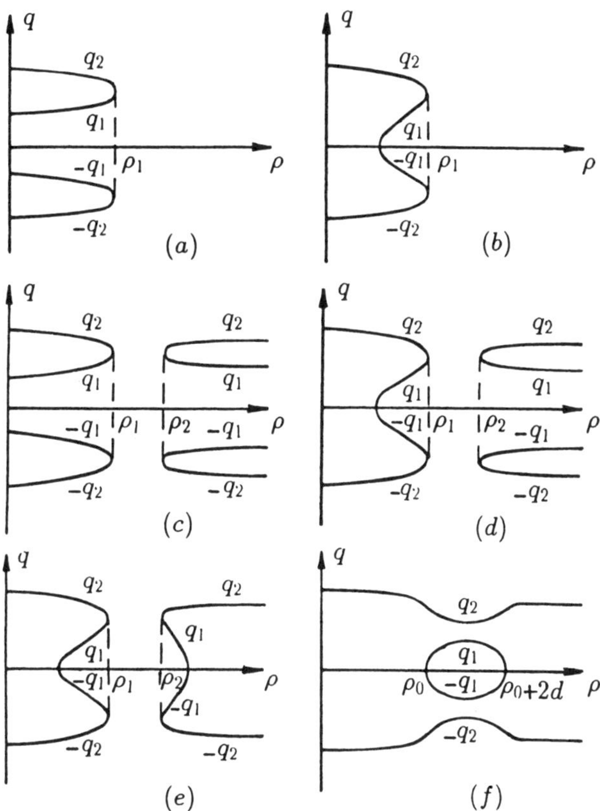

Figure 4.20. Plots of the transverse wavenumbers q_k as functions of the cylindrical coordinate ρ for duct with a nonmonotonic density variation. ρ_1 and ρ_2 stand for the branch points, and ρ_0 and $\rho_0 + 2d$ stand for the turning points; see text for discussion.

guided propagation along them. A good account of this topic is given by Strangeways (1982), whose work is strongly recommended.

It should be noted that although the above descriptions are complete enough to provide the background needed for the later work, there may be, of course, other types of ducts and guided waves (especially in the laboratory plasmas) which have not been mentioned in this chapter. Under the circumstances discussed above, only the so-called *forward modes* have been encountered in which the phase progress and power flow have the same direction. In general, when dealing with graded-profile ducts, within which the elements of dielectric tensor may change sign with ρ, the modes are possible in which the phase progress and power flow are oppositely directed. These are called the *backward modes*. Also, there may exist the *complex modes*. These are eigenmodes whose propagation constants are complex even in the absence of losses. An inspection of the dispersion equation shows that such complex solutions always appear in pairs. For example, if $p = p_n$ is a complex solution of the dispersion equation (4.71) for a non-absorbing duct, then $p = -p_n^*$ is also a solution. It will be explained later that such waves are excited in pairs, with their propagation constants $p = p_n$ and $p = -p_n^*$, but no energy flow can appear in the full field yielded by adding the two complex modes corresponding to each pair. Further discussion of such solutions need not be given since it would require us to depart even farther afield from the basic topic of this book.

Problems

4.1. Let $p_{m,n}$ be solutions of the dispersion equation (4.67) for a field-aligned cylindrical duct. Consider the situation when the static magnetic field is oppositely directed but the duct parameters are left unchanged. Noting that the dielectric tensor then becomes transposed, $\hat{\varepsilon}^{(T)}$, prove that solutions $p_{m,n}^{(T)}$ of the dispersion equation for this case are related to solutions $p_{m,n}$ by the relationship $p_{m,n}^{(T)} = p_{-m,n}$.

4.2. Show that when the plasma density in the duct tends to infinity, the dispersion equation (4.71) for the axisymmetric case reduces to

$$\frac{1}{n_1} \frac{K_1(S_1)}{S_1 \, K_0(S_1)} - \frac{1}{n_2} \frac{K_1(S_2)}{S_2 \, K_0(S_2)} = 0.$$

Verify that this equation corresponds to the case of an perfectly conducting cylinder of radius a. Prove that, for the frequency interval (2.71), the equation has a solution appropriate to a bound mode. Show that the propagation constant of this mode lies below $\mathcal{P}_c$ and varies with a between the limits $p = \varepsilon_a^{1/2}$ when $a \to 0$, and $p = \left[\varepsilon_a - \varepsilon_a^{1/2} g_a(\varepsilon_a - \eta_a)^{-1/2}\right]^{1/2}$ when $a \to \infty$. (Note carefully that here $\varepsilon_a > 0$, $g_a < 0$, and $\eta_a < 0$.)

4.3. Restore the calculations omitted in deriving formula (4.113).

4.4. Extend the analysis of possible types of modes, made in § 4.3 for whistler waves, to the frequency band $\omega < \Omega_{\mathrm{H}}$. Show that leaky modes guided along a density enhancement and bound modes guided along a density trough can be synthesized from Alfvén waves in that band. (Note that small terms of order ε/η and g/η must here be retained to obtain the desired result.)

4.5. Find the WKB solutions to coupled second-order equations (4.18) and (4.19) for a weakly nonuniform cylindrical duct with enhanced density in the frequency intervals (2.70) and (2.71). Show that the mode condition for determining real propagation constants $p = p'$ can be written

$$k_0 \int_0^{\rho_0} q_1(p, \rho)\, \mathrm{d}\rho = \pi n,$$

where $q_1(p, \rho) = q_1(p, N(\rho))$, ρ_0 is a turning point in which $q_1(p, \rho_0) = 0$, and n is a positive integer. Obtain the attenuation constants of leaky modes at frequencies (2.71). To do this, find the WKB solution for the quantity $F(\rho)$ in (4.149), substitute $F(\rho)$ into (4.157), and then use formulas (4.158) to (4.160).

Hint: For full details of the WKB solution for a cylindrical duct consult the work of Karpman and Kaufman (1982 b), which gives further references.

4.6. A radially nonuniform duct with a nonmonotonic profile arising due to thermal nonlinear effects was discussed in §4.9 under the condition that the density in its core is diminished. Sometimes the thermal-diffusion-driven redistribution of the plasma can give rise to ducts with enhanced density in the core, which is surrounded by an annular layer with decreased density, compared to the ambient value. Discuss the possible types of modes for such a duct in the frequency range (2.71).

Chapter 5

Integral Representation of Source-excited Fields on a Duct

5.1. Introduction

In the preceding discussions, we were dealing with source-free field solutions for ducts. In this chapter radiation from electromagnetic sources in the presence of a density duct will be considered in some detail. As before, the duct is aligned to be along an ambient static magnetic field. The aim is to establish some general principles, and so the mathematics in this chapter is simplified by assuming that the duct is a uniform, sharply bounded circular cylinder embedded in a uniform ambient magnetoplasma, though some general equations will also be valid for nonuniform ducts. The general treatment of the radiation problem for the case of a radially nonuniform duct as well as some discussion of the alternative field representations will be made in the next chapter.

To render our consideration more concrete, we shall here apply the formulation to radiation from given currents at frequencies belonging to the interval (2.71). The reasoning for this is similar to that in Chapter 3 (see § 3.3). It was found that in a homogeneous plasma, a uniform ring electric current would be very suitable for excitation of whistlers with the wavelengths corresponding to the branch q_1. It is intuitively clear that this source would do so when being immersed in the duct. One could therefore expect that a considerable amount of power radiated from a ring electric current would go to trapped waves. Conversely, linear electric currents and ring magnetic currents launch most of the radiation power into quasi-electrostatic whistler-mode waves corresponding to the escaping branch q_2.

In view of the above, when examining the effect of a duct on the radiation characteristics of given sources, we shall pay special attention to radiation from a ring of uniform electric current. For comparison, a ring of uniform magnetic current will simultaneously be considered. In what follows, the ring sources will be assumed to be concentric with the cylindrical plasma guide. Because of the azimuthally symmetric nature of the problem, the resultant field will not possess any variation with ϕ, and the derivative $\partial/\partial\phi$ may be set equal to zero.

Some generalization to the case of nonsymmetric sources, when $\partial/\partial\phi \neq 0$, will be made in the next chapter.

5.2. The field equations

The general source-free field equations for the cylindrical geometry were given in § 4.2, where the transverse field components were expressed through the E_z and $\mathcal{H}_z$ components. When the azimuthally symmetric sources are included, it is much more convenient to represent the field in another manner, via the azimuthal electric and magnetic fields.

Thus, let the current densities be

$$\mathbf{J}^e(\mathbf{r}) = \hat{\boldsymbol{\phi}}_0\, J_\phi^e(\rho, z) = \hat{\boldsymbol{\phi}}_0\, Z_0^{-1}\, \mathcal{J}_\phi^e(\rho, z), \tag{5.1}$$

$$\mathbf{J}^m(\mathbf{r}) = \hat{\boldsymbol{\phi}}_0\, J_\phi^m(\rho, z). \tag{5.2}$$

Noting that the fields excited by these sources are independent of the azimuth about $\mathbf{B}_0$, Maxwell' equations (2.38) to (2.43) may then be written

$$E_\rho = \frac{\mathrm{i}}{k_0\varepsilon}\left(\frac{\partial \mathcal{H}_\phi}{\partial z} + k_0 g E_\phi\right). \tag{5.3}$$

$$E_\phi = -\frac{\mathrm{i}}{k_0\varepsilon}\left(k_0 g E_\rho + \frac{\partial \mathcal{H}_\rho}{\partial z} - \frac{\partial \mathcal{H}_z}{\partial \rho} - \mathcal{J}_\phi^e\right), \tag{5.4}$$

$$E_z = -\frac{\mathrm{i}}{k_0\eta}\frac{1}{\rho}\frac{\partial}{\partial \rho}(\rho\,\mathcal{H}_\phi), \tag{5.5}$$

$$\mathcal{H}_\rho = -\frac{\mathrm{i}}{k_0}\frac{\partial E_\phi}{\partial z}, \tag{5.6}$$

$$\mathcal{H}_\phi = \frac{\mathrm{i}}{k_0}\left(\frac{\partial E_\rho}{\partial z} - \frac{\partial E_z}{\partial \rho} + J_\phi^m\right), \tag{5.7}$$

$$\mathcal{H}_z = \frac{\mathrm{i}}{k_0}\frac{1}{\rho}\frac{\partial}{\partial \rho}(\rho\, E_\phi). \tag{5.8}$$

The components E_ρ, E_z, $\mathcal{H}_\rho$, and $\mathcal{H}_z$ can be eliminated to yield

$$\frac{\partial}{\partial \rho}\left[\frac{1}{\rho}\frac{\partial}{\partial \rho}(\rho\, E_\phi)\right] + \frac{\partial^2 E_\phi}{\partial z^2} + \frac{k_0^2}{\varepsilon}\left[(\varepsilon^2 - g^2)E_\phi - \frac{g}{k_0}\frac{\partial \mathcal{H}_\phi}{\partial z}\right] = \mathrm{i}k_0\mathcal{J}_\phi^e \tag{5.9}$$

and

$$\frac{\partial}{\partial \rho}\left[\frac{1}{\rho\eta}\frac{\partial}{\partial \rho}(\rho\,\mathcal{H}_\phi)\right] + \frac{\partial}{\partial z}\left(\frac{1}{\varepsilon}\frac{\partial \mathcal{H}_\phi}{\partial z}\right) + k_0^2\left[\mathcal{H}_\phi + \frac{1}{k_0}\frac{\partial}{\partial z}\left(\frac{g}{\varepsilon}E_\phi\right)\right] = \mathrm{i}k_0 J_\phi^m. \tag{5.10}$$

These equations are applicable for an arbitrary (but prescribed) azimuthally symmetric distribution of plasma density N in the duct. If the duct parameters are assumed not to vary in the z direction, the solutions of Equations (5.9) and (5.10) may be represented by

$$E_\phi(\rho, z) = \frac{k_0}{2\pi} \int_{-\infty}^{+\infty} E_\phi(\rho, p) \exp(-ik_0 pz)\, dp,$$

$$\mathcal{H}_\phi(\rho, z) = \frac{k_0}{2\pi} \int_{-\infty}^{+\infty} \mathcal{H}_\phi(\rho, p) \exp(-ik_0 pz)\, dp,$$

(5.11)

which, when substituted into (5.9) and (5.10), together with the corresponding Fourier integral representations for $\mathcal{J}_\phi^e(\rho, z)$ and $J_\phi^m(\rho, z)$, yield

$$\frac{\partial}{\partial \rho} \left[\frac{1}{\rho} \frac{\partial}{\partial \rho} \left(\rho E_\phi(\rho, p) \right) \right]$$
$$+ \frac{k_0^2}{\varepsilon} \left\{ \left[\varepsilon \left(\varepsilon - p^2 \right) - g^2 \right] E_\phi(\rho, p) + igp\mathcal{H}_\phi(\rho, p) \right\} = ik_0 \mathcal{J}_\phi^e(\rho, p) \quad (5.12)$$

and

$$\frac{\partial}{\partial \rho} \left[\frac{1}{\rho\eta} \frac{\partial}{\partial \rho} \left(\rho \mathcal{H}_\phi(\rho, p) \right) \right] + \frac{k_0^2}{\varepsilon} \left[(\varepsilon - p^2)\mathcal{H}_\phi(\rho, p) - igpE_\phi(\rho, p) \right] = ik_0 J_\phi^m(\rho, p). \quad (5.13)$$

Further, consideration pertaining to the Green's function technique will be applied to Equations (5.12) and (5.13).

Let for

$$\mathcal{J}_\phi^e(\rho, z) = Z_0 I_0^e\, \delta(\rho - b)\, \delta(z), \qquad (5.14)$$

$$J_\phi^m(\rho, z) = I_0^m\, \delta(\rho - b) \left[U(z + d) - U(z - d) \right] (2d)^{-1}. \qquad (5.15)$$

The one-dimensional Fourier transforms, in p-domain, of (5.14) and (5.15) are then

$$\mathcal{J}_\phi^e(\rho, p) = Z_0 I_0^e\, \delta(\rho - b), \qquad (5.16)$$

$$J_\phi^m(\rho, p) = I_0^m\, \delta(\rho - b) \frac{\sin(k_0 pd)}{k_0 pd}. \qquad (5.17)$$

A need for spreading the magnetic current distribution in the z direction is caused by the fact that in the frequency range (2.71) the total radiation power of this source tends to infinity if $d \to 0$, as in the case of a homogeneous plasma. Notice that the factor representing the current spreading with z is used now in the form somewhat different from that in § 3.6. This is done only to simplify the mathematics later on.

Now the expressions (5.16) and (5.17) are substituted into Equations (5.12) and (5.13), which then become the defining equations for the axisymmetric Green's function problem in ρ-domain. In order to eliminate some of the function-theoretic tedium associated with use of Green's functions in obtaining the field solutions, we shall not state the systematized Green's function technique for the problem under consideration. The interested reader can study Green's function representations for general anisotropic waveguides, obtained via abstract operator methods, in the work of Bresler and Marcuvitz (1956). Note that in the next chapter we shall proceed via some alternative technique which, imbued with a certain formal elegance, appears much more convenient for readily obtaining the general solution. For the special problem herein, we utilize a procedure which, though less general, yields the required result more directly and by conventional methods.

5.3. Fields of ring currents, for a uniform duct

Now determine the solutions of Equations (5.12) and (5.13) where the Fourier transforms of the electric and magnetic current distributions are given by (5.16) and (5.17). The required solutions must be regular at $\rho = 0$ and satisfy the radiation condition for $\rho \to \infty$. They may be found in an analytical form for a piecewise constant profile of the plasma density.

Consider the step profile given by (4.39). The independent solutions of Equations (5.12) and (5.13) without sources are then known and represented via the cylindrical functions (see § 4.2, section 4.2.3). This allows for employing the method of variation of parameters to find the general solutions individually in domains $\rho < a$ and $\rho > a$ if sources are included. At the surface $\rho = a$, the appropriate boundary conditions must then be satisfied. Another possibility would be to utilize the Fourier – Bessel transformation of both sides of Equations (5.12) and (5.13) to determine their solutions for each domain. It is, however, more instructive to construct the required solutions directly from the general source-free solutions, already known to us, in domains $0 < \rho < \min\{a,b\}$, $\min\{a,b\} < \rho < \max\{a,b\}$, $\max\{a,b\} < \rho < \infty$ and then take into account that components $E_{\phi,z}(\rho,p)$, $\mathcal{H}_{\phi,z}(\rho,p)$ must satisfy the continuity conditions such as (4.65) at $\rho = a$ and the following conditions at $\rho = b$:

$$E_\phi(b - 0, p) = E_\phi(b + 0, p),$$

$$E_z(b - 0, p) = E_z(b + 0, p) - I_0^m \, \frac{\sin(k_0 pd)}{k_0 pd} \,,$$

$$\mathcal{H}_\phi(b - 0, p) = \mathcal{H}_\phi(b + 0, p),$$

$$\mathcal{H}_z(b - 0, p) = \mathcal{H}_z(b + 0, p) + Z_0 I_0^e. \tag{5.18}$$

These relations provide the jumps of the appropriate field components when crossing the current surface $\rho = b$ (e.g., Felsen and Marcuvitz, 1973, Chapter 1, § 5, formula (32a)).

The required solutions for source-free domains would be, from (4.36) and (4.38): $\rho < \min\{a,b\}$,

$$E_\phi(\rho,p) = \mathrm{i}\sum_{k=1}^{2} b_k\, J_1\left(\tilde{Q}_k\,\frac{\rho}{a}\right), \quad \mathcal{H}_\phi(\rho,p) = -\sum_{k=1}^{2} b_k \tilde{n}_k\, J_1\left(\tilde{Q}_k\,\frac{\rho}{a}\right); \qquad (5.19)$$

$\min\{a,b\} < \rho < \max\{a,b\}$,

$$E_\phi(\rho,p) = \begin{cases} \mathrm{i}\sum_{i=1}^{2}\sum_{k=1}^{2} d_{ik}\, \tilde{Z}_1^{(i)}\left(\tilde{Q}_k\,\frac{\rho}{a}\right) & \text{if } a > b, \\[2ex] \mathrm{i}\sum_{i=1}^{2}\sum_{k=1}^{2} d_{ik}\, Z_1^{(i)}\left(S_k\,\frac{\rho}{a}\right) & \text{if } a < b, \end{cases}$$

$$(5.20)$$

$$\mathcal{H}_\phi(\rho,p) = \begin{cases} -\sum_{i=1}^{2}\sum_{k=1}^{2} d_{ik}\,\tilde{n}_k \tilde{Z}_1^{(i)}\left(\tilde{Q}_k\,\frac{\rho}{a}\right) & \text{if } a > b, \\[2ex] -\sum_{i=1}^{2}\sum_{k=1}^{2} d_{ik}\, n_k Z_1^{(i)}\left(S_k\,\frac{\rho}{a}\right) & \text{if } a < b; \end{cases}$$

$\rho > \max\{a,b\}$,

$$E_\phi(\rho,p) = \mathrm{i}\sum_{k=1}^{2} c_k\, K_1\left(S_k\,\frac{\rho}{a}\right), \quad \mathcal{H}_\phi(\rho,p) = -\sum_{k=1}^{2} c_k n_k\, K_1\left(S_k\,\frac{\rho}{a}\right). \qquad (5.21)$$

In the preceding, $\tilde{Z}_m^{(i)}$ stands for the Bessel function $J_m(\zeta)$ if $i = 1$ and $Y_m(\zeta)$ if $i = 2$, while $Z_m^{(i)}(\zeta)$ stands for the modified Bessel function $I_m(\zeta)$ if $i = 1$ and $K_m(\zeta)$ if $i = 2$; b_k, d_{ik} and c_k are coefficients yet to be determined; the other notations are the same as in (4.66). The remaining components of $\mathbf{E}(\rho,p)$ and $\mathcal{H}(\rho,p)$ can be obtained via the relations similar to (5.3) through (5.8) if the operator $\partial/\partial z$ is replaced by $-\mathrm{i}k_0 p$. Using expressions (5.19) to (5.21) and applying the prescribed field conditions at $\rho = a$ and $\rho = b$ for $E_{\phi,z}(\rho,p)$, $\mathcal{H}_{\phi,z}(\rho,p)$, it is possible to obtain eight simultaneous inhomogeneous equations in the eight quantities b_k, d_{ik}, and c_k. The resulting equations may be written in matrix notation, thus:

$$\mathbf{A}\cdot\mathbf{F} = \mathbf{S}, \qquad (5.22)$$

where $\mathbf{F}$ and $\mathbf{S}$ denote the column matrices, whose elements are

$$F_1 = b_1, \quad F_2 = b_2, \quad F_3 = d_{11}, \quad F_4 = d_{12},$$
$$F_5 = d_{21}, \quad F_6 = d_{22}, \quad F_7 = c_1, \quad F_8 = c_2, \qquad (5.23)$$

and

$$S_1 = S_3 = S_5 = S_6 = S_7 = S_8 = 0,$$
$$S_2 = \mathrm{i}k_0 a\hat{\eta}\, I_0^m\, \frac{\sin(k_0 pd)}{k_0 pd}, \quad S_4 = -k_0 a Z_0 I_0^e, \qquad (5.24)$$

where

$$\hat{\eta} = \begin{cases} \tilde{\eta} & \text{if } b < a, \\ \eta_a & \text{if } b > a. \end{cases}$$

In Equation (5.22), $\mathbf{A}$ stands for the 8×8 matrix which is written as follows.

For $b < a$:

$$\mathbf{A} = \tag{5.25}$$

$$\begin{pmatrix}
J_1(\hat{\xi}_1) & J_1(\hat{\xi}_2) & -J_1(\hat{\xi}_1) & -J_1(\hat{\xi}_2) & -Y_1(\hat{\xi}_1) & -Y_1(\hat{\xi}_2) & 0 & 0 \\
\tilde{d}_1 J_0(\hat{\xi}_1) & \tilde{d}_2 J_0(\hat{\xi}_2) & -\tilde{d}_1 J_0(\hat{\xi}_1) & -\tilde{d}_2 J_0(\hat{\xi}_2) & -\tilde{d}_1 Y_0(\hat{\xi}_1) & -\tilde{d}_2 Y_0(\hat{\xi}_2) & 0 & 0 \\
\tilde{n}_1 J_1(\hat{\xi}_1) & \tilde{n}_2 J_1(\hat{\xi}_2) & -\tilde{n}_1 J_1(\hat{\xi}_1) & -\tilde{n}_2 J_1(\hat{\xi}_2) & -\tilde{n}_1 Y_1(\hat{\xi}_1) & -\tilde{n}_2 Y_1(\hat{\xi}_2) & 0 & 0 \\
\tilde{\xi}_1 J_0(\hat{\xi}_1) & \tilde{\xi}_2 J_0(\hat{\xi}_2) & -\tilde{\xi}_1 J_0(\hat{\xi}_1) & -\tilde{\xi}_2 J_0(\hat{\xi}_2) & -\tilde{\xi}_1 Y_0(\hat{\xi}_1) & -\tilde{\xi}_2 Y_0(\hat{\xi}_2) & 0 & 0 \\
0 & 0 & J_1(\tilde{\xi}_1) & J_1(\tilde{\xi}_2) & Y_1(\tilde{\xi}_1) & Y_1(\tilde{\xi}_2) & -K_1(S_1) & -K_1(S_2) \\
0 & 0 & \dfrac{\tilde{d}_1}{\tilde{\eta}} J_0(\tilde{\xi}_1) & \dfrac{\tilde{d}_2}{\tilde{\eta}} J_0(\tilde{\xi}_2) & \dfrac{\tilde{d}_1}{\tilde{\eta}} Y_0(\tilde{\xi}_1) & \dfrac{\tilde{d}_2}{\tilde{\eta}} Y_0(\tilde{\xi}_2) & \dfrac{d_1}{\eta_a} K_0(S_1) & \dfrac{d_2}{\eta_a} K_0(S_2) \\
0 & 0 & \tilde{n}_1 J_1(\tilde{\xi}_1) & \tilde{n}_2 J_1(\tilde{\xi}_2) & \tilde{n}_1 Y_1(\tilde{\xi}_1) & \tilde{n}_2 Y_1(\tilde{\xi}_2) & -n_1 K_1(S_1) & -n_2 K_1(S_2) \\
0 & 0 & \tilde{\xi}_1 J_0(\tilde{\xi}_1) & \tilde{\xi}_2 J_0(\tilde{\xi}_2) & \tilde{\xi}_1 Y_0(\tilde{\xi}_1) & \tilde{\xi}_2 Y_0(\tilde{\xi}_2) & S_1 K_0(S_1) & S_2 K_0(S_2)
\end{pmatrix}$$

where $\hat{\xi}_k = \tilde{\xi}_k \dfrac{b}{a} = \tilde{Q}_k \dfrac{b}{a}$, $\tilde{d}_1 = \tilde{n}_1 \tilde{Q}_1$, $\tilde{d}_2 = \tilde{n}_2 \tilde{Q}_2$, $d_1 = n_1 S_1$, $d_2 = n_2 S_2$.

For $b > a$:

$$\mathbf{A} = \tag{5.26}$$

$$\begin{pmatrix}
0 & 0 & I_1(\hat{S}_1) & I_1(\hat{S}_2) & K_1(\hat{S}_1) & K_1(\hat{S}_2) & -K_1(\hat{S}_1) & -K_1(\hat{S}_2) \\
0 & 0 & d_1 I_0(\hat{S}_1) & d_2 I_0(\hat{S}_2) & -d_1 K_0(\hat{S}_1) & -d_2 K_0(\hat{S}_2) & d_1 K_0(\hat{S}_1) & d_2 K_0(\hat{S}_2) \\
0 & 0 & n_1 I_1(\hat{S}_1) & n_2 I_1(\hat{S}_2) & n_1 K_1(\hat{S}_1) & n_2 K_1(\hat{S}_2) & -n_1 K_1(\hat{S}_1) & -n_2 K_1(\hat{S}_2) \\
0 & 0 & S_1 I_0(\hat{S}_1) & S_2 I_0(\hat{S}_2) & -S_1 K_0(\hat{S}_1) & -S_2 K_0(\hat{S}_2) & S_1 K_0(\hat{S}_1) & S_2 K_0(\hat{S}_2) \\
J_1(\tilde{\xi}_1) & J_1(\tilde{\xi}_2) & -I_1(S_1) & -I_1(S_2) & -K_1(S_1) & -K_1(S_2) & 0 & 0 \\
\dfrac{\tilde{d}_1}{\tilde{\eta}} J_0(\tilde{\xi}_1) & \dfrac{\tilde{d}_2}{\tilde{\eta}} J_0(\tilde{\xi}_2) & -\dfrac{d_1}{\eta_a} I_0(S_1) & -\dfrac{d_2}{\eta_a} I_0(S_2) & \dfrac{d_1}{\eta_a} K_0(S_1) & \dfrac{d_2}{\eta_a} K_0(S_2) & 0 & 0 \\
\tilde{n}_1 J_1(\tilde{\xi}_1) & \tilde{n}_2 J_1(\tilde{\xi}_2) & -n_1 I_1(S_1) & -n_2 I_1(S_2) & -n_1 K_1(S_1) & -n_2 K_1(S_2) & 0 & 0 \\
\tilde{\xi}_1 J_0(\tilde{\xi}_1) & \tilde{\xi}_2 J_0(\tilde{\xi}_2) & -S_1 I_0(S_1) & -S_2 I_0(S_2) & S_1 K_0(S_1) & S_2 K_0(S_2) & 0 & 0
\end{pmatrix}$$

where $\hat{S}_k = S_k \dfrac{b}{a}$ and the other notations are as in (5.25).

Upon solving Equation (5.22) for $\mathbf{F}$, the elements of $\mathbf{F}$ are substituted into (5.19) to (5.21) to yield the resulting expressions for $E_\phi(\rho, p)$ and $\mathcal{H}_\phi(\rho, p)$. After a considerable amount of algebraic manipulation, the solution can be shown to be as follows (see Zaboronkova *et al.*, 1993 c, 1997).*

---

* The algebra yielding the resulting form of the solution is effected by a straightforward procedure using the well-known expressions for the Wronskians of cylindrical functions (Abramowitz and Stegun, 1964). It is worth mentioning that the treatment is much more easily handled by the use of symbolic manipulation languages of the computer algebra.

For $b < a$:

$\rho < a$,

$$E_\phi(\rho, p) = \mathrm{i} \sum_{k=1}^{2} \left[B_k\, J_1\left(\tilde{Q}_k\, \frac{\rho}{a}\right) - \tilde{f}_k\, \tilde{G}_k^{(1)}(\rho)\, U(\rho - b) \right],$$

$$\mathcal{H}_\phi(\rho, p) = - \sum_{k=1}^{2} \tilde{n}_k \left[B_k\, J_1\left(\tilde{Q}_k\, \frac{\rho}{a}\right) - \tilde{f}_k\, \tilde{G}_k^{(1)}(\rho)\, U(\rho - b) \right];$$

(5.27)

$\rho > a$,

$$E_\phi(\rho, p) = \mathrm{i} \sum_{k=1}^{2} C_k\, K_1\left(S_k\, \frac{\rho}{a}\right),$$

$$\mathcal{H}_\phi(\rho, p) = - \sum_{k=1}^{2} n_k\, C_k\, K_1\left(S_k\, \frac{\rho}{a}\right).$$

(5.28)

For $b > a$:

$\rho < a$,

$$E_\phi(\rho, p) = \mathrm{i} \sum_{k=1}^{2} B_k\, J_1\left(\tilde{Q}_k\, \frac{\rho}{a}\right),$$

$$\mathcal{H}_\phi(\rho, p) = - \sum_{k=1}^{2} \tilde{n}_k B_k\, J_1\left(\tilde{Q}_k\, \frac{\rho}{a}\right);$$

(5.29)

$\rho > a$,

$$E_\phi(\rho, p) = \mathrm{i} \sum_{k=1}^{2} \left\{ C_k\, K_1\left(S_k\, \frac{\rho}{a}\right) + f_k\, G_k^{(1)}(\rho)\left[1 - U(\rho - b)\right] \right\},$$

$$\mathcal{H}_\phi(\rho, p) = - \sum_{k=1}^{2} n_k \left\{ C_k\, K_1\left(S_k\, \frac{\rho}{a}\right) + f_k\, G_k^{(1)}(\rho)\left[1 - U(\rho - b)\right] \right\}.$$

(5.30)

In the preceding expressions,

$$\tilde{f}_{1,2} = k_0 b\, \frac{\tilde{n}_{2,1} Z_0 I_0^e + \mathrm{i}\tilde{\eta} \mathrm{D} I_0^m}{\tilde{n}_{2,1} - \tilde{n}_{1,2}}, \quad f_{1,2} = k_0 b\, \frac{n_{2,1} Z_0 I_0^e + \mathrm{i}\eta_a \mathrm{D} I_0^m}{n_{2,1} - n_{1,2}}, \quad \mathrm{D} = \frac{\sin(k_0 p d)}{k_0 p d},$$

$$\tilde{G}_k^{(\ell)}(\rho) = \frac{\pi}{2} \tilde{Q}_k^{1-\ell} \left[J_\ell\left(\tilde{Q}_k\, \frac{\rho}{a}\right) Y_1\left(\tilde{Q}_k\, \frac{b}{a}\right) - J_1\left(\tilde{Q}_k\, \frac{b}{a}\right) Y_\ell\left(\tilde{Q}_k\, \frac{\rho}{a}\right) \right],$$

$$G_k^{(\ell)}(\rho) = S_k^{1-\ell} \left[K_\ell\left(S_k\, \frac{\rho}{a}\right) I_1\left(S_k\, \frac{b}{a}\right) + (-1)^\ell K_1\left(S_k\, \frac{b}{a}\right) I_\ell\left(S_k\, \frac{\rho}{a}\right) \right], \qquad (5.31)$$

and U is the Heaviside function. Since the field must decay with ρ in the outer region, it is necessary to require $\mathrm{Re}\,(s_k) > 0$ ($k = 1, 2$). An equivalent form of the solution, which is sometimes more convenient, is obtained by using the identity

$$K_m\left(S_k\,\frac{\rho}{a}\right) = \frac{\pi}{2}\,(-\mathrm{i})^{m+1} H_m^{(2)}\left(Q_k\,\frac{\rho}{a}\right), \tag{5.32}$$

where $Q_k = k_0 a q_k$, $S_k = k_0 a s_k$, $q_k = \mathrm{i} e^{-\mathrm{i}\pi} s_k \equiv q_k(p, N_a)$, and $k = 1, 2$. The required condition for q_k is then specified as

$$\mathrm{Im}\,(q_k) < 0. \tag{5.33}$$

In effect, this is the radiation condition to the problem defined by Equations (5.12) and (5.13) for the Fourier transforms $\mathbf{E}(\rho, p)$, $\mathcal{H}(\rho, p)$. It should be noted that the event when $\mathrm{Im}\,(q_k) \to 0$, which is possible in a loss-free medium for some values of p, must be considered as the limiting case corresponding to minute losses.

The coefficients B_k and C_k in expressions (5.27) to (5.30) are written as follows. For $b < a$:

$$\begin{aligned}
B_{1,2} &= \frac{B_{1,2}^{(\Delta)}(p)}{\Delta(p)} = \frac{1}{\Delta(p)}\left\{\tilde{f}_{1,2}\left[\left(\tilde{D}_{2,1}\,J(\tilde{Q}_{2,1}) + K(S_1)\,K(S_2)\right)\tilde{G}_{1,2}^{(0)}(a)\right.\right. \\
&\quad + \left.\left(\frac{\tilde{\eta}}{\eta_a}\,J(\tilde{Q}_{2,1}) + \tilde{D}_{1,2}\right)\tilde{G}_{1,2}^{(1)}(a)\right] - \tilde{f}_{2,1}\,\tilde{F}_{2,1}\,\frac{J_1(\tilde{Q}_{2,1}\,b/a)}{\tilde{Q}_{2,1}\,J_0(\tilde{Q}_{2,1})}\right\} \\
&\quad \times \left[\tilde{Q}_{1,2}\,J_0(\tilde{Q}_{1,2})\right]^{-1},
\end{aligned} \tag{5.34}$$

$$\begin{aligned}
C_{1,2} &= \frac{C_{1,2}^{(\Delta)}(p)}{\Delta(p)} = \frac{1}{\Delta(p)}\left\{\tilde{f}_1\left[L_{21,11}\,J(\tilde{Q}_2) - \frac{M_{12,11}}{M_0}\,K(S_{2,1})\right]\frac{J_1(\tilde{Q}_1\,b/a)}{\tilde{Q}_1\,J_0(\tilde{Q}_1)}\right. \\
&\quad + \left.\tilde{f}_2\left[L_{22,12}\,J(\tilde{Q}_1) - \frac{M_{22,21}}{M_0}\,K(S_{2,1})\right]\frac{J_1(\tilde{Q}_2\,b/a)}{\tilde{Q}_2\,J_0(\tilde{Q}_2)}\right\} \\
&\quad \times \left[(n_{1,2} - n_{2,1})\,S_{1,2}\,K_0(S_{1,2})\right]^{-1}.
\end{aligned}$$

For $b > a$:

$$\begin{aligned}
B_{1,2} &= \frac{B_{1,2}^{(\Delta)}(p)}{\Delta(p)} = \frac{1}{\Delta(p)}\left\{f_1\left[\frac{\tilde{\eta}M_{21,11}}{\eta_a\,M_0}\,J(\tilde{Q}_{2,1}) - L_{12,11}\,K(S_2)\right]\frac{K_1(S_1\,b/a)}{S_1\,K_0(S_1)}\right. \\
&\quad + \left.f_2\left[\frac{\tilde{\eta}M_{22,12}}{\eta_a\,M_0}\,J(\tilde{Q}_{2,1}) - L_{22,21}\,K(S_1)\right]\frac{K_1(S_2\,b/a)}{S_2\,K_0(S_2)}\right\} \\
&\quad \times \left[(\tilde{n}_{1,2} - \tilde{n}_{2,1})\,\tilde{Q}_{1,2}\,J_0(\tilde{Q}_{1,2})\right]^{-1}, \tag{5.35}
\end{aligned}$$

$$
\begin{aligned}
C_{1,2} &= \frac{C_{1,2}^{(\Delta)}(p)}{\Delta(p)} = -\frac{1}{\Delta(p)} \left\{ f_{1,2} \left[\left(D_{2,1}\,K(S_{2,1}) + \frac{\tilde{\eta}}{\eta_a}\,J(\tilde{Q}_1)\,J(\tilde{Q}_2) \right) G_{1,2}^{(0)}(a) \right.\right.\\
&\quad + \left.\left. \left(K(S_{2,1}) + D_{1,2} \right) G_{1,2}^{(1)}(a) \right] + f_{2,1}\,F_{2,1}\,\frac{K_1(S_{2,1}\,b/a)}{S_{2,1}\,K_0(S_{2,1})} \right\} \left[S_{1,2}\,K_0(S_{1,2}) \right]^{-1}
\end{aligned}
$$

In the preceding expressions,

$$
\Delta(p) = \tilde{D}_1\,J(\tilde{Q}_1) + \tilde{D}_2\,J(\tilde{Q}_2) + \frac{\tilde{\eta}}{\eta_a}\,J(\tilde{Q}_1)\,J(\tilde{Q}_2) + K(S_1)\,K(S_2) \tag{5.36}
$$

and

$$
\begin{aligned}
\tilde{D}_1 &= M_{12}L_{12}K(S_2) - M_{11}L_{22}K(S_1), & \tilde{D}_2 &= M_{21}L_{21}K(S_1) - M_{22}L_{11}K(S_2),\\
D_1 &= M_{21}L_{21}J(\tilde{Q}_2) - M_{11}L_{22}J(\tilde{Q}_1), & D_2 &= M_{12}L_{12}J(\tilde{Q}_1) - M_{22}L_{11}J(\tilde{Q}_2),\\
\tilde{F}_1 &= M_{12}L_{11}K(S_2) - M_{11}L_{21}K(S_1), & \tilde{F}_2 &= M_{21}L_{22}K(S_1) - M_{22}L_{12}K(S_2),\\
F_1 &= M_{21}L_{11}J(\tilde{Q}_2) - M_{11}L_{12}J(\tilde{Q}_1), & F_2 &= M_{12}L_{22}J(\tilde{Q}_1) - M_{22}L_{21}J(\tilde{Q}_2),\\
J(\tilde{Q}_k) &= \frac{J_1(\tilde{Q}_k)}{\tilde{Q}_k\,J_0(\tilde{Q}_k)}, & K(S_k) &= \frac{K_1(S_k)}{S_k\,K_0(S_k)},
\end{aligned} \tag{5.37}
$$

where the quantities M_{ij} and L_{ij} are defined by (4.69).

The solution for $E_\phi(\rho,p)$, $\mathcal{H}_\phi(\rho,p)$ given by formulas (5.27) to (5.31), together with (5.34) to (5.37), yields the total field for the problem with an excitation of a uniform cylindrical duct by ring electric and magnetic currents. The solution obtained may be applied to both crests and troughs of ionization, and to arbitrary values of parameters of the plasma as long as it is still describable by a dielectric tensor in the general form (2.19).

Note that when $a = b$, the solution reduces to the most simple form:

$\rho < a$,

$$
E_\phi(\rho,p) = i \sum_{k=1}^{2} B_k\,J_1\left(\tilde{Q}_k\,\frac{\rho}{a} \right), \quad \mathcal{H}_\phi(\rho,p) = -\sum_{k=1}^{2} \tilde{n}_k B_k\,J_1\left(\tilde{Q}_k\,\frac{\rho}{a} \right); \tag{5.38}
$$

$\rho > a$,

$$
E_\phi(\rho,p) = i \sum_{k=1}^{2} C_k\,K_1\left(S_k\,\frac{\rho}{a} \right), \quad \mathcal{H}_\phi(\rho,p) = -\sum_{k=1}^{2} n_k C_k\,K_1\left(S_k\,\frac{\rho}{a} \right);
$$

where the coefficients B_k and C_k may be found by proceeding to the limit $b \to a - 0$ in (5.34), or to the limit $b \to a + 0$ in (5.35).

The integrands in (5.11) have poles where

$$
\Delta(p) = 0 \tag{5.39}
$$

(see expressions (5.34) and (5.35) for the coefficients B_k and C_k to be substituted into (5.27) to (5.30)). But this is exactly the mode dispersion equation (4.71), and we thus come to the well-known and very important result that the poles are associated with waveguide modes. To obtain a representation which would exhibit such modes explicitly, it is necessary to make a suitable deformation of the path of integration in the complex p plane and perform residue evaluation at the poles. Such a deformation must, of course, take into account all the singularities of the integrands in (5.11). When these integrands, together with expressions (5.27) to (5.31), and (5.34) to (5.37), are examined, we see that the only singularities present are poles due to zeros of $\Delta(p)$, i.e., roots of (5.39), and branch cuts due to branch points of the functions $q_{1,2}(p, N)$ defined by (4.40), which are involved in the integrands of (5.11). It is necessary therefore to determine the position and shape of the branch cuts for these functions to make the integrands in (5.11) single-valued. This will be done in some detail in the following section. The well-known mathematical techniques of contour integration using multi-sheeted Riemann surfaces will be widely employed in the later work. The reader unfamiliar with this topic should consult standard texts such as Copson (1935) and Whittaker and Watson (1969). Many useful examples of application of contour integral methods can be found in treatises by Brekhovskikh (1960) and Felsen and Marcuvitz (1973).

5.4. Analytic properties of the functions $q_{1,2}$

The general expression for $q_{1,2}(p, N)$, for density N, may be written, from (2.64):

$$q_k(p, N) = \left\{ \left[\varepsilon^2 - g^2 + \varepsilon\eta - (\eta + \varepsilon) p^2 + (-1)^k R_q(p, N) \right] / 2\varepsilon \right\}^{1/2}, \quad k = 1, 2, \quad (5.40)$$

where

$$R_q(p, N) = \left\{ (\eta - \varepsilon)^2 p^4 + 2 \left[g^2(\eta + \varepsilon) - \varepsilon (\eta - \varepsilon)^2 \right] p^2 + (\varepsilon^2 - g^2 - \varepsilon\eta)^2 \right\}^{1/2}, \quad (5.41)$$

the notations of (4.40) being understood.

The function $R_q(p, N)$ defined by (5.41) is a two-valued function of p. It has branch points $p = \pm P_{b,c}$ where $R_q = 0$, and hence $q_1 = q_2$. These points are given by

$$p = \pm P_{b,c} = \pm \left\{ \varepsilon - (\eta + \varepsilon) \frac{g^2}{(\eta - \varepsilon)^2} + \frac{2\chi_{b,c}}{(\eta - \varepsilon)^2} \left[\varepsilon\eta g^2 \left(g^2 - (\eta - \varepsilon)^2 \right) \right]^{1/2} \right\}^{1/2}, \quad (5.42)$$

where $\chi_b = -\chi_c = -1$. To make the function $R_q(p, N)$ single-valued, it is convenient to choose the square root in (5.41) to have a positive real part, that is $\text{Re}\,(R_q) > 0$. Thence the branch cuts for R_q are to be drawn along the lines defined by the relation $\text{Re}\,(R_q) = 0$.

The functions $q_{1,2}(p, N)$, apart from the branch points (5.42), have the branch points $p = \pm P_{\rm o}(N) = \mp(\varepsilon + g)^{1/2}$ and $p = \pm P_{\rm x}(N) = \pm(\varepsilon - g)^{1/2}$ where $q_k = 0$. Since the branch cuts for $q_{1,2} = q_{1,2}(p, N_a)$ are due to the radiation condition (5.33) insuring decay of the integrands in (5.11) as $\rho \to \infty$, it is necessary to choose the branch cuts to be the lines defined by the relation $\mathrm{Im}\,(q_k) = 0$. As an example, this choice is exhibited for the frequency interval (2.71) in Figure 5.1, which shows the *"proper"* sheet of the multi-sheeted Riemann p plane. The sheet presented is defined by stipulating that simultaneously $\mathrm{Im}\,(q_{1,2}) < 0$ and $\mathrm{Re}\,(R_q) > 0$ in it. In Figure 5.1 the branch cuts I, II, and III correspond to the relations $\mathrm{Im}\,(q_1) = 0$, $\mathrm{Im}\,(q_2) = 0$, and $\mathrm{Re}\,(R_q) = 0$, respectively. There are also shown dashed lines IV and V whose equations are $\mathrm{Re}\,(q_1) = 0$ and $\mathrm{Re}\,(q_2) = 0$, respectively. On crossing one of the dashed lines marked by IV, therefore, $\mathrm{Re}\,(q_1)$ changes sign; so does $\mathrm{Re}\,(q_2)$, but when crossing one of the dashed lines marked by V.

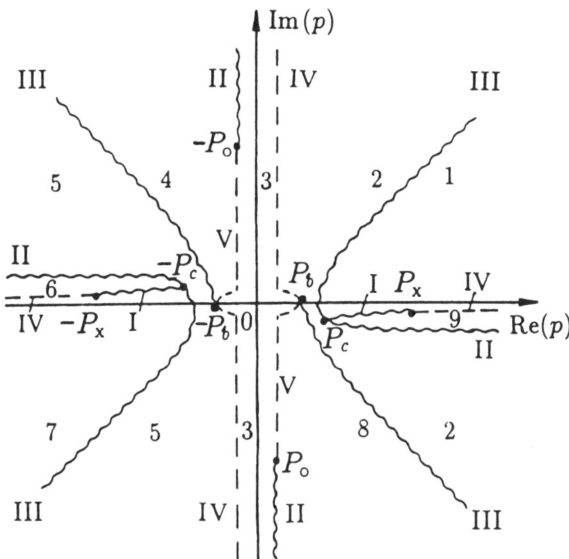

Figure 5.1. Location of branch singularities in the complex p plane (not to scale). $\mathrm{Re}\,(q_1) > 0$ in domains 1, 3, 4, 7, 8 and $\mathrm{Re}\,(q_1) < 0$ in domains 2, 5, 6, 9; $\mathrm{Re}\,(q_2) > 0$ in domains 2, 3, 5 and $\mathrm{Re}\,(q_2) < 0$ in domains 1, 4, 6–9.

It should be noted that in the case of a lossless plasma the branch points $\pm P_{b,c}$, $\pm P_{\rm x}$ lie on the real p axis, and $\pm P_{\rm o}$ on the imaginary p axis; the branch cuts I and II, but not III, and the dashed lines IV and V tend to merge with the corresponding segments of one of the p axes. In Figure 5.1 the branch cuts are illustrated for the case when a small amount of damping is introduced by collisions, so that the branch singularities are displaced slightly off the p axes. Clearly the mutual disposition of these singularities must be taken into account when proceeding to the loss-free limit.

We note that the above analysis is entirely analogous to Walker's discussion (1971), except that Walker adopted a convention for identifying the branches of q_k which was different from that used here.

5.5. Fields of ring currents, for a uniform duct (continued)

Consider the integral representation (5.11) of a source-excited field in more detail. For the reasons given in §§ 5.3 and 5.4, the integration along the real p axis, in (5.11), is carried out in the proper Reimann sheet chosen so that the inequalities $\mathrm{Im}\,(q_k) < 0$ and $\mathrm{Re}\,(R_q) > 0$ are valid in it for the functions $q_k(p, N_a)$ and $R_q(p, N_a)$ relating to the *ambient* plasma. Recall that the special notations were introduced in § 4.3 for the quantities P_x and P_c appropriate to the ambient plasma, namely $\mathcal{P}$ and $\mathcal{P}_c$. In this section, we shall however use the notations $P_{o,x} = P_{o,x}(N_a)$ and $P_c = P_c(N_a)$ for the ambient plasma.

The integrands in (5.11) remain continuous on crossing the branch cuts III (see Figure 5.1), though q_1 and q_2 obviously undergo a discontinuity across those cuts. This is explained for the following reasons. The quantity q_1 on one side of each cut III and the quantity q_2 on its other side do coincide. Furthermore, the integrands do not change when the simultaneous replacements $q_1 \to q_2$ and $q_2 \to q_1$ are made. Hence the corresponding integrands allow analytic continuation across the lines defined by $\mathrm{Re}\,(R_q) = 0$, and we have to take into account only the branch cuts I and II, as shown in Figure 5.2. Also, for each of the remaining branch cuts, the side where $\mathrm{Re}\,(q_k) < 0$ merges with the unshown branch cut for the Hankel functions involved in the integrands via (5.32). One should bear this in mind when crossing the branch cut to leave the proper sheet and appear on some other sheet. We note that no branch cuts are needed for $\tilde{q}_k = q_k(p, \tilde{N})$ since the integrands in (5.11) are even functions of $\tilde{q}_1$ and $\tilde{q}_2$.

Turning now to the pole singularities, it was noted above that all of these poles are obtained from equation (5.39). Hence the poles would be at points $p = p_n$ which are propagation constants of modes. It was shown in §§ 4.3 and 4.4 that for the frequencies specified by (2.71), the propagation constants of eigenmodes in loss-free ducts are located on the real p axis in the domain $p_n < P_c(N_a) = \mathcal{P}_c$. If a small amount of damping is introduced by collisions, the poles are displaced off the axis and their disposition becomes as in Figure 5.2. It is important that all the poles which account for the eigenmodes of a duct locate on the proper sheet of the p plane. These poles are marked as crosses at the points p_n, p_{-n} in Figure 5.2. Here

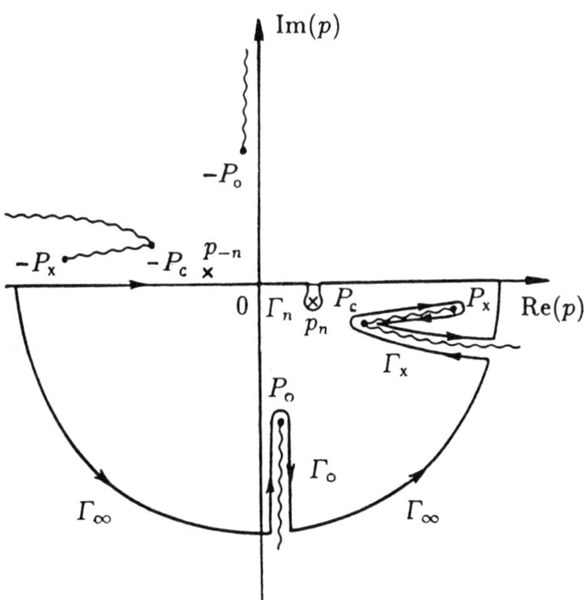

Figure 5.2. Singularities and paths of integration in the complex p plane (not to scale). The poles referring to eigenmodes are denoted by crosses.

we adopt a notation $p_{-n} = -p_n$, assuming that for n positive, $\text{Im}(p_n) < 0$. The poles for which $\text{Re}(p_n) \cdot \text{Im}(p_n) < 0$, correspond to forward eigenmodes.

Note that, in general, backward eigenmodes and complex eigenmodes may exist (see §4.9). For a loss-free duct, the propagation constants $p = p_n$ of the backward eigenmodes are also purely real. If minute losses were included, a small imaginary part of p_n would appear, but now the relation $\text{Re}(p_n) \cdot \text{Im}(p_n) > 0$ would be valid. The poles corresponding to the complex eigenmodes are complex even in the absence of losses and appear in pairs, p_n and $-p_n^*$; clearly p_{-n} and $-p_{-n}^*$ are then poles as well. The analysis of backward-mode poles and complex-mode poles is the same as for forward-mode poles and the details need not be given. In the present analysis, we confine ourselves to the case depicted in Figure 5.2, which is typical of a density trough capable of supporting bound modes (see §4.4, section 4.4.5).

After locating all of the singularities, it is now possible to deform the path of integration in order to obtain explicitly the contributions to the total field from individual modes. Referring to (5.11) it is evident that any deformation in the p plane must be carried out in the lower half of the plane for $z > 0$ (or, more precisely, for $z > d$ if $d \neq 0$, as in (5.15)) and in the upper half for $z < 0$ ($z < -d$) in order to ensure regularity at infinity for arbitrary values of z and ρ. In what follows we shall consider the case of positive z.

Determine the contribution from one mode, say, of number n. To do this, we distort the integration contour so as to coincide with that marked as Γ_n in Figure 5.2. By Cauchy's theorem of residues, the E_ϕ component is then given by

$$
\begin{aligned}
E_\phi(\rho, z) \;=\; & -\mathrm{i}k_0 \left(\left.\frac{\mathrm{d}\Delta}{\mathrm{d}p}\right|_{p=p_n}\right)^{-1} E_\phi^{(\Delta)}(\rho, p_n)\exp(-\mathrm{i}k_0 p_n z) \\
& + \frac{k_0}{2\pi}\int_{\Gamma_n} E_\phi(\rho, p)\exp(-\mathrm{i}k_0 p z)\,\mathrm{d}p,
\end{aligned}
\tag{5.43}
$$

where

$$
E_\phi^{(\Delta)}(\rho, p_n) = \begin{cases}
\mathrm{i}\sum_{k=1}^{2} B_k^{(\Delta)}(p_n)\, J_1\!\left(\tilde{Q}_k(p_n)\dfrac{\rho}{a}\right) & \text{if } \rho < a, \\[2ex]
\mathrm{i}\sum_{k=1}^{2} C_k^{(\Delta)}(p_n)\, K_1\!\left(S_k(p_n)\dfrac{\rho}{a}\right) & \text{if } \rho > a.
\end{cases}
\tag{5.44}
$$

The first term in (5.43) represents the residue, reversed in sign, of the integrand at the pole $p = p_n$. This term expresses the contribution from the nth eigenmode to the total field. The remaining field components can readily be represented in a similar manner.

Note that upon substituting $p = p_n$ into the expression for $B_k^{(\Delta)}(p)$ in the case $b < a$ (see (5.34)) one can arrive at the more simple formula

$$
\begin{aligned}
B_{1,2}^{(\Delta)}(p_n) \;=\; & \frac{1}{\tilde{Q}_{1,2}\, J_0(\tilde{Q}_{1,2})}\left\{\tilde{f}_{1,2}\left[\tilde{D}_{1,2} + \frac{\tilde{\eta}}{\eta a}\, J(\tilde{Q}_{2,1})\right]\frac{J_1(\tilde{Q}_{1,2}\,b/a)}{\tilde{Q}_{1,2}\, J_0(\tilde{Q}_{1,2})}\right. \\
& \left.\left. - \;\tilde{f}_{2,1}\,\tilde{F}_{2,1}\,\frac{J_1(\tilde{Q}_{2,1}\,b/a)}{\tilde{Q}_{2,1}\, J_0(\tilde{Q}_{2,1})}\right\}\right|_{p=p_n}.
\end{aligned}
\tag{5.45}
$$

The similar procedure for $C_k^{(\Delta)}$ in the case $b > a$ (see (5.35)) yields

$$
\begin{aligned}
C_{1,2}^{(\Delta)}(p_n) \;=\; & \frac{1}{S_{1,2}\, K_0(S_{1,2})}\left\{f_{1,2}\left[D_{1,2} + K(S_{2,1})\right]\frac{K_1(S_{1,2}\,b/a)}{S_{1,2}\, K_0(S_{1,2})}\right. \\
& \left.\left. - \;f_{2,1}\,F_{2,1}\,\frac{K_1(S_{2,1}\,b/a)}{S_{2,1}\, K_0(S_{2,1})}\right\}\right|_{p=p_n}.
\end{aligned}
\tag{5.46}
$$

In obtaining (5.45) and (5.46) we made use of the fact that $\Delta(p_n) = 0$ (see (5.36) and (5.39)) and took into account the expressions for the Wronskians of the Bessel functions $J_\ell(\zeta)$ and $Y_\ell(\zeta)$, and of the modified Bessel functions $I_\ell(\zeta)$ and $K_\ell(\zeta)$.

To obtain a representation exhibiting contributions from all the eigenmodes explicitly, the path of integration must be deformed to a semicircle Γ_∞ at infinity, which dose not contribute anything to the integral since the integrand becomes infinitely attenuated. The contribution to the integral is then due to residue contributions arising from the presence of poles, and due to integrations around branch cuts, that is along the contours Γ_{o} and Γ_{x}, as shown in Figure 5.2. Hence, from (5.11), one arrives at

$$
\begin{aligned}
\mathbf{E}(\rho, z) &= \sum_n (-\mathrm{i}k_0) \left(\left. \frac{\mathrm{d}\Delta}{\mathrm{d}p} \right|_{p=p_n} \right)^{-1} \mathbf{E}^{(\Delta)}(\rho, p_n) \exp(-\mathrm{i}k_0 p_n z) \\
&\quad + \frac{k_0}{2\pi} \int_{\Gamma_{\mathrm{o,x}}} \mathbf{E}(\rho, p) \exp(-\mathrm{i}k_0 p z)\, \mathrm{d}p,
\end{aligned}
$$

$$
\begin{aligned}
\boldsymbol{\mathcal{H}}(\rho, z) &= \sum_n (-\mathrm{i}k_0) \left(\left. \frac{\mathrm{d}\Delta}{\mathrm{d}p} \right|_{p=p_n} \right)^{-1} \boldsymbol{\mathcal{H}}^{(\Delta)}(\rho, p_n) \exp(-\mathrm{i}k_0 p_n z) \\
&\quad + \frac{k_0}{2\pi} \int_{\Gamma_{\mathrm{o,x}}} \boldsymbol{\mathcal{H}}(\rho, p) \exp(-\mathrm{i}k_0 p z)\, \mathrm{d}p,
\end{aligned}
\tag{5.47}
$$

where

$$
\mathbf{E}^{(\Delta)}(\rho, p) = \Delta(p)\, \mathbf{E}(\rho, p), \quad \boldsymbol{\mathcal{H}}^{(\Delta)}(\rho, p) = \Delta(p)\, \boldsymbol{\mathcal{H}}(\rho, p),
\tag{5.48}
$$

in which the 'n' summation coves all the poles located in the lower half of the complex p plane and the terms under the summation sign represent the fields of individual eigenmodes.

Expressions (5.47) give a spectral representation of the source-excited field. The eigenmodes, that is the *discrete* spectral contributions, are now clearly exhibited by the residue series*, while the remaining contribution is due to integrals over the *continuous* spectral values of p corresponding to the branch cuts. Thus, only the singularities located in the proper (spectral) sheet of the p plane account for the eigenmodes (i.e. waves of a discrete spectrum), called also the proper (bound) modes (see § 4.3). Besides these spectral poles, there may be poles in the *improper* (nonspectral) sheets of the plane. No eigenmodes may be associated with such improper (nonspectral) poles which do not comply with the regularity condition for $\mathbf{E}(\rho, p)$, $\boldsymbol{\mathcal{H}}(\rho, p)$ at infinity ($\rho \to +\infty$). These poles account for leaky modes (see § 4.3), which can heavily affect the radiation field in some regions of the observation space. This aspect, however, will be treated in detail in Chapter 6. The present discussion is only concerned with obtaining a representation which would exhibit leaky-mode contributions in the residue form. The discussion is made for the important special case of a density enhancement, assuming that the wave frequency belongs to range (2.71).

---

* It is evident from the above analysis that if the complex eigenmodes were here, the two complex poles, either p_n and $-p_n^*$ (for positive z) or p_{-n} and $-p_{-n}^*$ (for negative z), would be included simultaneously by the deformed path of integration. Thus they would contribute residues in pairs. In contrast to this, for the propagating eigenmodes only one of the two poles, either p_n (for positive z) or p_{-n} (for negative z) contributes a residue.

Recall that under the above stated circumstances it is possible that leaky modes occur without eigenmodes. In this case, the leaky-mode poles, that is roots $p = p_\nu$ of Equation (5.39), lie in the improper sheet defined by the inequalities $\text{Im}\,(q_1) < 0$, $\text{Im}\,(q_2) > 0$, and $\text{Re}\,(R_q) > 0$. The leaky-mode poles are complex even in a loss-free case (see §4.4) and their location in the p plane is as shown in Figure 5.3. These are marked as circles in order to be not confused with poles in the proper sheet.

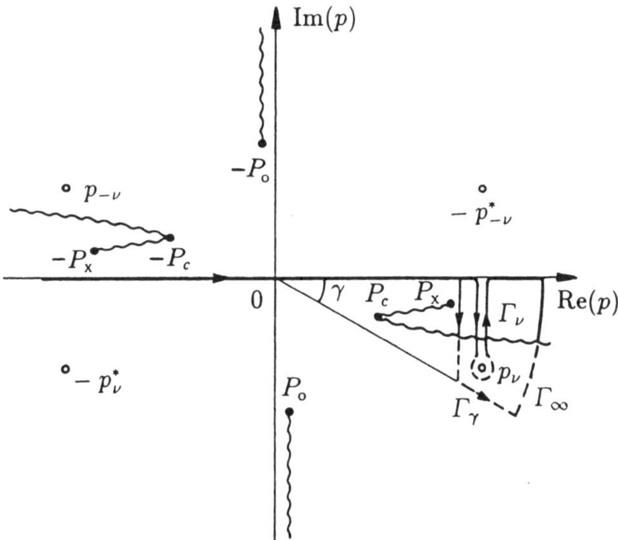

Figure 5.3. Singularities and paths of integration in the complex p plane (not to scale). The dashed lines are in improper sheet in which leaky-mode poles are located.

An inspection of Equation (5.39), together with (5.36), reveals that if p_ν is a solution, then $-p_\nu^*$, $p_{-\nu} \equiv -p_\nu$ and $-p_{-\nu}^*$ are also solutions of (5.39). In the improper sheet shown in Figure 5.3, the pole p_ν corresponds to the forward-propagating leaky mode of order ν, while the pole $p_{-\nu}$ corresponds to the backward-propagating leaky mode of the same order. The poles $-p_{-\nu}^*$ and $-p_\nu^*$ correspond to modes by which energy is being fed into the duct. We shall not discuss the latter set of poles, though they may be important when we consider how a mode may be launched in a duct from sources outside.

Bearing in mind the location of all the nonspectral singularities, we now distort the initial integration path in (5.11) to coincide with a contour Γ_ν, as shown in Figure 5.3. This yields the leaky-mode contributions in the form of a residue series. Notice that integration contours are indicated partly as solid lines and partly as dashed lines. Only the solid lines are in the proper sheet of the multi-sheeted Riemann p

plane. The dashed lines are in the improper sheet where nonspectral poles are located. It is clear that the solid line, on crossing the branch cut II, continues as the dashed line into the improper sheet where the relations $-3\pi/2 < \arg[q_2(p_\nu)] < -\pi$ are evidently valid. This, in particular, implies that the second kind of the Hankel function $H_m^{(2)}(k_0 q_2 \rho)$, in accordance with (3.54), can be replaced by its first kind, so that $H_m^{(2)}(k_0 q_2(p_\nu)\rho) = -e^{im\pi} H_m^{(1)}(k_0 e^{i\pi} q_2(p_\nu)\rho)$, where $e^{i\pi} q_2(p_\nu)$ satisfies the relations $-\pi/2 < \arg\left[e^{i\pi} q_2(p_\nu)\right] < 0$. Recall that just the first kind of the Hankel function was used in Chapter 4 to represent the escaping part of the field in such a leaky mode; its "trapped" part is described by $H_m^{(2)}(k_0 q_1(p_\nu)\rho)$ which can be replaced, via (5.32), by the more convenient modified Bessel function of the second kind $K_m(k_0\, s_1(p_\nu)\, \rho)$. It is now evident that the described deformation of the path of integration will furnish the representation which is obtainable formally from (5.43) by making the replacements $p_n \to p_\nu$ and $\Gamma_n \to \Gamma_\nu$. Thus the leaky-mode pole p_ν intercepted during the path deformation yields, for $z > 0$, the leaky-mode field in the form of a residue contribution.

A further analysis shows that contributions from the leaky modes whose propagation constants lie in the range (4.75), can all be represented by the residue series. Before proceeding in this way, recall that the relation $p_\nu'' \ll p_\nu'$ was established for these modes in §§ 4.4 and 4.8. It is therefore possible to introduce an angle

$$\bar{\gamma} = \max_{\nu}\left\{ \arctan\left(-\operatorname{Im}\left[q_2(p_\nu)\right]/\operatorname{Re}\left[q_2(p_\nu)\right]\right)\right\} \tag{5.49}$$

which will satisfy the inequality $\bar{\gamma} \ll \pi$. Now we can distort the path of integration so as to run in the improper sheet, for $\operatorname{Re}(p) > P_{\mathrm{x}}(N_a)$, along the line which makes an angle γ with the real p axis so that $\bar{\gamma} < \gamma \ll \pi$. The new contour, marked as Γ_γ in Figure 5.3, is supplemented by an arc Γ_∞ at infinity. It is a simple matter to show that the dashed part of Γ_∞ contributes nothing to the result of integration if

$$z > d + (\varrho - a)(-\eta_a/\varepsilon_a)^{1/2} \tag{5.50}$$

where

$$\varrho = \begin{cases} \varrho_0 & \text{if } \rho < \varrho_0 \\ \rho & \text{if } \rho > \varrho_0 \end{cases}, \qquad \varrho_0 = \max\{a, b\}. \tag{5.51}$$

Obviously, the quantity d in (5.50) must be set equal to zero when the source (5.14) is used. Further, the remainder of Γ_∞ gives no contribution as it is located in the proper sheet where the integrands vanish for $p \to \infty$. Thus we get

$$\mathbf{E}(\rho, z) = \sum_{\nu}(-ik_0)\left(\frac{d\Delta}{dp}\bigg|_{p=p_\nu}\right)^{-1} \mathbf{E}^{(\Delta)}(\rho, p_\nu)\exp(-ik_0 p_\nu z)$$
$$+ \frac{k_0}{2\pi}\int_{\Gamma_\gamma} \mathbf{E}(\rho, p)\exp(-ik_0 p z)\, dp, \tag{5.52}$$

$$
\begin{aligned}
\mathcal{H}(\rho, z) \;=\;& \sum_{\nu}(-\mathrm{i}k_0)\left(\frac{\mathrm{d}\Delta}{\mathrm{d}p}\bigg|_{p=p_\nu}\right)^{-1}\mathcal{H}^{(\Delta)}(\rho, p_\nu)\exp(-\mathrm{i}k_0 p_\nu z)\\
&+\;\frac{k_0}{2\pi}\int_{\Gamma_\gamma}\mathcal{H}(\rho, p)\exp(-\mathrm{i}k_0 p z)\,\mathrm{d}p,
\end{aligned}
$$

where the 'ν' summation covers all the improper poles intercepted during the path deformation; the vector quantities $\mathbf{E}^{(\Delta)}(\rho, p)$ and $\mathcal{H}^{(\Delta)}(\rho, p)$ are defined by (5.48). A similar representation can be obtained for the region

$$
z < -d - (\varrho - a)(-\eta_a/\varepsilon_a)^{1/2} \tag{5.53}
$$

by suitably deforming the path of integration in the second quadrant of the complex p plane.

In view of the close similarity between the residue series in (5.47) and (5.52), one can expect that it is possible to work with leaky modes as with proper modes in mathematical developments. This topic is discussed in § 6.7.

Thus, we have shown how an exact solution for the problem of a given source in the presence of a cylindrical duct can be obtained via a radial-transmission formulation. The solution has the form of an integral representation which may be transformed into the spectral (modal) form interpretable as a z-transmission representation.

The above theory has only dealt with the general physical nature of the various source-excited modal fields. The study has been made in connection with the problem of radiation from the ring sources carrying the simplest azimuthally symmetric current distributions. This work could be continued. For example, once the axisymmetric Green's function problem in ρ-domain is solved, the fields can readily be found when symmetric annular currents are distributed over the ρ coordinate. By analogy with the above treatment, excitation of a duct by nonsymmetric annular currents is easily handled. In the next chapter the theory will be extended so as to suffice for the complete analysis of radiation from arbitrary sources in the presence of a duct.

It is worth mentioning that there is a remarkable parallel between the nature of spectral and nonspectral waves for magnetized ducts and isotropic guiding structures and that there are numerous works on various open isotropic waveguides, involving plasma waveguides (for example, Marcuvitz, 1956; Brekhovskikh, 1960; Budden, 1961 b; Wait, 1962, 1966; Tamir and Oliner, 1963; Felsen and Marcuvitz, 1966, 1973). Several close studies on excitation of cylindrical axially-magnetized plasma structures should also be mentioned, such as the works of Seshadri and Yip (1966), Yip and Seshadri (1967), and Vorob'ev and Rukhadze (1994).

Problems

5.1. Let the current density be specified as

$$\mathbf{J}^{e,m}(\rho,\phi,z) = \hat{\phi}_0 \, I_\phi^{e,m}(\phi)\, \delta(\rho - b)\, [U(z+d) - U(z-d)](2d)^{-1}.$$

Find the source-excited field on a uniform cylindrical duct of radius a by the use of the following field representation:

$$\mathbf{E}(\rho,\phi,z) = \sum_{m=-\infty}^{+\infty} e^{-im\phi}\, \frac{k_0}{2\pi} \int_{-\infty}^{+\infty} \mathbf{E}_m(\rho,p)\, e^{-ik_0 pz} \mathrm{d}p,$$

$$\mathcal{H}(\rho,\phi,z) = \sum_{m=-\infty}^{+\infty} e^{-im\phi}\, \frac{k_0}{2\pi} \int_{-\infty}^{+\infty} \mathcal{H}_m(\rho,p)\, e^{-ik_0 pz} \mathrm{d}p.$$

Hint: Utilize a similar representation for the current density and employ the boundary conditions at $\rho = a$ and $\rho = b$ for the field quantities $\mathbf{E}_m(\rho,p)$ and $\mathcal{H}_m(\rho,p)$.

5.2. Determine the position and shape of the branch cuts defined by $\mathrm{Re}\,(R_q) = 0$ and $\mathrm{Im}\,(q_k) = 0$ for the frequency intervals (2.70) and (2.72). Compare the results with those obtained in § 5.4 for the frequency interval (2.71).

Chapter 6

Modal Representation of Source-excited Fields on a Duct

6.1. Introduction. The boundary-value problem for a duct

In this chapter we deal with a different field representation from that used earlier. It expresses the source-excited field on a duct as a superposition of the fields in modes propagating along the duct axis. Very often such a *modal representation* turns out more convenient than the equivalent integral representation obtained in Chapter 5 with the use of the Green's function approach. The term "modes" in this context refers to transverse field configurations which individually satisfy the source-free field equations and the required boundary conditions (involving some conditions at the singular point, $\rho = \infty$) and contain the longitudinal coordinate z only through the factor e^{-ik_0pz} where p is a normalized axial wavenumber. The field distributions in modes are defined by the vector functions commonly known as the eigenfunctions of the duct. The purpose of this chapter is to lay the foundations of the modal, or eigenfunction, approach, by deriving the basic properties of the eigenfunctions and showing how they are used to represent the field on a duct in a magnetized plasma.

As before, the density N is here assumed to be a constant, N_a, in the outer region of a duct $(\rho > a)$ and varying with ρ in its core $(\rho < a)$. In the core, $N(\rho)$ is allowed to be a piecewise smooth function having jump-like discontinuities at some points $\rho = \rho_i$ where $i = 1, \ldots, n$ and $0 < \rho_1 < \ldots < \rho_n \leq a$. To avoid complicated effects associated with the plasma resonance and the hybrid resonances, we shall also assume that even if the elements of the plasma dielectric tensor $\hat{\varepsilon}(\rho)$ can change sign, this occurs only at the points $\rho = \rho_i$ of the jump of $N(\rho)$.

Since the plasma density does not vary with z, the duct is translationally invariant. The electric and magnetic field in source-free regions of the duct are then expressible in the form

$$\mathbf{E} = \sum_{s=(-)}^{(+)} \sum_{\alpha=o}^{x} \mathbf{E}_{s,\alpha}(\rho,\phi,q)\exp\left[-ik_0 p_{s,\alpha}(q)\,z\right],$$

$$\mathcal{H} = \sum_{s=(-)}^{(+)} \sum_{\alpha=o}^{x} \mathcal{H}_{s,\alpha}(\rho,\phi,q)\exp\left[-ik_0 p_{s,\alpha}(q)\,z\right],$$

$$(6.1)$$

where the subscript $s = (+)$ is used for *forward-propagating waves* which travel in the positive z-direction, and the subscript $s = (-)$ for *backward-propagating waves* which travel in the negative z-direction. The functions $p_{s,\alpha}(q)$ obey the conventions $p_{+,\alpha}(q) = p_\alpha(q)$ and $p_{-,\alpha}(q) = -p_\alpha(q)$, whence $p_{-,\alpha}(q) = -p_{+,\alpha}(q)$ and, accordingly, $p_{-s,\alpha}(q) = -p_{s,\alpha}(q)$. The functions $p_\alpha(q)$ are defined in turn by

$$p_\alpha(q) = \left[\varepsilon_a - \frac{1}{2}\left(1 + \frac{\varepsilon_a}{\eta_a}\right)q^2 + \chi_\alpha\, R_p(q)\right]^{1/2},\qquad (6.2)$$

in which

$$R_p(q) = \left[\frac{1}{4}\left(1 - \frac{\varepsilon_a}{\eta_a}\right)^2 q^4 - \frac{g_a^2}{\eta_a}q^2 + g_a^2\right]^{1/2}\qquad (6.3)$$

and $\chi_o = -\chi_x = -1$, where the conditions $\mathrm{Im}\,(p_\alpha) < 0$ and $\mathrm{Re}\,(R_p) > 0$ are assumed to be satisfied. It is evident from comparison (6.2) and (6.3) with (2.63) that the functions $p_o(q)$ and $p_x(q)$ now describe the axial wavenumbers versus the transverse wavenumber q for the "ordinary" ($\alpha = o$) and "extraordinary" ($\alpha = x$) characteristic modes of a *homogeneous ambient magnetoplasma with the density N_a*. The vector functions $\mathbf{E}_{s,\alpha}(\rho,\phi,q)$ and $\mathcal{H}_{s,\alpha}(\rho,\phi,q)$ can be written

$$\mathbf{E}_{s,\alpha}(\rho,\phi,q) = \sum_{m=-\infty}^{\infty} \mathbf{E}_{m,s,\alpha}(\rho,q)\exp(-im\phi),$$

$$\mathcal{H}_{s,\alpha}(\rho,\phi,q) = \sum_{m=-\infty}^{\infty} \mathcal{H}_{m,s,\alpha}(\rho,q)\exp(-im\phi),$$

$$(6.4)$$

with m being the azimuthal index.

The field quantities $\mathbf{E}_{m,s,\alpha}(\rho,q)$ and $\mathcal{H}_{m,s,\alpha}(\rho,q)$ in both the duct core and the outer regions are governed by the equations coinciding in form with Equations (4.14) and (4.15), in which one must now take $p = p_{s,\alpha}(q)$ and the relevant dielectric-tensor components, and replace the quantities E_z and $\mathcal{H}_z$ by the corresponding components of $\mathbf{E}_{m,s,\alpha}$ and $\mathcal{H}_{m,s,\alpha}$. By making the same replacements in formulas (4.10) to (4.13), it is possible to deduce the relationships between the transverse and longitudinal components of $\mathbf{E}_{m,s,\alpha}$ and $\mathcal{H}_{m,s,\alpha}$.

In the later work, relationships between the fields of forward- and backward-propagating waves will be used. To obtain these relationships, we decompose the field vectors $\mathbf{E}_{m,s,\alpha}$ and $\boldsymbol{\mathcal{H}}_{m,s,\alpha}$ into transverse and longitudinal parts, denoted by subscripts $\perp$ and z, so that

$$\mathbf{E}_{m,s,\alpha} = \mathbf{E}_{\perp;m,s,\alpha} + \hat{\mathbf{z}}_0\, E_{z;m,s,\alpha},$$

$$\boldsymbol{\mathcal{H}}_{m,s,\alpha} = \boldsymbol{\mathcal{H}}_{\perp;m,s,\alpha} + \hat{\mathbf{z}}_0\, \mathcal{H}_{z;m,s,\alpha}. \tag{6.5}$$

Then we can deduce from (4.10) to (4.15) that the backward-propagating fields are simply related to the forward-propagating fields. There are two possibilities, either

$$\mathbf{E}_{m,-,\alpha} = -\mathbf{E}_{\perp;m,+,\alpha} + \hat{\mathbf{z}}_0\, E_{z;m,+,\alpha},$$

$$\boldsymbol{\mathcal{H}}_{m,-,\alpha} = \boldsymbol{\mathcal{H}}_{\perp;m,+,\alpha} - \hat{\mathbf{z}}_0\, \mathcal{H}_{z;m,+,\alpha} \tag{6.6}$$

or

$$\mathbf{E}_{m,-,\alpha} = \mathbf{E}_{\perp;m,+,\alpha} - \hat{\mathbf{z}}_0\, E_{z;m,+,\alpha},$$

$$\boldsymbol{\mathcal{H}}_{m,-,\alpha} = -\boldsymbol{\mathcal{H}}_{\perp;m,+,\alpha} + \hat{\mathbf{z}}_0\, \mathcal{H}_{z;m,+,\alpha}. \tag{6.7}$$

We adopt the latter convention throughout.

To determine the complete eigenfunction set suitable for representing the total field on the duct, one must find the solutions of the differential equations governing $\mathbf{E}_{m,s,\alpha}$ and $\boldsymbol{\mathcal{H}}_{m,s,\alpha}$. The required solutions must be regular at $\rho = 0$ and satisfy the *boundedness conditions*

$$\rho^{1/2}|\mathbf{E}_{m,s,\alpha}(\rho,q)| < R_{m,\alpha}^{(1)}, \quad \rho^{1/2}|\boldsymbol{\mathcal{H}}_{m,s,\alpha}(\rho,q)| < R_{m,\alpha}^{(2)} \tag{6.8}$$

in the limit when $\rho \to \infty$, where $R_{m,\alpha}^{(1,2)}$ are constants. We shall verify that summing all the particular solutions which obey (6.8) yields the total field satisfying the usual radiation condition at infinity. It should be noted that the inequalities (6.8) are entirely equivalent to a similar condition suggested by Shevchenko (1966, 1971 a) to obtain the eigenfunction set of an open isotropic waveguide. The alternative method of determining the eigenfunctions of such a waveguide, based on the use of the characteristic Green's function, was developed by Marcuvitz (1956) and Felsen and Marcuvitz (1966, 1973). Although their approach could be generalized to the case of a magnetized duct, the direct solution of the boundary-value problem with the conditions (6.8) at the singular point (infinity) would here seem more tractable.

6.2. The eigenfunction set for the nonuniform duct

In the preceding section, we set up the singular boundary-value problem in the domain $0 \le \rho < \infty$. The appropriate transverse wavenumbers q may be considered as eigenvalues of this problem. Therefore our next steps will consist first in obtaining the eigenvalue spectrum and the eigenfunction set and then representing the total field in the form of an eigenfunction expansion.

6.2.1. Some mathematical developments

In the *outer region* $\rho > a$, where the plasma density is constant, the general solutions of the source-free field equations can be represented in the analytic form (see § 4.2, section 4.2.2). The longitudinal electric and magnetic components can be written, from (4.29),

$$
\begin{aligned}
E_{z;m,s,\alpha}(\rho,q) &= \frac{i}{\eta_a}\left\{ n^{(1)}_{s,\alpha}q\left[C^{(1)}_{m,s,\alpha}(q)\,H^{(1)}_m(k_0 q\rho) + C^{(2)}_{m,s,\alpha}(q)\,H^{(2)}_m(k_0 q\rho)\right]\right. \\
&\quad \left. + n^{(2)}_{s,\alpha}q_\alpha\left[D_{m,s,\alpha}(q)\,H^{(1)}_m(k_0 q_\alpha\rho) + C_{m,s,\alpha}(q)\,H^{(2)}_m(k_0 q_\alpha\rho)\right]\right\},
\end{aligned}
$$

$$(6.9)$$

$$
\begin{aligned}
\mathcal{H}_{z;m,s,\alpha}(\rho,q) &= -q\left[C^{(1)}_{m,s,\alpha}(q)\,H^{(1)}_m(k_0 q\rho) + C^{(2)}_{m,s,\alpha}(q)\,H^{(2)}_m(k_0 q\rho)\right] \\
&\quad - q_\alpha\left[D_{m,s,\alpha}(q)\,H^{(1)}_m(k_0 q\rho) + C_{m,s,\alpha}(q)\,H^{(2)}_m(k_0 q_\alpha\rho)\right],
\end{aligned}
$$

where $C^{(1,2)}_{m,s,\alpha}$, $C_{m,s,\alpha}$, and $D_{m,s,\alpha}$ are coefficients yet to be determined. Other notations in (6.9) are given by

$$
n^{(1)}_{s,\alpha}(q) = -\varepsilon_a\left[q^2 + p^2_\alpha(q) + \frac{g_a^2}{\varepsilon_a} - \varepsilon_a\right] \Big/ g_a\, p_\alpha(q), \tag{6.10}
$$

$$
n^{(2)}_{s,\alpha}(q) = -\varepsilon_a\left[q^2_\alpha(q) + p^2_\alpha(q) + \frac{g_a^2}{\varepsilon_a} - \varepsilon_a\right] \Big/ g_a\, p_\alpha(q), \tag{6.11}
$$

and

$$
q_\alpha(q) = \left[\varepsilon_a - p^2_\alpha(q) - \frac{g_a}{\varepsilon_a}\left(g_a - \frac{\eta_a\, p_{s,\alpha}(q)}{n^{(1)}_{s,\alpha}(q)}\right)\right]^{1/2}. \tag{6.12}
$$

For definiteness, the *condition* $\mathrm{Im}\,(q_\alpha) < 0$ *will henceforth be assumed.* The remaining electric and magnetic components can be obtained with the help of expressions (4.10) to (4.13). Unlike the isotropic medium, the "auxiliary" normalized transverse wavenumbers $q_{o,x}(q)$ appear in (6.9). These quantities give the same values of the axial wavenumbers $p_{o,x}$ as does the transverse wavenumber q. This can be clarified as follows. It is seen from (2.64) that there are always two different values q_1 and q_2 of the transverse wavenumber, which correspond to the same fixed value of p. In the ambient medium, the three quantities q_1, q_2, and p satisfy the identity

$$\left(p^2 + q_1^2 + \frac{g_a^2}{\varepsilon_a} - \varepsilon_a\right)\left(p^2 + q_2^2 + \frac{g_a^2}{\varepsilon_a} - \varepsilon_a\right) = -\eta_a\, g_a^2\, p^2/\varepsilon_a^2, \qquad (6.13)$$

which follows directly from the relation (4.57). Now let $p = p_\alpha(q)$ in (6.13). Then, setting one of the quantities $q_{1,2}$ equal to q, and another to q_α, we immediately arrive at the formula (6.12) for $q_\alpha(q)$. Notice that the appearance of "auxiliary" transverse wavenumbers is connected only with the anisotropic properties of a magnetized plasma.

In the *core* $\rho < a$, where the plasma density varies, the above analysis cannot be used, and a more complicated development must be employed. Before we go on to discuss a procedure of finding the field solutions in the core, we shall obtain some general relationships which will be needed in our further treatment. We turn directly to Maxwell's equations

$$\nabla \times \mathbf{E}_{\mathrm{I}} = -\mathrm{i}k_0\boldsymbol{\mathcal{H}}_{\mathrm{I}}, \qquad (6.14)$$

$$\nabla \times \boldsymbol{\mathcal{H}}_{\mathrm{I}} = \mathrm{i}k_0\hat{\varepsilon}_{\mathrm{I}} \cdot \mathbf{E}_{\mathrm{I}}, \qquad (6.15)$$

$$\nabla \times \mathbf{E}_{\mathrm{II}} = -\mathrm{i}k_0\boldsymbol{\mathcal{H}}_{\mathrm{II}}, \qquad (6.16)$$

$$\nabla \times \boldsymbol{\mathcal{H}}_{\mathrm{II}} = \mathrm{i}k_0\hat{\varepsilon}_{\mathrm{II}} \cdot \mathbf{E}_{\mathrm{II}} \qquad (6.17)$$

written for the fields $\mathbf{E}_{\mathrm{I}}$, $\boldsymbol{\mathcal{H}}_{\mathrm{I}}$ and $\mathbf{E}_{\mathrm{II}}$, $\boldsymbol{\mathcal{H}}_{\mathrm{II}}$ which are taken (at the same frequency) in two different media described by dielectric tensors $\hat{\varepsilon}_{\mathrm{I}}$ and $\hat{\varepsilon}_{\mathrm{II}}$. Making the dot products of (6.14), (6.15), (6.16), and (6.17) with $\boldsymbol{\mathcal{H}}_{\mathrm{II}}$, $\mathbf{E}_{\mathrm{II}}$, $-\boldsymbol{\mathcal{H}}_{\mathrm{I}}$, and $-\mathbf{E}_{\mathrm{I}}$ respectively, and adding the resulting equations term by term, we obtain

$$\nabla \cdot (\mathbf{E}_{\mathrm{I}} \times \boldsymbol{\mathcal{H}}_{\mathrm{II}} - \mathbf{E}_{\mathrm{II}} \times \boldsymbol{\mathcal{H}}_{\mathrm{I}}) = \mathrm{i}k_0\,(\mathbf{E}_{\mathrm{II}} \cdot \hat{\varepsilon}_{\mathrm{I}} \cdot \mathbf{E}_{\mathrm{I}} - \mathbf{E}_{\mathrm{I}} \cdot \hat{\varepsilon}_{\mathrm{II}} \cdot \mathbf{E}_{\mathrm{II}}). \qquad (6.18)$$

By applying the divergence theorem in the two-dimensional form, we deduce

$$\int_\Sigma \nabla \cdot (\mathbf{E}_{\mathrm{I}} \times \boldsymbol{\mathcal{H}}_{\mathrm{II}} - \mathbf{E}_{\mathrm{II}} \times \boldsymbol{\mathcal{H}}_{\mathrm{I}})\, \mathrm{d}\sigma = \frac{\partial}{\partial z}\int_\Sigma (\mathbf{E}_{\mathrm{I}} \times \boldsymbol{\mathcal{H}}_{\mathrm{II}} - \mathbf{E}_{\mathrm{II}} \times \boldsymbol{\mathcal{H}}_{\mathrm{I}}) \cdot \hat{\mathbf{z}}_0\, \mathrm{d}\sigma$$

$$+ \oint_L (\mathbf{E}_{\mathrm{I}} \times \boldsymbol{\mathcal{H}}_{\mathrm{II}} - \mathbf{E}_{\mathrm{II}} \times \boldsymbol{\mathcal{H}}_{\mathrm{I}}) \cdot \hat{\mathbf{n}}\, \mathrm{d}\ell, \qquad (6.19)$$

where Σ is an arbitrary cross-sectional area of the duct. The line integral on the right of (6.19) is along the boundary L of Σ, and $\hat{\mathbf{n}}$ is the unit outward normal on L in the plane of Σ.

Now let the medium I be a magnetoplasma with the density profile $N(\rho)$ such as in the duct, so that $\hat{\varepsilon}_I = \hat{\varepsilon}$. As $\mathbf{E}_I$, $\mathcal{H}_I$ we may then take the field

$$\mathbf{E}_I = \mathbf{E}_{m,s,\alpha}(\rho, q) \exp[-im\phi - ik_0\, p_{s,\alpha}(q)\, z],$$

$$\mathcal{H}_I = \mathcal{H}_{m,s,\alpha}(\rho, q) \exp[-im\phi - ik_0\, p_{s,\alpha}(q)\, z] \tag{6.20}$$

governed by the general source-free differential equations. Let the medium II be a uniform magnetoplasma with the constant (ambient) density N_a and the *oppositely directed ambient static magnetic field* $(-\mathbf{B}_0)$. For such a medium, commonly called "transposed", one must write $\hat{\varepsilon}_{II} = \hat{\varepsilon}_a^{(T)}$ where $\hat{\varepsilon}_a^{(T)}$ is the transposed dielectric tensor defined by $\varepsilon_{a,ij}^{(T)} = \varepsilon_{a,ji}$. Let the field in the medium II be

$$\mathbf{E}_{II} = \overline{\mathbf{E}}_{\tilde{m},\tilde{s},\tilde{\alpha}}^{(T)}(\rho, q) \exp[-i\tilde{m}\phi - ik_0\, p_{\tilde{s},\tilde{\alpha}}(q)\, z],$$

$$\mathcal{H}_{II} = \overline{\mathcal{H}}_{\tilde{m},\tilde{s},\tilde{\alpha}}^{(T)}(\rho, q) \exp[-i\tilde{m}\phi - ik_0\, p_{\tilde{s},\tilde{\alpha}}(q)\, z], \tag{6.21}$$

where we put $\tilde{m} = -m$, $\tilde{s} = -s$, $\tilde{\alpha} = \alpha$ and take the same value of q as for the field I. The vector quantities $\overline{\mathbf{E}}_{\tilde{m},\tilde{s},\tilde{\alpha}}^{(T)}(\rho, q)$ and $\overline{\mathcal{H}}_{\tilde{m},\tilde{s},\tilde{\alpha}}^{(T)}(\rho, q)$ can readily be found from the general field expressions of section 4.2.2 (§ 4.2) by making the replacements $m \to \tilde{m}$, $p \to p_{\tilde{s},\tilde{\alpha}}(q)$, $n_{1,2} \to -n_{\tilde{s},\tilde{\alpha}}^{(1,2)}$, $q_1 \to q$, $q_2 \to q_{\tilde{\alpha}}(q)$, and $(\varepsilon, g, \eta) \to (\varepsilon_a, -g_a, \eta_a)$. Next, we substitute (6.20) and (6.21) into (6.18), and perform integration of (6.18) over the cross-sectional area Σ bounded by the lines $\tilde{\rho} = \rho$ and $\tilde{\rho} = \rho_0$. Using the divergence theorem (6.19) and passing to the limit $\rho_0 \to 0$, we deduce after some rearrangement

$$\tilde{\rho} \left[\mathbf{E}_{m,s,\alpha}(\tilde{\rho}, q) \times \overline{\mathcal{H}}_{-m,-s,\alpha}^{(T)}(\tilde{\rho}, q) - \overline{\mathbf{E}}_{-m,-s,\alpha}^{(T)}(\tilde{\rho}, q) \times \mathcal{H}_{m,s,\alpha}(\tilde{\rho}, q) \right] \cdot \hat{\boldsymbol{\rho}}_0 \Bigg|_{\tilde{\rho}=0}^{\tilde{\rho}=\rho}$$

$$= ik_0 \int_0^\rho \overline{\mathbf{E}}_{-m,-s,\alpha}^{(T)}(\tilde{\rho}, q) \cdot [\hat{\varepsilon}(\tilde{\rho}) - \hat{\varepsilon}_a] \cdot \mathbf{E}_{m,s,\alpha}(\tilde{\rho}, q)\, \tilde{\rho}\, d\tilde{\rho}. \tag{6.22}$$

In obtaining (6.22) we made use of the fact that the tangential components of $\mathbf{E}_{m,s,\alpha}(\tilde{\rho}, q)$ and $\mathcal{H}_{m,s,\alpha}(\tilde{\rho}, q)$ must be continuous at all cylindrical surfaces $\tilde{\rho} = \rho_i$ of the jump of $N(\rho)$, whose radii ρ_i lie in the variability range $0 < \tilde{\rho} < \rho$ of the integration variable $\tilde{\rho}$. The relation (6.22) is the starting point for much of the later work in this chapter.

6.2.2. Field on the source-free duct

First consider the *azimuthally symmetric* case. To simplify the notation, the azimuthal index m which is now equal to zero will be omitted throughout. The analysis for the more complicated nonsymmetric case, when $m \neq 0$, will be given later. Turning to the relationship (6.22), we note that, for $m = 0$, it is more convenient to employ the azimuthal components $E_{\phi;s,\alpha}(\tilde{\rho}, q)$ and $\mathcal{H}_{\phi;s,\alpha}(\tilde{\rho}, q)$ instead of the longitudinal components in order to represent the field $\mathbf{E}_{s,\alpha}(\tilde{\rho}, q)$, $\mathcal{H}_{s,\alpha}(\tilde{\rho}, q)$. The components $E_{\phi;s,\alpha}(\rho, q)$ and $\mathcal{H}_{\phi;s,\alpha}(\rho, q)$ satisfy the differential equations

$$\frac{\partial}{\partial \rho}\left[\frac{1}{\rho}\frac{\partial}{\partial \rho}\left(\rho\, E_{\phi;s,\alpha}\right)\right] = -k_0^2 \varepsilon^{-1}\left\{\left[\varepsilon\left(\varepsilon - p_\alpha^2\right) - g^2\right] E_{\phi;s,\alpha} + i p_{s,\alpha} g \mathcal{H}_{\phi;s,\alpha}\right\}, \quad (6.23)$$

$$\frac{\partial}{\partial \rho}\left[\frac{1}{\eta\rho}\frac{\partial}{\partial \rho}\left(\rho\, \mathcal{H}_{\phi;s,\alpha}\right)\right] = -k_0^2 \varepsilon^{-1}\left[\left(\varepsilon - p_\alpha^2\right)\mathcal{H}_{\phi;s,\alpha} - i p_{s,\alpha} g E_{\phi;s,\alpha}\right] \quad (6.24)$$

(compare (5.12) and (5.13)), which are much simpler than the general field equations (4.14) and (4.15). The corresponding longitudinal components are then

$$E_{z;s,\alpha}(\tilde{\rho}, q) = -\frac{i}{k_0 \eta \tilde{\rho}}\frac{\partial}{\partial \tilde{\rho}}\left[\tilde{\rho}\,\mathcal{H}_{\phi;s,\alpha}(\tilde{\rho}, q)\right],$$

$$\mathcal{H}_{z;s,\alpha}(\tilde{\rho}, q) = \frac{i}{k_0 \tilde{\rho}}\frac{\partial}{\partial \tilde{\rho}}\left[\tilde{\rho}\, E_{\phi;s,\alpha}(\tilde{\rho}, q)\right]. \tag{6.25}$$

It is evident from the condition of regularity at $\rho = 0$ for the fields of the duct that

$$E_{\phi;s,\alpha}(0, q) = 0, \quad \mathcal{H}_{\phi;s,\alpha}(0, q) = 0. \tag{6.26}$$

Conversely, the components (6.25) are nonzero at $\rho = 0$. We introduce quantities

$$u_{s,\alpha}(q) = \lim_{\tilde{\rho}\to 0}\frac{(-i)}{k_0 \tilde{\rho}}\frac{\partial}{\partial \tilde{\rho}}\left[\tilde{\rho}\, E_{\phi;s,\alpha}(\tilde{\rho}, q)\right],$$

$$v_{s,\alpha}(q) = \lim_{\tilde{\rho}\to 0}\frac{1}{k_0 \tilde{\rho}}\frac{\partial}{\partial \tilde{\rho}}\left[\tilde{\rho}\,\mathcal{H}_{\phi;s,\alpha}(\tilde{\rho}, q)\right], \tag{6.27}$$

which are simply related to the longitudinal components (6.25) at $\tilde{\rho} = 0$ and convenient for the later work. In order to obtain the required form of the axisymmetric solution we take some auxiliary field $\overline{\mathbf{E}}_{-s,\alpha}^{(T)}(\tilde{\rho}, q)$, $\overline{\mathcal{H}}_{-s,\alpha}^{(T)}(\tilde{\rho}, q)$ with its azimuthal components

$$\overline{E}_{\phi;-s,\alpha}^{(T)}(\tilde{\rho}, q) = i\left\{A_{s,\alpha}^{(1)}\left[J_1(k_0 q\rho)\, Y_1(k_0 q\tilde{\rho}) - Y_1(k_0 q\rho)\, J_1(k_0 q\tilde{\rho})\right]\right.$$
$$\left. + A_{s,\alpha}^{(2)}\left[J_1(k_0 q_\alpha\rho)\, Y_1(k_0 q_\alpha\tilde{\rho}) - Y_1(k_0 q_\alpha\rho)\, J_1(k_0 q_\alpha\tilde{\rho})\right]\right\},$$

$$\tag{6.28}$$

$$\overline{\mathcal{H}}_{\phi;-s,\alpha}^{(T)}(\tilde{\rho}, q) = -\left\{A_{s,\alpha}^{(1)}\, n_{s,\alpha}^{(1)}\left[J_1(k_0 q\rho)\, Y_1(k_0 q\tilde{\rho}) - Y_1(k_0 q\rho)\, J_1(k_0 q\tilde{\rho})\right]\right.$$
$$\left. + A_{s,\alpha}^{(2)}\, n_{s,\alpha}^{(2)}\left[J_1(k_0 q_\alpha\rho)\, Y_1(k_0 q_\alpha\tilde{\rho}) - Y_1(k_0 q_\alpha\rho)\, J_1(k_0 q_\alpha\tilde{\rho})\right]\right\},$$

where $A_{s,\alpha}^{(1,2)}$ are arbitrary constants and the quantity ρ is the same as the upper limit in (6.22). In (6.28) we also make use of the fact that $q_\alpha^{(T)} = q_\alpha$ and $\left(n_{-s,\alpha}^{(1,2)}\right)^{(T)} = n_{s,\alpha}^{(1,2)}$. The longitudinal components $\overline{E}_{z;\,-s,\alpha}^{(T)}$ and $\overline{\mathcal{H}}_{z;\,-s,\alpha}^{(T)}$ are readily found by analogy with (6.25). Although the components (6.28) are unbounded at $\tilde{\rho} = 0$, this is unimportant because they pertain to a purely auxiliary, not physically realizable, field.

Now it is a simple matter to express the quantities $E_{\phi;\,s,\alpha}(\rho, q)$ and $\mathcal{H}_{\phi;\,s,\alpha}(\rho, q)$ in an implicit form. Using (6.25) to (6.28), along with formulas

$$\lim_{\zeta \to 0} \zeta\, Y_0(\zeta) = 0, \quad \lim_{\zeta \to 0} \zeta\, Y_1(\zeta) = -\frac{2}{\pi},$$

and setting $A_{s,\alpha}^{(1,2)} = n_{s,\alpha}^{(2,1)}\left(n_{s,\alpha}^{(2,1)} - n_{s,\alpha}^{(1,2)}\right)^{-1}$, we obtain from (6.22) that

$$
\begin{aligned}
E_{\phi;\,s,\alpha}(\rho, q) &= \frac{i n_{s,\alpha}^{(2)}}{2\left(n_{s,\alpha}^{(2)} - n_{s,\alpha}^{(1)}\right)} \left[G_{s,\alpha}^{(1)}(\rho, q)\, H_1^{(1)}(k_0 q \rho) + G_{s,\alpha}^{(2)}(\rho, q)\, H_1^{(2)}(k_0 q \rho)\right] \\
&+ \frac{i n_{s,\alpha}^{(1)}}{2\left(n_{s,\alpha}^{(1)} - n_{s,\alpha}^{(2)}\right)} \left[F_{s,\alpha}(\rho, q)\, H_1^{(1)}(k_0 q_\alpha \rho) + G_{s,\alpha}(\rho, q)\, H_1^{(2)}(k_0 q_\alpha \rho)\right].
\end{aligned} \tag{6.29}
$$

For $A_{s,\alpha}^{(1)} = -A_{s,\alpha}^{(2)} = \left(n_{s,\alpha}^{(2)} - n_{s,\alpha}^{(1)}\right)^{-1}$, a similar process yields

$$
\begin{aligned}
\mathcal{H}_{\phi;\,s,\alpha}(\rho, q) &= \frac{\eta_a}{2\left(n_{s,\alpha}^{(2)} - n_{s,\alpha}^{(1)}\right)} \left[G_{s,\alpha}^{(1)}(\rho, q)\, H_1^{(1)}(k_0 q \rho) + G_{s,\alpha}^{(2)}(\rho, q)\, H_1^{(2)}(k_0 q \rho) \right. \\
&\left. \quad - F_{s,\alpha}(\rho, q)\, H_1^{(1)}(k_0 q_\alpha \rho) - G_{s,\alpha}(\rho, q)\, H_1^{(2)}(k_0 q_\alpha \rho)\right].
\end{aligned} \tag{6.30}
$$

In the preceding expressions,

$$
\begin{aligned}
G_{s,\alpha}^{(1,2)} &= \frac{w_{s,\alpha}^{(1)}(q)}{q} - \chi^{(1,2)} \int_0^\rho f_{s,\alpha}^{(1)}(\tilde{\rho}, q)\, H_1^{(2,1)}(k_0 q \tilde{\rho})\, \tilde{\rho}\, \mathrm{d}\tilde{\rho}, \\
G_{s,\alpha}(\rho, q) &= \frac{w_{s,\alpha}^{(2)}(q)}{q_a} - \int_0^\rho f_{s,\alpha}^{(2)}(\tilde{\rho}, q)\, H_1^{(1)}(k_0 q_\alpha \tilde{\rho})\, \tilde{\rho}\, \mathrm{d}\tilde{\rho}, \\
F_{s,\alpha}(\rho, q) &= \frac{w_{s,\alpha}^{(2)}(q)}{q_\alpha} + \int_0^\rho f_{s,\alpha}^{(2)}(\tilde{\rho}, q)\, H_1^{(2)}(k_0 q_\alpha \tilde{\rho})\, \tilde{\rho}\, \mathrm{d}\tilde{\rho},
\end{aligned} \tag{6.31}
$$

where

$$w_{s,\alpha}^{(1,2)}(q) = u_{s,\alpha}(q) - \eta_a^{-1}\, n_{s,\alpha}^{(1,2)}\, v_{s,\alpha}(q),$$

$$
\begin{aligned}
f_{s,\alpha}^{(1,2)}(\rho,q) = \frac{\pi k_0^2}{2}\Bigg\{ & \left[\Delta\varepsilon\left(1 + \frac{g}{\varepsilon\,\varepsilon_a}\left(p_{s,\alpha}\, n_{s,\alpha}^{(1,2)} + g_a\right)\right)\right. \\
& - \Delta g\left(\frac{1}{\varepsilon_a}\, p_{s,\alpha}\, n_{s,\alpha}^{(1,2)} + \frac{g}{\varepsilon} + \frac{g_a}{\varepsilon_a}\right) - \Delta\eta\, p_{s,\alpha}\, n_{s,\alpha}^{(1,2)}\, \frac{g}{\eta_a\varepsilon}\Bigg]\, E_{\phi;\,s,\alpha}(\rho,q) \\
& - \mathrm{i}\left[\Delta\varepsilon\,\frac{p_{s,\alpha}}{\varepsilon}\left(\frac{1}{\varepsilon_a}\, p_{s,\alpha}\, n_{s,\alpha}^{(1,2)} + \frac{g_a}{\varepsilon_a}\right) - \Delta g\,\frac{p_{s,\alpha}}{\varepsilon} - \Delta\eta\,\frac{n_{s,\alpha}^{(1,2)}}{\eta_a\varepsilon}\,(p_\alpha^2 - \varepsilon)\right]\mathcal{H}_{\phi;\,s,\alpha}(\rho,q) \\
& + \mathrm{i}\,\frac{n_{s,\alpha}^{(1,2)}}{k_0^2\,\eta\,\eta_a}\,\frac{\mathrm{d}\eta}{\mathrm{d}\rho}\,\frac{1}{\rho}\,\frac{\partial}{\partial\rho}\left[\rho\,\mathcal{H}_{\phi;\,s,\alpha}(\rho,q)\right]\Bigg\},
\end{aligned}
$$

$$\Delta\varepsilon = \varepsilon(\rho) - \varepsilon_a, \quad \Delta g = g(\rho) - g_a, \quad \Delta\eta = \eta(\rho) - \eta_a, \quad \chi^{(1)} = -\chi^{(2)} = -1.$$

In obtaining (6.31) and (6.32) use was made of the expression for the Wronskian of Bessel functions $J_m(\zeta)$ and $Y_m(\zeta)$, and of the identity

$$
\begin{aligned}
\int_0^\rho \frac{\partial}{\partial\tilde\rho}&\left[\tilde\rho\,\overline{\mathcal{H}}_{\phi;\,-s,\alpha}^{(T)}(\tilde\rho,q)\right]\frac{\Delta\eta}{\eta}\,\frac{1}{\tilde\rho}\,\frac{\partial}{\partial\tilde\rho}\left[\tilde\rho\mathcal{H}_{\phi;\,s,\alpha}(\tilde\rho,q)\right]\mathrm{d}\tilde\rho \\
&= \frac{\Delta\eta(\tilde\rho)}{\eta(\tilde\rho)}\,\frac{\partial}{\partial\tilde\rho}\left[\tilde\rho\,\mathcal{H}_{\phi;\,s,\alpha}(\tilde\rho,q)\right]\overline{\mathcal{H}}_{\phi;\,-s,\alpha}^{(T)}(\tilde\rho,q)\Bigg|_{\tilde\rho=0}^{\tilde\rho=\rho} \\
&\quad - \int_0^\rho \left\{\Delta\eta(\tilde\rho)\,\frac{\partial}{\partial\tilde\rho}\left[\frac{1}{\eta\tilde\rho}\,\frac{\partial}{\partial\tilde\rho}\left(\tilde\rho\,\mathcal{H}_{\phi;\,s,\alpha}(\tilde\rho,q)\right)\right]\right. \\
&\quad\quad \left. + \frac{\mathrm{d}\eta}{\mathrm{d}\tilde\rho}\,\frac{1}{\eta\tilde\rho}\,\frac{\partial}{\partial\tilde\rho}\left[\tilde\rho\mathcal{H}_{\phi;\,s,\alpha}(\tilde\rho,q)\right]\right\}\overline{\mathcal{H}}_{\phi;\,-s,\alpha}^{(T)}(\tilde\rho,q)\,\tilde\rho\,\mathrm{d}\tilde\rho,
\end{aligned}
\tag{6.33}
$$

where the first term in the braces under the sign of the last integral was replaced by the right side of Equation (6.24) to yield the resulting form for $f_{s,\alpha}^{(1,2)}(\rho,q)$.

In the outer region $\rho > a$, where $f_{s,\alpha}^{(1,2)}(\rho,q) = 0$ and $n_{s,\alpha}^{(1)}\, n_{s,\alpha}^{(2)} = -\eta_a$, formulas (6.29) and (6.30) give the customary form for the azimuthal components

$$
\begin{aligned}
E_{\phi;\,s,\alpha}(\rho,q) &= \mathrm{i}\Big[C_{s,\alpha}^{(1)}(q)\,H_1^{(1)}(k_0 q\rho) + C_{s,\alpha}^{(2)}(q)\,H_1^{(2)}(k_0 q\rho) \\
&\quad\quad + D_{s,\alpha}(q)\,H_1^{(1)}(k_0 q_\alpha\rho) + C_{s,\alpha}(q)\,H_1^{(2)}(k_0 q_\alpha\rho)\Big],
\end{aligned}
$$

$$
\begin{aligned}
\mathcal{H}_{\phi;\,s,\alpha}(\rho,q) &= -n_{s,\alpha}^{(1)}\Big[C_{s,\alpha}^{(1)}(q)\,H_1^{(1)}(k_0 q\rho) + C_{s,\alpha}^{(2)}(q)\,H_1^{(2)}(k_0 q\rho)\Big] \\
&\quad - n_{s,\alpha}^{(2)}\Big[D_{s,\alpha}(q)\,H_1^{(1)}(k_0 q_\alpha\rho) + C_{s,\alpha}(q)\,H_1^{(2)}(k_0 q_\alpha\rho)\Big],
\end{aligned}
\tag{6.34}
$$

with

$$C_{s,\alpha}^{(1,2)}(q) = \frac{n_{s,\alpha}^{(2)}}{2\left(n_{s,\alpha}^{(2)} - n_{s,\alpha}^{(1)}\right)} \left[\frac{w_{s,\alpha}^{(1)}(q)}{q} - \chi^{(1,2)} \int_0^a f_{s,\alpha}^{(1)}(\tilde{\rho}, q)\, H_1^{(2,1)}(k_0 q \tilde{\rho})\, \tilde{\rho}\, \mathrm{d}\tilde{\rho}\right],$$

$$C_{s,\alpha}(q) = \frac{n_{s,\alpha}^{(1)}}{2\left(n_{s,\alpha}^{(1)} - n_{s,\alpha}^{(2)}\right)} \left[\frac{w_{s,\alpha}^{(2)}(q)}{q_\alpha} - \int_0^a f_{s,\alpha}^{(2)}(\tilde{\rho}, q)\, H_1^{(1)}(k_0 q_\alpha \tilde{\rho})\, \tilde{\rho}\, \mathrm{d}\tilde{\rho}\right], \quad (6.35)$$

$$D_{s,\alpha}(q) = \frac{n_{s,\alpha}^{(1)}}{2\left(n_{s,\alpha}^{(1)} - n_{s,\alpha}^{(2)}\right)} \left[\frac{w_{s,\alpha}^{(2)}(q)}{q_\alpha} + \int_0^a f_{s,\alpha}^{(2)}(\tilde{\rho}, q)\, H_1^{(2)}(k_0 q_\alpha \tilde{\rho})\, \tilde{\rho}\, \mathrm{d}\tilde{\rho}\right].$$

Note that the result (6.34) is entirely consistent with the previous expressions (6.9) for the longitudinal components, where one must set $m = 0$ for the azimuthally symmetric case.

Next, we have to extend the above theory to the general *nonsymmetric* case when $m \neq 0$. This can be made without difficulty but the mathematics is tedious. Therefore, we shall only list the salient steps which lead to the resulting expressions and then present results in the form convenient for the future analysis.

When $\partial/\partial\phi \neq 0$ and, accordingly, $m \neq 0$, it is more suitable to employ the longitudinal components throughout and deal with the homogeneous field equations written in terms of these components (see Equations (4.14) and (4.15)). If the fields of the duct are regular at $\rho = 0$, then, for $m = \pm 1, \pm 2, \ldots$, the components $E_{z;m,s,\alpha}(\rho, q)$ and $\mathcal{H}_{z;m,s,\alpha}(\rho, q)$ behave like $\sim \rho^{|m|}$ as $\rho \to 0$, whence

$$E_{z;m,s,\alpha}(0, q) = 0, \quad \mathcal{H}_{z;m,s,\alpha}(0, q) = 0,$$

$$u_{m,s,\alpha}(q) = \lim_{\tilde{\rho}\to 0} \frac{1}{k_0^{|m|}\, \tilde{\rho}^{|m|-1}} \frac{\partial}{\partial\tilde{\rho}}\, E_{z;m,s,\alpha}(\tilde{\rho}, q) \neq 0, \quad (6.36)$$

$$v_{m,s,\alpha}(q) = \lim_{\tilde{\rho}\to 0} \frac{1}{k_0^{|m|}\, \tilde{\rho}^{|m|-1}} \frac{\partial}{\partial\tilde{\rho}}\, \mathcal{H}_{z;m,s,\alpha}(\tilde{\rho}, q) \neq 0.$$

To obtain a required field representation we again apply the vector relation (6.22). The auxiliary field $\overline{\mathbf{E}}_{-m,-s,\alpha}^{(\mathrm{T})}(\tilde{\rho}, q)$, $\overline{\mathcal{H}}_{-m,-s,\alpha}^{(\mathrm{T})}(\tilde{\rho}, q)$ to be used in (6.22) must now be defined via

$$\overline{E}_{z;-m,-s,\alpha}^{(\mathrm{T})}(\tilde{\rho}, q) = \frac{\mathrm{i}}{\eta_a}\left\{ A_{m,s,\alpha}^{(1)}\, n_{s,\alpha}^{(1)}\, q\left[J_m(k_0 q\rho)\, Y_m(k_0 q\tilde{\rho}) - Y_m(k_0 q\rho)\, J_m(k_0 q\tilde{\rho})\right]\right.$$

$$\left. + A_{m,s,\alpha}^{(2)}\, n_{s,\alpha}^{(2)}\, q_\alpha\left[J_m(k_0 q_\alpha\rho)\, Y_m(k_0 q_\alpha\tilde{\rho}) - Y_m(k_0 q_\alpha\rho)\, J_m(k_0 q_\alpha\tilde{\rho})\right]\right\},$$

$$(6.37)$$

$$\overline{\mathcal{H}}_{z;-m,-s,\alpha}^{(\mathrm{T})}(\tilde{\rho}, q) = -\left\{ A_{m,s,\alpha}^{(1)}\, q\left[J_m(k_0 q\rho)\, Y_m(k_0 q\tilde{\rho}) - Y_m(k_0 q\rho)\, J_m(k_0 q\tilde{\rho})\right]\right.$$

$$\left. + A_{m,s,\alpha}^{(2)}\, q_\alpha\left[J_m(k_0 q_\alpha\rho)\, Y_m(k_0 q_\alpha\tilde{\rho}) - Y_m(k_0 q_\alpha\rho)\, J_m(k_0 q_\alpha\tilde{\rho})\right]\right\}.$$

In the above, we made use of the fact that

$$J_{-m}(\zeta) = (-1)^m \, J_m(\zeta), \quad Y_{-m}(\zeta) = (-1)^m \, Y_m(\zeta)$$

if $m = 1, 2, \ldots$. Further, we shall take into account the relation

$$\lim_{\zeta \to 0} \zeta^{|m|} \, Y_m(\zeta) = -\frac{\sigma_m}{\pi} \, ,$$

where

$$\sigma_m = \begin{cases} 2^m \, (m-1)! & \text{if } m = 1, 2, \ldots, \\ (-1)^m \, 2^{|m|} \, (|m| - 1)! & \text{if } m = -1, -2, \ldots . \end{cases} \tag{6.38}$$

Next, bearing (6.36) and (6.37) in mind, we express the transverse components of the fields occurring in (6.22) via the corresponding longitudinal components, with the help of the general formulas (4.10) to (4.13), and then make the operations indicated in (6.22). To obtain a required representation for $E_{z\,;\,m,s,\alpha}(\rho, q)$ we set

$$A^{(1)}_{m,s,\alpha} = \frac{q}{n^{(2)}_{s,\alpha} - n^{(1)}_{s,\alpha}} \frac{1}{\left(p_\alpha^2 - \varepsilon_a\right)^2 - g_a^2} \, , \qquad A^{(2)}_{m,s,\alpha} = \frac{q_\alpha}{n^{(1)}_{s,\alpha} - n^{(2)}_{s,\alpha}} \frac{1}{\left(p_\alpha^2 - \varepsilon_a\right)^2 - g_a^2} \, ,$$

while for $\mathcal{H}_{z\,;\,m,s,\alpha}(\rho, q)$,

$$A^{(1)}_{m,s,\alpha} = \frac{q\,n^{(2)}_{s,\alpha}}{n^{(2)}_{s,\alpha} - n^{(1)}_{s,\alpha}} \frac{1}{\left(p_\alpha^2 - \varepsilon_a\right)^2 - g_a^2} \, , \qquad A^{(2)}_{m,s,\alpha} = \frac{q_\alpha\,n^{(1)}_{s,\alpha}}{n^{(1)}_{s,\alpha} - n^{(2)}_{s,\alpha}} \frac{1}{\left(p_\alpha^2 - \varepsilon_a\right)^2 - g_a^2} \, .$$

The resulting expressions for $E_{z\,;\,m,s,\alpha}(\rho, q)$ and $\mathcal{H}_{z\,;\,m,s,\alpha}(\rho, q)$ in the outer region then reduce to the usual representation (6.9), with the coefficients $C^{(1,2)}_{m,s,\alpha}$, $C_{m,s,\alpha}$, $D_{m,s,\alpha}$ given by

$$C^{(1,2)}_{m,s,\alpha}(q) = \frac{q\,n^{(2)}_{s,\alpha}}{2\left(n^{(2)}_{s,\alpha} - n^{(1)}_{s,\alpha}\right)} \left[\frac{w^{(1)}_{m,s,\alpha}(q)}{q^{|m|}} - \chi^{(1,2)} \int_0^a f^{(1)}_{m,s,\alpha}(\tilde{\rho}, q) \, H^{(2,1)}_m(k_0 q \tilde{\rho}) \, \tilde{\rho} \, \mathrm{d}\tilde{\rho} \right],$$

$$C_{m,s,\alpha}(q) = \frac{q_\alpha\,n^{(1)}_{s,\alpha}}{2\left(n^{(1)}_{s,\alpha} - n^{(2)}_{s,\alpha}\right)} \left[\frac{w^{(2)}_{m,s,\alpha}(q)}{q_\alpha^{|m|}} - \int_0^a f^{(2)}_{m,s,\alpha}(\tilde{\rho}, q) \, H^{(1)}_m(k_0 q_\alpha \tilde{\rho}) \, \tilde{\rho} \, \mathrm{d}\tilde{\rho} \right], \tag{6.39}$$

$$D_{m,s,\alpha}(q) = \frac{q_\alpha\,n^{(1)}_{s,\alpha}}{2\left(n^{(1)}_{s,\alpha} - n^{(2)}_{s,\alpha}\right)} \left[\frac{w^{(2)}_{m,s,\alpha}(q)}{q_\alpha^{|m|}} + \int_0^a f^{(2)}_{m,s,\alpha}(\tilde{\rho}, q) \, H^{(2)}_m(k_0 q_\alpha \tilde{\rho}) \, \tilde{\rho} \, \mathrm{d}\tilde{\rho} \right].$$

In the preceding expressions,

$$\begin{aligned} w^{(1,2)}_{m,s,\alpha}(q) &= \frac{\mathrm{i}}{2} \sigma_m \, \beta(0) \Bigg\{ \left[p_{s,\alpha} \left(\beta_a \Delta^{(1,2)}_1(0) - g(0) \right) \right. \\ &\quad + \left. \frac{g^2(0) - (\varepsilon(0) - p_\alpha^2)\,\varepsilon(0)}{\eta_a} \, n^{(1,2)}_{s,\alpha} \right] u_{m,s,\alpha}(q) \\ &\quad - \mathrm{i} \left[\beta_a \Delta^{(1,2)}_2(0) + p_\alpha^2 - \varepsilon(0) - \frac{g(0)}{\eta_a} \, p_{s,\alpha} \, n^{(1,2)}_{s,\alpha} \right] v_{m,s,\alpha}(q) \Bigg\}, \end{aligned} \tag{6.40}$$

where

$$\beta = \left[\left(p_\alpha^2 - \varepsilon\right)^2 - g^2\right]^{-1}, \quad \beta_a = \left[\left(p_\alpha^2 - \varepsilon_a\right)^2 - g_a^2\right]^{-1},$$

$$\Delta_{1,2}^{(1,2)} = -\Delta\varepsilon\left[ga_{1,2}^{(1,2)} + (\varepsilon - p_\alpha^2)\,a_{2,1}^{(1,2)}\right] + \Delta g\left[ga_{2,1}^{(1,2)} + (\varepsilon - p_\alpha^2)\,a_{1,2}^{(1,2)}\right], \quad (6.41)$$

$$a_1^{(1,2)} = \frac{g_a}{\eta_a}\,p_{s,\alpha}\,n_{s,\alpha}^{(1,2)} + \varepsilon_a - p_\alpha^2, \quad a_2^{(1,2)} = \frac{\varepsilon_a - p_\alpha^2}{\eta_a}\,p_{s,\alpha}\,n_{s,\alpha}^{(1,2)} + g_a.$$

The functions $f_{m,s,\alpha}^{(1,2)}$ can be found to be

$$\begin{aligned}
f_{m,s,\alpha}^{(1,2)}(\rho, q) = \frac{\pi k_0^2}{2}\Bigg\{&\left[\frac{n_{s,\alpha}^{(1,2)}}{\eta_a}\,\Delta\eta + \frac{m\,p_{s,\alpha}\,\beta_a}{k_0^2\rho}\,\frac{\partial}{\partial\rho}\left(\beta\Delta_2^{(1,2)}\right)\right] E_{z;m,s,\alpha}(\rho, q)\\
&+ k_0^{-2}\,\beta_a\,p_{s,\alpha}\left[\frac{\partial}{\partial\rho}\left(\beta\Delta_1^{(1,2)}\right)\frac{\partial}{\partial\rho}\,E_{z;m,s,\alpha}(\rho, q) + \beta\Delta_1^{(1,2)}\,\hat{L}_m\,E_{z;m,s,\alpha}(\rho, q)\right]\\
&- \mathrm{i}\,\frac{m\,\beta_a}{k_0^2\rho}\,\frac{\partial}{\partial\rho}\left(\beta\Delta_1^{(1,2)}\right)\mathcal{H}_{z;m,s,a}(\rho, q)\\
&- \mathrm{i}k_0^{-2}\beta_a\left[\frac{\partial}{\partial\rho}\left(\beta\Delta_2^{(1,2)}\right)\frac{\partial}{\partial\rho}\,\mathcal{H}_{z;m,s,\alpha}(\rho, q) + \beta\Delta_2^{(1,2)}\,\hat{L}_m\,\mathcal{H}_{z;m,s,\alpha}(\rho, q)\right]\Bigg\}, \quad (6.42)
\end{aligned}$$

in which the quantities $\hat{L}_m\,E_{z;m,s,\alpha}$ and $\hat{L}_m\,\mathcal{H}_{z;m,s,\alpha}$ have the same meaning as in Equations (4.14) and (4.15), and can readily be expressed with the help of those equations via the components $E_{z;m,s,\alpha}$, $\mathcal{H}_{z;m,s,\alpha}$ and their first derivatives with respect to ρ (one should only replace the total differentiaton sign $\mathrm{d}/\mathrm{d}\rho$ by the partial differentiation sign $\partial/\partial\rho$ in (4.14) and (4.15) throughout). The other notations in formulas (6.39) to (6.42) are as in (6.32). The results presented above for $m = \pm 1, \pm 2, \ldots$ complement the preceding theory formulated under the assumption that $m = 0$.

Thus, we have obtained the formal representations of the field components in the outer region of the duct. These representations enable one to link the field solutions in the nonuniform duct core and in the uniform outer region, where they are expressible in terms of the Hankel functions with the coefficients $C_{m,s,\alpha}^{(1,2)}$, $C_{m,s,\alpha}$ and $D_{m,s,\alpha}$. The expressions (6.35) and (6.39) for these coefficients are in form suitable for numerical calculation (note: it must be remembered that the index $m = 0$ has been omitted in (6.35) for the notational simplicity). If the quantities $u_{m,s,\alpha}(q)$ and $v_{m,s,\alpha}(q)$, which define the solution behavior in the neighborhood of the point $\rho = 0$, are specified for a given q, the source-free field equations (4.14) and (4.15), or Equations (6.23) and (6.24) for $m = 0$, can be solved by means of numerical integration. When the integration over the nonuniform part of the duct (i. e., the range $0 < \rho \leq a$) is complete the coefficients $C_{m,s,\alpha}^{(1,2)}$, $C_{m,s,\alpha}$, and $D_{m,s,\alpha}$ can be found for the chosen q. Since these coefficients define the behavior of the field at infinity, they can be used for determination of the values of q which provide the fulfillment of conditions (6.8). These values produce the eigenvalue spectrum, and the corresponding solutions are then the eigenfunctions of the duct.

6.2.3. The eigenfunction expansion

Turning now to the general expressions (6.9) written for the outer region, it was noted above that the quantity q_α is complex and defined so as to satisfy the condition $\mathrm{Im}\,(q_\alpha) < 0.^*$ Because of this fact, the function $H_m^{(1)}(k_0\,q_\alpha\,\rho)$ in (6.9) tends to infinity when $\rho \to \infty$, so that the prescribed conditions (6.8) will be violated. To avoid this, it is necessary to require

$$D_{m,s,\alpha}(q) = 0 \tag{6.43}$$

that can always be ensured by the suitable choice of ratios $u_{m,s,o}(q)/v_{m,s,o}(q)$ and $u_{m,s,x}(q)/v_{m,s,x}(q)$ for any preassigned value of q. *The fulfillment of (6.43) will henceforth be assumed throughout.* Then the field representation in the outer region $\rho > a$ goes over to

$$
\begin{aligned}
E_{z;m,s,\alpha}(\rho,q) &= \frac{i}{\eta_a}\left\{ n_{s,\alpha}^{(1)}\,q\left[C_{m,s,\alpha}^{(1)}(q)\,H_m^{(1)}(k_0 q\rho) + C_{m,s,\alpha}^{(2)}(q)\,H_m^{(2)}(k_0 q\rho)\right]\right.\\
&\qquad\left. + n_{s,\alpha}^{(2)}\,q_\alpha\,C_{m,s,\alpha}(q)\,H_m^{(2)}(k_0\,q_\alpha\,\rho)\right\},
\end{aligned}
$$

$$\tag{6.44}$$

$$
\begin{aligned}
\mathcal{H}_{z;m,s,\alpha}(\rho,q) &= -q\left[C_{m,s,\alpha}^{(1)}(q)\,H_m^{(1)}(k_0 q\rho) + C_{m,s,\alpha}^{(2)}(q)\,H_m^{(2)}(k_0 q\rho)\right]\\
&\qquad - q_\alpha\,C_{m,s,\alpha}(q)\,H_m^{(2)}(k_0\,q_\alpha\,\rho),
\end{aligned}
$$

where $m = 0, \pm 1, \ldots$. If we wish to find the corresponding solutions everywhere, the procedure we must follow for fixed subscripts m, s, α and a given value of q is to set either $u_{m,s,\alpha}$ or $v_{m,s,\alpha}$ equal to an arbitrary constant, say unity, and then compute the coefficient $D_{m,s,\alpha}$ for various values of either $v_{m,s,\alpha}$ or $u_{m,s,\alpha}$, respectively, in order to provide a zero of $D_{m,s,\alpha}$. On finding the value of $u_{m,s,\alpha}/v_{m,s,\alpha}$ which ensures (6.43), the field components in the region $0 < \rho \le a$ and the coefficients $C_{m,s,\alpha}^{(1,2)}$ and $C_{m,s,\alpha}$ are then computed. The problem thus becomes one of analyzing the behavior of $C_{m,s,\alpha}^{(1,2)}$ and $C_{m,s,\alpha}$ regarded as functions of q.

Some general remarks can be made concerning the above coefficients $C_{m,s,\alpha}^{(1,2)}$ and $C_{m,s,\alpha}$. It can be inferred directly from source-free field equations (4.14) and (4.15) that, in the core, our solutions $E_{z;m,s,\alpha}(\rho,q)$ and $\mathcal{H}_{z;m,s,\alpha}(\rho,q)$ depend on q^2, so that $E_{z;m,s,\alpha}(\rho,-q) = E_{z;m,s,\alpha}(\rho,q)$ and $\mathcal{H}_{z;m,s,\alpha}(\rho,-q) = \mathcal{H}_{z;m,s,\alpha}(\rho,q)$ for

---

* In some special cases the auxiliary transverse wavenumber q_α may become real in a loss-free plasma. For such a case, it is sufficient to include minute losses and then employ the formalism stated below. In the resulting formulas, of course, losses may be set equal to zero.

$\rho \leq a$. By making use of the evenness of functions $p_\alpha(q)$, $q_\alpha(q)$, and $n_{s,\alpha}^{(1,2)}(q)$ we then establish that the quantities $w_{m,s,\alpha}^{(1,2)}$ and $f_{m,s,\alpha}^{(1,2)}$, occurring in (6.35) and (6.39), are also even functions of q. With this point in mind, we apply the circuital relations (3.54) for the Hankel functions, along with the relation

$$J_m\left(e^{\mp i\pi}\,\zeta\right) = (-1)^m\,J_m(\zeta) \tag{6.45}$$

for the Bessel functions, to get

$$\begin{aligned}
C_{m,s,\alpha}^{(1)}(e^{-i\pi}\,q) &= (-1)^{m+1}\,C_{m,s,\alpha}^{(2)}(q),\\
C_{m,s,\alpha}^{(2)}(e^{i\pi}\,q) &= (-1)^{m+1}\,C_{m,s,\alpha}^{(1)}(q),\\
\Delta C_{m,s,a}(e^{\mp i\pi}\,q) &= (-1)^{m+1}\,\Delta C_{m,s,\alpha}(q),
\end{aligned} \tag{6.46}$$

where

$$\Delta C_{m,s,\alpha}(q) = C_{m,s,\alpha}^{(2)}(q) - C_{m,s,\alpha}^{(1)}(q) \tag{6.47}$$

and $m = 0,\pm 1,\pm 2,\dots$. Here it must be remembered that the Hankel functions $H_m^{(1,2)}(k_0 q\rho)$ involved in the expressions (6.35) and (6.39), as well as in the field representations (6.44) for the outer region, have a branch point at $q = 0$, with the branch cut passing from this point along the negative real q axis. Therefore, it is necessary to distinguish the negative quantities $e^{-i\pi}\,|q|$ and $e^{i\pi}\,|q|$ lying on different sides of the branch cut. Making the rearrangements

$$\begin{aligned}
C_{m,s,\alpha}^{(1)}&(e^{-i\pi}\,q)\,H_m^{(1)}(k_0\,e^{-i\pi}\,q\rho) + C_{m,s,\alpha}^{(2)}(e^{-i\pi}\,q)\,H_m^{(2)}(k_0\,e^{-i\pi}\,q\rho)\\
&= 2C_{m,s,\alpha}^{(1)}(e^{-i\pi}\,q)\,J_m(k_0\,e^{-i\pi}\,q\rho) + \Delta C_{m,s,\alpha}(e^{-i\pi}\,q)\,H_m^{(2)}(k_0\,e^{-i\pi}\,q\rho)
\end{aligned} \tag{6.48}$$

and

$$\begin{aligned}
C_{m,s,\alpha}^{(1)}&(e^{i\pi}\,q)\,H_m^{(1)}(k_0\,e^{i\pi}\,q\rho) + C_{m,s,\alpha}^{(2)}(e^{i\pi}\,q)\,H_m^{(2)}(k_0\,e^{i\pi}\,q\rho)\\
&= -\Delta C_{m,s,\alpha}(e^{i\pi}\,q)\,H_m^{(1)}(k_0\,e^{i\pi}\,q\rho) + 2C_{m,s,\alpha}^{(2)}(e^{i\pi}\,q)\,J_m(k_0\,e^{i\pi}\,q\rho)
\end{aligned} \tag{6.49}$$

in (6.44) and using (3.54) togeter with (6.45) to (6.47), we deduce that, in the outer region,

$$E_{z;\,m,s,\alpha}(\rho, e^{\mp i\pi}\,q) = E_{z;\,m,s,\alpha}(\rho, q),$$

$$\mathcal{H}_{z;\,m,s,\alpha}(\rho, e^{\mp i\pi}\,q) = \mathcal{H}_{z;\,m,s,\alpha}(\rho, q).$$

It is evident that similar relations can be found for the remaining field components, and we thus have

$$\begin{aligned}
\mathbf{E}_{m,s,\alpha}(\rho, e^{\mp i\pi}\,q) &= \mathbf{E}_{m,s,\alpha}(\rho, q),\\
\boldsymbol{\mathcal{H}}_{m,s,\alpha}(\rho, e^{\mp i\pi}\,q) &= \boldsymbol{\mathcal{H}}_{m,s,\alpha}(\rho, q).
\end{aligned} \tag{6.50}$$

It remains to determine the eigenvalues q ensuring the fulfillment of conditions (6.8). It can easily be verified from (6.44) that the field solutions corresponding to real values of q do satisfy the inequalities (6.8). Since the foregoing analysis shows that the negative real values of q do not give new solutions compared with the case of positive real q, we may consider only the positive values of q. The corresponding vector eigenfunctions $\mathbf{E}_{m,s,\alpha}(\rho,q)$, $\boldsymbol{\mathcal{H}}_{m,s,\alpha}(\rho,q)$ then give the field solutions

$$
\begin{aligned}
\mathbf{E}_{m,s,\alpha}(\mathbf{r},q) &= \mathbf{E}_{m,s,\alpha}(\rho,q)\exp[-im\phi - ik_0 p_{s,\alpha}(q)\,z], \\
\boldsymbol{\mathcal{H}}_{m,s,\alpha}(\mathbf{r},q) &= \boldsymbol{\mathcal{H}}_{m,s,\alpha}(\rho,q)\exp[-im\phi - ik_0 p_{s,\alpha}(q)\,z].
\end{aligned}
\tag{6.51}
$$

These solutions represent the 'modes' appropriate to the continuous part $0 \leq q < \infty$ of the eigenvalue spectrum and are commonly called the *improper modes*. The improper modes do not satisfy the radiation condition at infinity and, in contrast to the bound (proper) modes, cannot be individually excited by given sources. In what follows, it will be shown that only by superposing a continuum of improper modes can the physical field satisfying the radiation condition be yielded.

Along with real q, some discrete complex values of q can provide the solutions obeying conditions (6.8). Indeed, complex values $q = q_{m,n}$ which are roots of the equation

$$
C^{(1)}_{m,s,\alpha}(q) = 0,
\tag{6.52}
$$

and complex values $q = \tilde{q}_{m,n}$ which are roots of the equation

$$
C^{(2)}_{m,s,\alpha}(q) = 0,
\tag{6.53}
$$

give solutions satisfying (6.8) if $\mathrm{Im}(q_{m,n}) < 0$ and $\mathrm{Im}(\tilde{q}_{m,n}) > 0$ $(n = 1, 2, \ldots)$. From (6.46) one sees that $\tilde{q}_{m,n} = e^{i\pi}\, q_{m,n}$, so that

$$
\begin{aligned}
\mathbf{E}_{m,s,\alpha}(\rho,\tilde{q}_{m,n}) &= \mathbf{E}_{m,s,\alpha}(\rho,q_{m,n}), \\
\boldsymbol{\mathcal{H}}_{m,s,\alpha}(\rho,\tilde{q}_{m,n}) &= \boldsymbol{\mathcal{H}}_{m,s,\alpha}(\rho,q_{m,n}).
\end{aligned}
\tag{6.54}
$$

Bearing these relations in mind, we shall hereafter consider only the roots $q = q_{m,n}$ with $\mathrm{Im}(q_{m,n}) < 0$.

As a practical matter, these roots are found numerically by using the field equations in the non-dimensionalized form. Let coefficients $D_{m,s,\alpha}(q)$ and $C^{(1)}_{m,s,\alpha}(q)$ be denoted as $d^{(u)}_{m,s,\alpha}(q)$ and $c^{(u)}_{m,s,\alpha}(q)$ in the case when $u_{m,s,\alpha}(q) = 1$, $v_{m,s,\alpha}(q) = 0$, and as $d^{(v)}_{m,s,\alpha}(q)$ and $c^{(v)}_{m,s,\alpha}(q)$ in the case when $u_{m,s,\alpha}(q) = 0$, $v_{m,s,\alpha}(q) = 1$. The quantities $q_{n,m}$ satisfying Equations (6.52) and (6.43) simultaneously are then roots of the equation

$$
\begin{vmatrix}
d^{(u)}_{m,s,\alpha}(q) & d^{(v)}_{m,s,\alpha}(q) \\[2mm]
c^{(u)}_{m,s,\alpha}(q) & c^{(v)}_{m,s,\alpha}(q)
\end{vmatrix}
= d^{(u)}_{m,s,\alpha}(q)\, c^{(v)}_{m,s,\alpha}(q) - d^{(v)}_{m,s,\alpha}(q)\, c^{(u)}_{m,s,\alpha}(q) = 0.
$$

It is evident that the roots $q_{m,n}$ *correspond to proper modes*, that is, bound modes of a duct. Strictly speaking, if the quantity $q_{m,n}$ giving a proper mode with the propagation constant $p_\alpha(q_{m,n})$ is the solution of the eigenvalue equation (6.52), then the quantity $q_\alpha(q_{m,n})$ is also the solution of the same equation. Since $q_{m,n}$ and $q_\alpha(q_{m,n})$ correspond to the same eigenmode, there is an ambiguity in what quantity, $q_{m,n}$ or $q_\alpha(q_{m,n})$, may be chosen as the proper eigenvalue. To avoid this ambiguity, we shall adopt the following convention for the unique choice of the eigenvalue $q_{m,n}$:

$$|\operatorname{Im}(q_{m,n})| < \left|\operatorname{Im}\left(q_\alpha(q_{m,n})\right)\right|. \tag{6.55}$$

The label α ensuring the fulfillment of this convention will be denoted by $\hat{\alpha}$, so that $p_{m,n} = p_{\hat{\alpha}}(q_{m,n})$, a notation $p_{m,-n} = -p_{m,n}$ being understood. The corresponding vector eigenfunctions

$$\mathbf{E}_{m,\pm n}(\rho) = \mathbf{E}_{m,\pm,\hat{\alpha}}(\rho, q_{m,n}), \quad \mathcal{H}_{m,\pm n}(\rho) = \mathcal{H}_{m,\pm,\hat{\alpha}}(\rho, q_{m,n}) \tag{6.56}$$

then give the fields of proper modes, hereinafter denoted as

$$\begin{aligned}
\mathbf{E}_{m,\pm n}(\mathbf{r}) &= \mathbf{E}_{m,\pm n}(\rho)\exp[-im\phi \mp ik_0\,p_{m,n}\,z], \\
\mathcal{H}_{m,\pm n}(\mathbf{r}) &= \mathcal{H}_{m,\pm n}(\rho)\exp[-im\phi \mp ik_0\,p_{m,n}\,z].
\end{aligned} \tag{6.57}$$

Thus, we can assert that the eigenvalue spectrum consists of the discrete part $q = q_{m,n}$ and the continuous part $0 \le q < \infty$. Summation (integration) over all these eigenvalues yields the following eigenfunction expansion for the total field:

$$\mathbf{E}(\mathbf{r}) = \sum_{(-)}^{(+)} \sum_{m=-\infty}^{\infty} \left[\sum_n a_{m,n}\,\mathbf{E}_{m,n}(\mathbf{r}) + \sum_{\alpha=o}^{x} \int_0^\infty a_{m,s,\alpha}(q)\,\mathbf{E}_{m,s,\alpha}(\mathbf{r},q)\,\mathrm{d}q\right],$$

$$\mathcal{H}(\mathbf{r}) = \sum_{(-)}^{(+)} \sum_{m=-\infty}^{\infty} \left[\sum_n a_{m,n}\,\mathcal{H}_{m,n}(\mathbf{r}) + \sum_{\alpha=o}^{x} \int_0^\infty a_{m,s,\alpha}(q)\,\mathcal{H}_{m,s,\alpha}(\mathbf{r},q)\,\mathrm{d}q\right] \tag{6.58}$$

(Zaboronkova *et al.*, 1993 c; Kondrat'ev *et al.*, 1994, 1996). Here the summation marked by the sign $\sum_n$ is taken over the discrete eigenvalues, and the integration is extended over the continuous part of the eigenvalue spectrum; the sign $\sum_{(-)}^{(+)}$ denotes the summation over the forward-propagating waves ($n, s = $ ' $+$ ') and the backward-propagating waves ($-n, s = $ ' $-$ '). The quantities $a_{m,n}$ and $a_{m,s,\alpha}(q)$ are called the *excitation coefficients* (excitation factors) of modes belonging, respectively, to the discrete and continuous parts of the eigenvalue spectrum.

It should be noted that the eigenfunction expansion (6.58) is a generalization of the well-known modal field-representation for open isotropic waveguides (e. g.,

Shevchenko 1966, 1971 a; Snyder and Love, 1983) to the case of ducts in magnetized plasmas.

6.3. The eigenfunction set for the uniform duct

As mentioned in the preceding sections, for the general case, one has to resort to numerical integration of the source-free field equations in order to find the eigenfunctions of a duct. There is, however, a special case when the eigenfunctions can be represented in terms of known functions. This is possible if the plasma density N takes a constant value, $\tilde{N}$, in the core. Although the eigenfunctions for this case can be deduced from the general formulation of § 6.2, a simpler way is to obtain them using the general field solutions in a homogeneous plasma and applying the boundary conditions at $\rho = a$, along with the requisite conditions at $\rho = 0$ and $\rho \to \infty$. We show this for the axisymmetric field configurations when $m = 0$, omitting, for brevity, the index m.

In the inner region of the duct, the axisymmetric solution, regular at $\rho = 0$, can be written, from (4.58):

$$E_{\phi;\,s,\alpha}(\rho, q) = \mathrm{i} \sum_{k=1}^{2} B_{s,\alpha}^{(k)}(q)\, J_1(k_0\, \tilde{q}_\alpha^{(k)}\, \rho),$$

$$\mathcal{H}_{\phi;\,s,\alpha}(\rho, q) = -\sum_{k=1}^{2} B_{s,\alpha}^{(k)}(q)\, \tilde{n}_{s,\alpha}^{(k)}\, J_1(k_0\, \tilde{q}_\alpha^{(k)}\, \rho),$$

(6.59)

where

$$\tilde{n}_{s,\alpha}^{(k)} = -\tilde{\varepsilon} \left[\left(\tilde{q}_\alpha^{(k)} \right)^2 + p_\alpha^2(q) + \frac{\tilde{g}^2}{\tilde{\varepsilon}} - \tilde{\varepsilon} \right] \Big/ \tilde{g}\, p_{s,\alpha}(q), \quad \tilde{q}_\alpha^{(k)} = q_k(p_\alpha(q), \tilde{N}),$$

and the other notations are the same as in (4.40) and (4.47). In the outer region, the solution is written thus:

$$E_{\phi;\,s,\alpha}(\rho, q) = \mathrm{i} \left[\sum_{k=1}^{2} C_{s,\alpha}^{(k)}(q)\, H_1^{(k)}(k_0 q\rho) + C_{s,\alpha}(q)\, H_1^{(2)}(k_0\, q_\alpha\, \rho) \right],$$

$$\mathcal{H}_{\phi;\,s,\alpha}(\rho, q) = -\left[\sum_{k=1}^{2} C_{s,\alpha}^{(k)}(q)\, n_{s,\alpha}^{(1)}\, H_1^{(k)}(k_0 q\rho) + C_{s,\alpha}(q)\, n_{s,\alpha}^{(2)}\, H_1^{(2)}(k_0\, q_\alpha\, \rho) \right]. \quad (6.60)$$

The remaining field components are readily found by analogy with (4.58) and (4.59). The above expressions automatically provide the required values of ratios $u_{s,\alpha}/v_{s,\alpha}$ for $\alpha = $ 'o', 'x', which assure the condition (6.43) for $D_{s,\alpha}(q)$. In the case of a nonuniform duct, this condition is satisfied through the special procedure of numerical finding $u_{s,\alpha}(q)/v_{s,\alpha}(q)$, as discussed in § 6.2.

By applying boundary conditions such as (4.65), we arrive at four simultaneous equations in the five quantities $B_{s,\alpha}^{(1)}$, $B_{s,\alpha}^{(2)}$, $C_{s,\alpha}^{(1)}$, $C_{s,\alpha}^{(2)}$, and $C_{s,\alpha}$. The equations can be written concisely in matrix form thus:

$$
\begin{pmatrix}
J_1(\tilde{Q}_\alpha^{(1)}) & J_1(\tilde{Q}_\alpha^{(2)}) & -H_1^{(2)}(Q) & -H_1^{(2)}(Q_\alpha) \\[2ex]
\tilde{n}_{s,\alpha}^{(1)}\tilde{Q}_\alpha^{(1)} J_0(\tilde{Q}_\alpha^{(1)}) & \tilde{n}_{s,\alpha}^{(2)}\tilde{Q}_\alpha^{(2)} J_0(\tilde{Q}_\alpha^{(2)}) & -\dfrac{\tilde{\eta}}{\eta_a}\, n_{s,\alpha}^{(1)} Q H_0^{(2)}(Q) & -\dfrac{\tilde{\eta}}{\eta_a}\, n_{s,\alpha}^{(2)} Q_\alpha H_0^{(2)}(Q_\alpha) \\[2ex]
\tilde{n}_{s,\alpha}^{(1)} J_1(\tilde{Q}_\alpha^{(1)}) & \tilde{n}_{s,\alpha}^{(2)} J_1(\tilde{Q}_\alpha^{(2)}) & -n_{s,\alpha}^{(1)} H_1^{(2)}(Q) & -n_{s,\alpha}^{(2)} H_1^{(2)}(Q_\alpha) \\[2ex]
\tilde{Q}_\alpha^{(1)} J_0(\tilde{Q}_\alpha^{(1)}) & \tilde{Q}_\alpha^{(2)} J_0(\tilde{Q}_\alpha^{(2)}) & -Q H_0^{(2)}(Q) & -Q_\alpha H_0^{(2)}(Q_\alpha)
\end{pmatrix}
$$

$$
\cdot
\begin{pmatrix}
B_{s,\alpha}^{(1)} \\[1.5ex]
B_{s,\alpha}^{(2)} \\[1.5ex]
C_{s,\alpha}^{(2)} \\[1.5ex]
C_{s,\alpha}
\end{pmatrix}
= C_{s,\alpha}^{(1)}
\begin{pmatrix}
H_1^{(1)}(Q) \\[1.5ex]
\dfrac{\tilde{\eta}}{\eta_a}\, n_{s,\alpha}^{(1)} Q H_0^{(1)}(Q) \\[1.5ex]
n_{s,\alpha}^{(1)} H_1^{(1)}(Q) \\[1.5ex]
Q H_0^{(1)}(Q)
\end{pmatrix},
\tag{6.61}
$$

where

$$
\tilde{Q}_\alpha^{(1,2)} = k_0 a\, \tilde{q}_\alpha^{(1,2)}, \quad Q = k_0 a q, \quad Q_\alpha = k_0 a q_\alpha.
$$

It is clear that one of the above five quantities, $C_{s,\alpha}^{(1)}$ say, may be set equal to an arbitrary factor independent of ρ. This is because the eigenfunctions are defined up to such a factor. Now let $\tilde{\Delta}$ be a determinant of the matrix on the left side of Equation (6.61). One can readily verify that the equation $\tilde{\Delta} = 0$ reduces exactly to the dispersion equation (4.71) for the axisymmetric modes of the uniform duct. Therefore it is convenient to choose of the above-mentioned arbitrary factor in the field solution so as to ensure the following normalization:

$$
C_{s,\alpha}^{(1)}(q) = -\tilde{\Delta}\left[(n_{s,\alpha}^{(1)} - n_{s,\alpha}^{(2)})(\tilde{n}_{s,\alpha}^{(1)} - \tilde{n}_{s,\alpha}^{(2)})\, \tilde{Q}_\alpha^{(1)} J_0(\tilde{Q}_\alpha^{(1)})\, \tilde{Q}_\alpha^{(2)} J_0(\tilde{Q}_\alpha^{(2)})\right]^{-1}.
$$

We now see that a zero of $\tilde{\Delta}$ immediately provides the fulfillment of the mode-determining equation (6.52), as required. Upon calculating $\tilde{\Delta}$, the coefficients $B_{s,\alpha}^{(1,2)}$, $C_{s,\alpha}^{(1,2)}$, and $C_{s,\alpha}$ after some algebra can be written (Kondrat'ev *et al.*, 1994, 1996):

$$
\begin{aligned}
B_{s,\alpha}^{(1,2)}(q) = {} & 4\mathrm{i}\left[(\tilde{n}_{s,\alpha}^{(2,1)} - n_{s,\alpha}^{(1)})\frac{\tilde{\eta}}{\eta_a} J(\tilde{Q}_\alpha^{(2,1)}) - \left(\tilde{n}_{s,\alpha}^{(2,1)} - \frac{\tilde{\eta}}{\eta_a} n_{s,\alpha}^{(1)}\right) H^{(2)}(Q_\alpha)\right] \\
& \times\; Q_\alpha H_0^{(2)}(Q_\alpha)\left[\pi\,(\tilde{n}_{s,\alpha}^{(1,2)} - \tilde{n}_{s,\alpha}^{(2,1)})\, \tilde{Q}_\alpha^{(1,2)} J_0(\tilde{Q}_\alpha^{(1,2)})\right]^{-1},
\end{aligned}
\tag{6.62}
$$

$$
\begin{aligned}
C_{s,\alpha}^{(1,2)}(q) \;=\; &\Big\{ [\mathcal{M}_{12}\,\mathcal{L}_{12}\,H^{(2)}(Q_\alpha) - \mathcal{M}_{11}\,\mathcal{L}_{22}\,H^{(2,1)}(Q)]\,J(\tilde{Q}_\alpha^{(1)}) \\
&+ \; [\mathcal{M}_{21}\,\mathcal{L}_{21}\,H^{(2,1)}(Q) - \mathcal{M}_{22}\,\mathcal{L}_{11}\,H^{(2)}(Q_\alpha)]\,J(\tilde{Q}_\alpha^{(2)}) \\
&- \; \tilde{\eta}\,\eta_a^{-1}\,J(\tilde{Q}_\alpha^{(1)})\,J(\tilde{Q}_\alpha^{(2)}) - H^{(2,1)}(Q)\,H^{(2)}(Q_\alpha) \Big\} \\
&\times \; \chi^{(1,2)}\,Q\,H_0^{(2,1)}(Q)\,Q_\alpha\,H_0^{(2)}(Q_\alpha),
\end{aligned}
\tag{6.63}
$$

$$
C_{s,\alpha}(q) = 4\mathrm{i}\pi^{-1}\Big[\mathcal{M}_{21}\,\mathcal{L}_{11}\,J(\tilde{Q}_\alpha^{(2)}) - \mathcal{M}_{11}\,\mathcal{L}_{12}\,J(\tilde{Q}_\alpha^{(1)}) \Big],
\tag{6.64}
$$

where

$$
\begin{aligned}
\mathcal{M}_{ij} &= \mathcal{M}_0(\tilde{n}_{s,\alpha}^{(i)} - n_{s,\alpha}^{(j)}), \quad \mathcal{M}_0 = (n_{s,\alpha}^{(1)} - n_{s,\alpha}^{(2)})^{-1}(\tilde{n}_{s,\alpha}^{(1)} - \tilde{n}_{s,\alpha}^{(2)})^{-1}, \\
\mathcal{L}_{ij} &= \frac{\tilde{\eta}}{\eta_a}\,n_{s,\alpha}^{(i)} - \tilde{n}_{s,\alpha}^{(j)}, \quad i,j = 1,2, \\
\chi^{(1)} &= -1, \quad \chi^{(2)} = 1, \\
J(\zeta) &= \frac{J_1(\zeta)}{\zeta\,J_0(\zeta)}, \quad H^{(k)}(\zeta) = \frac{H_1^{(k)}(\zeta)}{\zeta\,H_0^{(k)}(\zeta)}.
\end{aligned}
\tag{6.65}
$$

Once the expressions for $B_{s,\alpha}^{(1,2)}$, $C_{s,\alpha}^{(1,2)}$, and $C_{s,\alpha}$ are known, the eigenfunctions of a uniform duct are fully determined. It is important that the proper modes are obtained again with the help of Equation (6.52) which yields all of the discrete eigenvalues. The continuous improper modes correspond, as before, to the continuum $0 \le q < \infty$.

The theory of this section can be formulated without the simplification $m = 0$, but the algebra is more complicated. Obtaining the expressions for the eigenfunctions of a uniform duct in the case when $m \ne 0$, may be left as an exercise for the reader.

6.4. Mode orthogonality

It can be shown that the modes of the discrete and continuous spectrum satisfy the following *orthogonality conditions*:

$$
\begin{aligned}
J_{\tilde{m},\tilde{n}}^{m,n} \;&=\; \int_0^{2\pi} \mathrm{d}\phi \int_0^{\infty} \Big(\mathbf{E}_{m,n}(\mathbf{r}) \times \mathcal{H}_{\tilde{m},\tilde{n}}^{(\mathrm{T})}(\mathbf{r}) - \mathbf{E}_{\tilde{m},\tilde{n}}^{(\mathrm{T})}(\mathbf{r}) \times \mathcal{H}_{m,n}(\mathbf{r}) \Big) \cdot \hat{\mathbf{z}}_0 \rho\, \mathrm{d}\rho \\
&=\; N_{m,n}\,\delta_{m,-\tilde{m}}\,\delta_{n,-\tilde{n}},
\end{aligned}
\tag{6.66}
$$

$$
J_{\tilde{m},\tilde{n}}^{m,s,\alpha} = \int_0^{2\pi} \mathrm{d}\phi \int_0^{\infty} \Big(\mathbf{E}_{m,s,\alpha}(\mathbf{r},q) \times \mathcal{H}_{\tilde{m},\tilde{n}}^{(\mathrm{T})}(\mathbf{r}) - \mathbf{E}_{\tilde{m},\tilde{n}}^{(\mathrm{T})}(\mathbf{r}) \times \mathcal{H}_{m,s,\alpha}(\mathbf{r},q) \Big) \cdot \hat{\mathbf{z}}_0 \rho\, \mathrm{d}\rho = 0,
\tag{6.67}
$$

$$J_{\tilde{m},\tilde{s},\tilde{\alpha}}^{m,s,\alpha} = \int_0^{2\pi} \mathrm{d}\phi \int_0^{\infty} \left(\mathbf{E}_{m,s,\alpha}(\mathbf{r},q) \times \mathcal{H}_{\tilde{m},\tilde{s},\tilde{\alpha}}^{(\mathrm{T})}(\mathbf{r},\tilde{q}) \right. \tag{6.68}$$

$$\left. - \mathbf{E}_{\tilde{m},\tilde{s},\tilde{\alpha}}^{(\mathrm{T})}(\mathbf{r},\tilde{q}) \times \mathcal{H}_{m,s,\alpha}(\mathbf{r},q) \right) \cdot \hat{\mathbf{z}}_0 \rho\, \mathrm{d}\rho = N_{m,s,\alpha}(q)\, \delta(q-\tilde{q})\, \delta_{m,-\tilde{m}}\, \delta_{s,-\tilde{s}}\, \delta_{\alpha,\tilde{\alpha}},$$

where

$$N_{m,s,\alpha}(q) \;=\; \sigma_s\, N_{m,\alpha}(q) = -\sigma_s\, \frac{16\pi}{k_0^2} \left(\frac{\mathrm{d}p_\alpha}{\mathrm{d}q} \right)^{-1}$$

$$\times \; \left(1 + \frac{n_{s,\alpha}^{(1)^2}}{\eta_a} \right) C_{m,s,\alpha}^{(1)}(q)\, C_{m,s,\alpha}^{(2)}(q), \quad \sigma_{\pm} = \pm 1, \tag{6.69}$$

$$N_{m,n} = \frac{1}{2\pi\mathrm{i}} \left. \frac{\mathrm{d}N_{m,\hat{\alpha}}}{\mathrm{d}q} \right|_{q=q_{m,n}}. \tag{6.70}$$

In the above, $N_{m,n}$ is the *normalization* for the forward-propagating proper mode with its indices m, n, and $N_{m,\alpha}(q)$ is the normalization for the forward-propagating improper mode with its indices m, α. The symbol "T" stands, as before, for the "transposed" medium, $\tilde{s} = (\pm)$, and $\tilde{\alpha} = $ 'o', 'x'. Recall that δ is the Dirac delta-function, and $\delta_{i,j}$ is the Kronecker delta, i.e. $\delta_{i,j} = 1$ if $i = j$ and $\delta_{i,j} = 0$ otherwise.

To prove the above orthogonality conditions, we apply the relation (6.18) where $\hat{\varepsilon}_{\mathrm{I}} = \hat{\varepsilon}$ and $\hat{\varepsilon}_{\mathrm{II}} = \hat{\varepsilon}^{(\mathrm{T})}$. In this case, the right side of (6.18) can be shown to be equal to zero, so that (6.18), with the help of (6.19), reduces to

$$\frac{\partial}{\partial z} \int_\Sigma \left(\mathbf{E}_{\mathrm{I}} \times \mathcal{H}_{\mathrm{II}}^{(\mathrm{T})} - \mathbf{E}_{\mathrm{II}}^{(\mathrm{T})} \times \mathcal{H}_{\mathrm{I}} \right) \cdot \hat{\mathbf{z}}_0\, \mathrm{d}\sigma$$

$$= - \oint_L \left(\mathbf{E}_{\mathrm{I}} \times \mathcal{H}_{\mathrm{II}}^{(\mathrm{T})} - \mathbf{E}_{\mathrm{II}}^{(\mathrm{T})} \times \mathcal{H}_{\mathrm{I}} \right) \cdot \hat{\mathbf{n}}\, \mathrm{d}\ell. \tag{6.71}$$

Setting $\mathbf{E}_{\mathrm{I}} = \mathbf{E}_{m,n}(\mathbf{r})$, $\mathcal{H}_{\mathrm{I}} = \mathcal{H}_{m,n}(\mathbf{r})$ and $\mathbf{E}_{\mathrm{II}}^{(\mathrm{T})} = \mathbf{E}_{\tilde{m},\tilde{n}}^{(\mathrm{T})}(\mathbf{r})$, $\mathcal{H}_{\mathrm{II}}^{(\mathrm{T})} = \mathcal{H}_{\tilde{m},\tilde{n}}^{(\mathrm{T})}(\mathbf{r})$, where

$$\mathbf{E}_{m,n}(\mathbf{r}) = \mathbf{E}_{m,n}(\rho) \exp(-\mathrm{i}m\phi - \mathrm{i}k_0\, p_{m,n}\, z),$$

$$\mathcal{H}_{m,n}(\mathbf{r}) = \mathcal{H}_{m,n}(\rho) \exp(-\mathrm{i}m\phi - \mathrm{i}k_0\, p_{m,n}\, z)$$

and

$$\mathbf{E}_{\tilde{m},\tilde{n}}^{(\mathrm{T})}(\mathbf{r}) = \mathbf{E}_{\tilde{m},\tilde{n}}^{(\mathrm{T})}(\rho) \exp(-\mathrm{i}\tilde{m}\phi - \mathrm{i}k_0\, p_{\tilde{m},\tilde{n}}^{(\mathrm{T})}\, z),$$

$$\mathcal{H}_{\tilde{m},\tilde{n}}^{(\mathrm{T})}(\mathbf{r}) = \mathcal{H}_{\tilde{m},\tilde{n}}^{(\mathrm{T})}(\rho) \exp(-\mathrm{i}\tilde{m}\phi - \mathrm{i}k_0\, p_{\tilde{m},\tilde{n}}^{(\mathrm{T})}\, z),$$

we perform integration in (6.71) over a cross-sectional circular area Σ of an infinite radius, making use of the fact that in the cylindrical coordinates $\mathrm{d}\sigma = \rho\, \mathrm{d}\rho\, \mathrm{d}\phi$. After some algebra we get

$$J_{\tilde{m},\tilde{n}}^{m,n} \;=\; \int_0^{2\pi} \exp\left[-\mathrm{i}\,(m+\tilde{m})\,\phi \right] \mathrm{d}\phi \, \lim_{\rho\to\infty} \frac{\exp[-\mathrm{i}k_0\,(p_{m,n} - p_{-\tilde{m},-\tilde{n}})\,z]}{\mathrm{i}k_0\,(p_{m,n} - p_{-\tilde{m},-\tilde{n}})}$$

$$\times \;\; \rho\Big(\mathbf{E}_{m,n}(\rho) \times \mathcal{H}_{\tilde{m},\tilde{n}}^{(\mathrm{T})}(\rho) - \mathbf{E}_{\tilde{m},\tilde{n}}^{(\mathrm{T})}(\rho) \times \mathcal{H}_{m,n}(\rho) \Big) \cdot \hat{\boldsymbol{\rho}}_0. \tag{6.72}$$

Here we made use of the fact that $p_{\tilde{m},\tilde{n}}^{(T)} = p_{-\tilde{m},\tilde{n}} = -p_{-\tilde{m},-\tilde{n}}$. These relationships, though intuitively clear, can be deduced from the dispersion equation for proper modes. Noting that

$$\int_0^{2\pi} \exp\left[-\,i\,(m + \tilde{m})\,\phi\right] d\phi = 2\pi\,\delta_{m,-\tilde{m}} \tag{6.73}$$

and observing that the fields of proper modes vanish with ρ, one sees that the right side of (6.72) equals zero if $p_{m,n} \neq p_{-\tilde{m},-\tilde{n}}$. In the absence of mode degeneration* this implies that $J_{\tilde{m},\tilde{n}}^{m,n} = 0$ for $m \neq -\tilde{m}$ and $n \neq -\tilde{n}$. By using (6.7) one finds that $N_{m,-n} = -N_{m,n}$. We have thus proved the relation (6.66), which expresses the orthogonality condition for proper modes.

The relation (6.67) may be established analogously if we set

$$\mathbf{E}_{\mathrm{I}} = \mathbf{E}_{m,s,\alpha}(\rho, q)\exp[-im\phi - ik_0\,p_{s,\alpha}(q)\,z],$$

$$\boldsymbol{\mathcal{H}}_{\mathrm{I}} = \boldsymbol{\mathcal{H}}_{m,s,\alpha}(\rho, q)\exp[-im\phi - ik_0\,p_{s,\alpha}(q)\,z]$$

and take the field $\mathbf{E}_{\mathrm{II}}^{(T)}$, $\boldsymbol{\mathcal{H}}_{\mathrm{II}}^{(T)}$ as that in the preceding derivation. Following the above procedure and making use of the fact that $p_{s,\alpha}(q) \neq p_{-\tilde{m},-\tilde{n}}$ for all *real values* of q, one can readily obtain the condition (6.67). Hence each proper mode is orthogonal to each of improper modes, i. e., modes of the continuous spectrum, and, therefore, to the total field radiating outwards.

Next, to derive the relation (6.68), we take the last form for $\mathbf{E}_{\mathrm{I}}$, $\boldsymbol{\mathcal{H}}_{\mathrm{I}}$ and set

$$\mathbf{E}_{\mathrm{II}}^{(T)} = \mathbf{E}_{\tilde{m},\tilde{s},\tilde{\alpha}}^{(T)}(\rho, \tilde{q})\exp[-i\tilde{m}\phi - ik_0\,p_{\tilde{s},\tilde{\alpha}}(\tilde{q})\,z],$$

$$\boldsymbol{\mathcal{H}}_{\mathrm{II}}^{(T)} = \boldsymbol{\mathcal{H}}_{\tilde{m},\tilde{s},\tilde{\alpha}}^{(T)}(\rho, \tilde{q})\exp[-i\tilde{m}\phi - ik_0\,p_{\tilde{s},\tilde{\alpha}}(\tilde{q})\,z],$$

and then resort to the same procedure. We thus find

$$J_{\tilde{m},\tilde{s},\tilde{\alpha}}^{m,s,\alpha} = \int_0^{2\pi} \exp\left[-\,i\,(m+\tilde{m})\,\phi\right] d\phi \ \lim_{\rho \to \infty} \frac{\exp\left\{-ik_0\left[p_{s,\alpha}(q) - p_{-\tilde{s},\tilde{\alpha}}(\tilde{q})\right]z\right\}}{ik_0\left[p_{s,\alpha}(q) - p_{-\tilde{s},\tilde{\alpha}}(\tilde{q})\right]}$$

$$\times \rho\left(\mathbf{E}_{m,s,\alpha}(\rho, q) \times \boldsymbol{\mathcal{H}}_{\tilde{m},\tilde{s},\tilde{\alpha}}^{(T)}(\rho, \tilde{q}) - \mathbf{E}_{\tilde{m},\tilde{s},\tilde{\alpha}}^{(T)}(\rho, \tilde{q}) \times \boldsymbol{\mathcal{H}}_{m,s,\alpha}(\rho, q)\right) \cdot \hat{\boldsymbol{\rho}}_0. \tag{6.74}$$

First, consider the case when the conditions $m = -\tilde{m}$, $s = -\tilde{s}$, and $\alpha = \tilde{\alpha}$ are simultaneously satisfied. By inspection of Equations (4.11), (4.13), (4.14), and (4.15) it is noted that the following convention can be adopted without loss of generality:

$$E_{\phi;\,-m,-s,\alpha}^{(T)}(\rho, q) = E_{\phi;\,m,s,\alpha}(\rho, q), \quad E_{z;\,-m,-s,\alpha}^{(T)}(\rho, q) = E_{z;\,m,s,\alpha}(\rho, q),$$

$$\mathcal{H}_{\phi;\,-m,-s,\alpha}^{(T)}(\rho, q) = \mathcal{H}_{\phi;\,m,s,\alpha}(\rho, q), \quad \mathcal{H}_{z;\,-m,-s,\alpha}^{(T)}(\rho, q) = \mathcal{H}_{z;\,m,s,\alpha}(\rho, q). \tag{6.75}$$

---

* By mode degeneration, we mean a situation when two modes with different sets of indices m, n and, consequently, different field structures have the same propagation constants. For the degenerate modes, one needs an additional orthogonalization procedure.

Using the relationships (6.75) together with the large-argument approximations of the Hankel functions $H_m^{(1,2)}(k_0 q \rho)$ and $H_m^{(2)}(k_0 q_\alpha \rho)$, one obtains:

$$J_{-m,-s,\alpha}^{m,s,\alpha} = \lim_{\rho \to \infty} \frac{4(\tilde{q}-q)\exp\left\{-ik_0[p_{s,\alpha}(q)-p_{s,\alpha}(\tilde{q})]z\right\}}{k_0^2(q\,\tilde{q})^{1/2}\,[p_{s,\alpha}(q)-p_{s,\alpha}(\tilde{q})]}\left[1+\frac{1}{\eta_a}\,n_{s,\alpha}^{(1)}(q)\,n_{s,\alpha}^{(1)}(\tilde{q})\right]$$

$$\times\left\{(q+\tilde{q})\left[C_{m,s,\alpha}^{(1)}(q)\,C_{m,s,\alpha}^{(2)}(\tilde{q})+C_{m,s,\alpha}^{(1)}(\tilde{q})\,C_{m,s,\alpha}^{(2)}(q)\right]\frac{\sin\left(k_0\rho\,(q-\tilde{q})\right)}{q-\tilde{q}}\right.$$

$$-\,i\,(q+\tilde{q})\left[C_{m,s,\alpha}^{(1)}(q)\,C_{m,s,\alpha}^{(2)}(\tilde{q})-C_{m,s,\alpha}^{(1)}(\tilde{q})\,C_{m,s,\alpha}^{(2)}(q)\right]\frac{\cos\left(k_0\rho\,(q-\tilde{q})\right)}{q-\tilde{q}}$$

$$+\,i\,(-1)^m\left[C_{m,s,\alpha}^{(1)}(q)\,C_{m,s,\alpha}^{(1)}(\tilde{q})-C_{m,s,\alpha}^{(2)}(q)\,C_{m,s,\alpha}^{(2)}(\tilde{q})\right]\sin\left(k_0\rho\,(q+\tilde{q})\right)$$

$$\left.+\,(-1)^m\left[C_{m,s,\alpha}^{(1)}(q)\,C_{m,s,\alpha}^{(1)}(\tilde{q})+C_{m,s,\alpha}^{(2)}(q)\,C_{m,s,\alpha}^{(2)}(\tilde{q})\right]\cos\left(k_0\rho\,(q+\tilde{q})\right)\right\}. \quad (6.76)$$

In the above, use was made of the fact that the functions $H_n^{(2)}(k_0 q_\alpha \rho)$ vanish with ρ because of the condition $\mathrm{Im}\,(q_\alpha) < 0$. Passing, weakly, to the limit $\rho \to \infty$ in (6.76), as is made in the theory of distributions (Schwartz, 1950), and using the well-known relation

$$\delta(\xi) = \lim_{R\to\infty}\frac{\sin R\xi}{\pi\xi}, \qquad (6.77)$$

we obtain from (6.76) the orthogonality condition (6.68) for improper modes, with $N_{m,s,\alpha}(q)$ being given by (6.69). In the other possible cases, when either $s \neq -\tilde{s}$ and $\alpha \neq \tilde{\alpha}$ (where $p_{s,\alpha}(q) \neq p_{-\tilde{s},\tilde{\alpha}}(\tilde{q})$ for all real values of q and $\tilde{q}$) or $m \neq -\tilde{m}$, we have the trivial result for $J_{\tilde{m},\tilde{s},\tilde{\alpha}}^{m,s,\alpha}$, as expressed by (6.68).

We note that improper modes of the continuous spectrum possessing the same eigenvalues q and indices s, α but different azimuthal indices are degenerate as they have the same propagation constant $p = p_{s,\alpha}(q)$. Nevertheless, any two such modes are orthogonal due to the "angular" orthogonality (6.73).

It now remains to establish the relation (6.70) for $N_{m,n}$. The normalization $N_{m,n}$ can be represented by

$$\begin{aligned}N_{m,n} \;=\; & 2\pi\lim_{R\to\infty}\lim_{q\to q_{m,n}}\int_0^R\left(\mathbf{E}_{m,n}(\mathbf{r})\times\boldsymbol{\mathcal{H}}_{-m,-,\hat{\alpha}}^{(\mathrm{T})}(\mathbf{r},q)\right.\\ & -\;\mathbf{E}_{-m,-,\hat{\alpha}}^{(\mathrm{T})}(\mathbf{r},q)\times\boldsymbol{\mathcal{H}}_{m,n}(\mathbf{r})\Big)\cdot\hat{\mathbf{z}}_0\rho\,\mathrm{d}\rho.\end{aligned} \qquad (6.78)$$

Following the steps analogous to those that led to (6.72), and making use of the fact that $p_{m,n} = p_{\hat{\alpha}}(q_{m,n})$, we rewrite (6.78) as

$$N_{m,n} = 2\pi \lim_{R\to\infty} \lim_{q\to q_{m,n}} \frac{\exp\left\{-ik_0\left[p_{\hat{\alpha}}(q_{m,n}) - p_{\hat{\alpha}}(q)\right]z\right\}}{ik_0\left[p_{\hat{\alpha}}(q_{m,n}) - p_{\hat{\alpha}}(q)\right]}$$

$$\times \rho\left(\mathbf{E}_{m,n}(\rho)\times\mathcal{H}^{(\mathrm{T})}_{-m,-,\hat{\alpha}}(\rho,q) - \mathbf{E}^{(\mathrm{T})}_{-m,-,\hat{\alpha}}(\rho,q)\times\mathcal{H}_{m,n}(\rho)\right)\cdot\hat{\rho}_0 \Bigg|_{\rho=R}. \qquad (6.79)$$

Performing the operations indicated here and taking into account the relationships (6.75) and the chain of inequalities $\mathrm{Im}\, q_{\hat{\alpha}}(q_{m,n}) < \mathrm{Im}\, q_{m,n} < 0$ (see (6.55)), we find

$$N_{m,n} = \frac{8i}{k_0^2}\left(\frac{\mathrm{d}p_{\hat{\alpha}}}{\mathrm{d}q}\right)^{-1}\left(1 + \frac{n^{(1)^2}_{s,\hat{\alpha}}}{\eta_a}\right)\frac{\mathrm{d}C^{(1)}_{m,s,\hat{\alpha}}}{\mathrm{d}q}\, C^{(2)}_{m,s,\hat{\alpha}}\Bigg|_{q=q_{m,n}}. \qquad (6.80)$$

Using (6.52) and (6.69), we come from (6.80) to the final expression (6.70) for $N_{m,n}$.

The orthogonality conditions (6.66) to (6.68), along with relations (6.69) and (6.70), are valid for both absorbing and nonabsorbing ducts. These conditions have diverse application, ranging from the calculation of mode excitation coefficients due to current sources, to the formulation of coupled mode equations for axially nonuniform ducts.

In nonabsorbing ducts, for proper modes with *purely real* propagation constants, the so-called *power orthogonality* takes place. That is

$$P^{m,n}_{\tilde{m},\tilde{n}} = \frac{1}{2}\,\mathrm{Re}\int_0^{2\pi}\mathrm{d}\phi\int_0^\infty\left(\mathbf{E}_{m,n}(\mathbf{r})\times\mathcal{H}^*_{\tilde{m},\tilde{n}}(\mathbf{r}) + \mathbf{E}_{\tilde{m},\tilde{n}}(\mathbf{r})\times\mathcal{H}^*_{m,n}(\mathbf{r})\right)\cdot\hat{\mathbf{z}}_0\rho\,\mathrm{d}\rho = 0$$
$$(6.81)$$

for $m\neq\tilde{m}$, $n\neq\tilde{n}$.

This can be obtained if we first set $\hat{\varepsilon}_{\mathrm{I;II}} = \hat{\varepsilon}$ and $\mathbf{E}_{\mathrm{I;II}} = \mathbf{E}_{m,n;\,\tilde{m},\tilde{n}}(\mathbf{r})$, $\mathcal{H}_{\mathrm{I;II}} = \mathcal{H}_{m,n;\,\tilde{m},\tilde{n}}(\mathbf{r})$ in Equations (6.14) to (6.17), perform the operation of complex conjugation in equations for $\mathbf{E}_{\tilde{m},\tilde{n}}$, $\mathcal{H}_{\tilde{m},\tilde{n}}$ and then follow a procedure similar to that used in deriving (6.71) and (6.72). By taking into account the evident identity

$$\mathrm{Re}\left[\left(\mathbf{E}_{\tilde{m},\tilde{n}}(\mathbf{r})\times\mathcal{H}^*_{m,n}(\mathbf{r})\right)\cdot\hat{\mathbf{z}}_0\right] = \mathrm{Re}\left[\left(\mathbf{E}^*_{\tilde{m},\tilde{n}}(\mathbf{r})\times\mathcal{H}_{m,n}(\mathbf{r})\right)\cdot\hat{\mathbf{z}}_0\right],$$

it is readily found that

$$P^{m,n}_{\tilde{m},\tilde{n}} = \frac{1}{2}\lim_{\rho\to\infty}\mathrm{Re}\left\{\int_0^{2\pi}\frac{\rho}{ik_0\left(p_{m,n} - p_{\tilde{m},\tilde{n}}\right)}\right.$$

$$\left.\times\left(\mathbf{E}_{m,n}(\mathbf{r})\times\mathcal{H}^*_{\tilde{m},\tilde{n}}(\mathbf{r}) + \mathbf{E}^*_{\tilde{m},\tilde{n}}(\mathbf{r})\times\mathcal{H}_{m,n}(\mathbf{r})\right)\cdot\hat{\boldsymbol{\rho}}_0\,\mathrm{d}\phi\right\}. \qquad (6.82)$$

For the reasons discussed above, the expression on the right of (6.82) vanishes if $p_{m,n}\neq p_{\tilde{m},n}$ (i.e., $m\neq\tilde{m}$, $n=\tilde{n}$). Hence

$$P^{m,n}_{\tilde{m},\tilde{n}} = 2P_{m,n}\,\delta_{m,\tilde{m}}\,\delta_{n,\tilde{n}}, \qquad (6.83)$$

where $P_{m,n}$ is the power transported along the duct by a proper mode with its indices m, n. The result given by (6.83) allows for adding the "partial" powers transported by separate proper modes in order to find the total power flow associated with them.

6.5. Calculation of modal excitation coefficients

To determine the excitation coefficients of proper modes belonging to the discrete spectrum and of improper modes belonging to the continuous spectrum, we first establish one important integral relationship between two field solutions of inhomogeneous Maxwell's equations for the case of a gyrotropic medium, assuming that both fields as well as their sources vary with time with the same frequency. Let some electric and magnetic currents with their densities $\mathcal{J}_I^e$ and $\mathbf{J}_I^m$ are specified in a medium described by a dielectric tensor $\hat{\varepsilon}(\mathbf{r})$, and let $\mathbf{E}_I$, $\mathcal{H}_I$ be a source-excited field in this medium. Let some other current sources with their densities $\mathcal{J}_{II}^e$ and $\mathbf{J}_{II}^m$ are specified in a medium described by the transposed tensor $\hat{\varepsilon}^{(T)}(\mathbf{r})$, and let $\mathbf{E}_{II}^{(T)}$, $\mathcal{H}_{II}^{(T)}$ be a source-excited field there.

Then, by writing inhomogeneous Maxwell's equations for the two fields, $\mathbf{E}_I$, $\mathcal{H}_I$ and $\mathbf{E}_{II}^{(T)}$, $\mathcal{H}_{II}^{(T)}$, and making operations similar to those used for derivation of (6.18), it is found that

$$\nabla \cdot \left(\mathbf{E}_I \times \mathcal{H}_{II}^{(T)} - \mathbf{E}_{II}^{(T)} \times \mathcal{H}_I \right) = \mathcal{J}_I^e \cdot \mathbf{E}_{II}^{(T)} - \mathbf{J}_I^m \cdot \mathcal{H}_{II}^{(T)} - \mathcal{J}_{II}^e \cdot \mathbf{E}_I + \mathbf{J}_{II}^m \cdot \mathcal{H}_I. \quad (6.84)$$

This relationship is commonly known as the "transposed" *Lorentz's lemma*, in the differential form. The integral form of the lemma can be obtained by using the three-dimensional divergence theorem, yielding

$$\oint_\Sigma \left(\mathbf{E}_I \times \mathcal{H}_{II}^{(T)} - \mathbf{E}_{II}^{(T)} \times \mathcal{H}_I \right) \cdot \hat{\mathbf{n}}\, d\sigma$$
$$= \int_V \left(\mathcal{J}_I^e \cdot \mathbf{E}_{II}^{(T)} - \mathbf{J}_I^m \cdot \mathcal{H}_{II}^{(T)} - \mathcal{J}_{II}^e \cdot \mathbf{E}_I + \mathbf{J}_{II}^m \cdot \mathcal{H}_I \right) d\mathbf{r}, \quad (6.85)$$

where integration on the right applies to volume V bounded by the closed surface Σ, on which $\hat{\mathbf{n}}$ is the unit outward normal. For a non-gyrotropic medium, where $\hat{\varepsilon} = \hat{\varepsilon}^{(T)}$, the symbol "T" must be omitted, and this gives a formulation of the usual Lorentz's lemma (see Vaynshteyn, 1988).* The "transposed" Lorentz's lemma has diverse application for many problems of radiation (see, for example, Kondrat'ev and Talanov, 1965) and diffraction of electromagnetic waves in magnetized plasmas. We shall see below that it provides a simple method for obtaining the excitation coefficients $a_{m,n}$ and $a_{m,s,\alpha}(q)$ in the eigenfunction expansion (6.58).

---

* The draft version of a similar relationship was formulated for the first time by Lorentz (1896).

If the sources occupy a bounded region and the surface Σ is taken at infinity, then the surface integral in (6.85) vanishes, and we thus obtain the "transposed" form of the *reciprocity theorem*:

$$\int_V \left(\boldsymbol{\mathcal{J}}_{\mathrm{I}}^e \cdot \mathbf{E}_{\mathrm{II}}^{(\mathrm{T})} - \mathbf{J}_{\mathrm{I}}^m \cdot \boldsymbol{\mathcal{H}}_{\mathrm{II}}^{(\mathrm{T})} \right) \mathrm{d}\mathbf{r} = \int_V \left(\boldsymbol{\mathcal{J}}_{\mathrm{II}}^e \cdot \mathbf{E}_{\mathrm{I}} - \mathbf{J}_{\mathrm{II}}^m \cdot \boldsymbol{\mathcal{H}}_{\mathrm{I}} \right) \mathrm{d}\mathbf{r}. \tag{6.86}$$

Note that the standard reciprocity theorem, which follows from the usual Lorentz's lemma, is not true in a magnetoplasma because it is a non-reciprocal medium.

In the absence of sources, the Lorentz's lemma is formulated only in terms of the fields, thus:

$$\oint_\Sigma \left(\mathbf{E}_{\mathrm{I}} \times \boldsymbol{\mathcal{H}}_{\mathrm{II}}^{(\mathrm{T})} - \mathbf{E}_{\mathrm{II}}^{(\mathrm{T})} \times \boldsymbol{\mathcal{H}}_{\mathrm{I}} \right) \cdot \hat{\mathbf{n}} \, \mathrm{d}\sigma = 0.$$

This relation represented in somewhat another form was, in fact, used in derivation of the orthogonality conditions of § 6.4.

Now it is a simple matter to determine the modal excitation coefficients, using the "transposed" Lorentz's lemma. Consider a prescribed distribution of electric currents $\boldsymbol{\mathcal{J}}^e(\mathbf{r})$ and magnetic currents $\mathbf{J}^m(\mathbf{r})$, which occupy a volume V located between the planes $z = z_1$ and $z = z_2$ and bounded by the surface $\rho = b$. For definiteness, assume that $z_2 > z_1$. The total field everywhere outside the volume V satisfies the source-free Maxwell equations and, therefore, is expanded in proper and continuous improper modes of the duct, as expressed by the general expansion (6.58). When applying that expansion to the concrete regions outside the volume occupied by currents, one must take into account the radiation condition at $|z| \to \infty$, which requires that the energy transport be outgoing. Small collisional losses included, the above requirement means that the total field must be vanishing as $|z| \to \infty$. Then the field outside the source region can be written thus:

$z > z_2$,

$$\mathbf{E}(\mathbf{r}) = \sum_{m=-\infty}^{\infty} \left[\sum_n a_{m,n}\, \mathbf{E}_{m,n}(\mathbf{r}) + \sum_{\alpha=o}^{x} \int_0^\infty a_{m,+,\alpha}(q)\, \mathbf{E}_{m,+,\alpha}(\mathbf{r},q)\, \mathrm{d}q \right],$$

$$\boldsymbol{\mathcal{H}}(\mathbf{r}) = \sum_{m=-\infty}^{\infty} \left[\sum_n a_{m,n}\, \boldsymbol{\mathcal{H}}_{m,n}(\mathbf{r}) + \sum_{\alpha=o}^{x} \int_0^\infty a_{m,+,\alpha}(q)\, \boldsymbol{\mathcal{H}}_{m,+,\alpha}(\mathbf{r},q)\, \mathrm{d}q \right]; \tag{6.87}$$

$z < z_1$,

$$\mathbf{E}(\mathbf{r}) = \sum_{m=-\infty}^{\infty} \left[\sum_n a_{m,-n}\, \mathbf{E}_{m,-n}(\mathbf{r}) + \sum_{\alpha=o}^{x} \int_0^\infty a_{m,-,\alpha}(q)\, \mathbf{E}_{m,-,\alpha}(\mathbf{r},q)\, \mathrm{d}q \right],$$

$$\boldsymbol{\mathcal{H}}(\mathbf{r}) = \sum_{m=-\infty}^{\infty} \left[\sum_n a_{m,-n}\, \boldsymbol{\mathcal{H}}_{m,-n}(\mathbf{r}) + \sum_{\alpha=o}^{x} \int_0^\infty a_{m,-,\alpha}(q)\, \boldsymbol{\mathcal{H}}_{m,-,\alpha}(\mathbf{r},q)\, \mathrm{d}q \right]; \tag{6.88}$$

$z_1 < z < z_2, \quad \rho > b,$

$$
\begin{aligned}
\mathbf{E}(\mathbf{r}) &= \sum_{m=-\infty}^{\infty} \left\{ \sum_n \left[a_{m,n}(z)\,\mathbf{E}_{m,n}(\mathbf{r}) + a_{m,-n}(z)\,\mathbf{E}_{m,-n}(\mathbf{r}) \right] \right. \\
&\quad + \left. \sum_{\alpha=0}^{x} \int_0^\infty \left[a_{m,+,\alpha}(z,q)\,\mathbf{E}_{m,+,\alpha}(\mathbf{r},q) + a_{m,-,\alpha}(z,q)\,\mathbf{E}_{m,-,\alpha}(\mathbf{r},q) \right] dq \right\},
\end{aligned}
$$

$$(6.89)$$

$$
\begin{aligned}
\mathcal{H}(\mathbf{r}) &= \sum_{m=-\infty}^{\infty} \left\{ \sum_n \left[a_{m,n}(z)\,\mathcal{H}_{m,n}(\mathbf{r}) + a_{m,-n}(z)\,\mathcal{H}_{m,-n}(\mathbf{r}) \right] \right. \\
&\quad + \left. \sum_{\alpha=0}^{x} \int_0^\infty \left[a_{m,+,\alpha}(z,q)\,\mathcal{H}_{m,+,\alpha}(\mathbf{r},q) + a_{m,-,\alpha}(z,q)\,\mathcal{H}_{m,-,\alpha}(\mathbf{r},q) \right] dq \right\},
\end{aligned}
$$

where the excitation coefficients depend on z only within the range $z_1 < z < z_2$, and are z-independent elsewhere. For the region $z_1 < z < z_2$, $0 < \rho < b$, which is occupied by the current sources, the total field is represented in a more complicated manner, and this question will be discussed in some detail later.

The excitation coefficients in expansions (6.87) and (6.88) are readily obtained by using Lorentz's lemma in its integral form (6.85), where we take $\mathcal{J}_{\mathrm{I}}^e = \mathcal{J}^e(\mathbf{r})$ and $\mathbf{J}_{\mathrm{I}}^m = \mathbf{J}^m(\mathbf{r})$, but $\mathcal{J}_{\mathrm{II}}^e = \mathbf{J}_{\mathrm{II}}^m = 0$. Let a closed surface of integration Σ in (6.85) be composed of three surfaces, S_+, S_-, and S_R, defined by

$$
\begin{aligned}
S_+ &: \quad z = z_+, \quad 0 \le \rho \le R, \\
S_- &: \quad z = z_-, \quad 0 \le \rho \le R, \\
S_R &: \quad z_- < z < z_+, \quad \rho = R,
\end{aligned}
$$

$$(6.90)$$

where the constant quantities $z_\pm$ are such that $z_+ > z_2$ and $z_- < z_1$, and $R \to \infty$, so that the distant surface S_R is located at infinity. Then the field $\mathbf{E}_{\mathrm{I}}$, $\mathcal{H}_{\mathrm{I}}$ on the surfaces S_+, S_- and S_R is given by expansions (6.87), (6.88) and (6.89), respectively. As the field $\mathbf{E}_{\mathrm{II}}$, $\mathcal{H}_{\mathrm{II}}$, we may take the field of any mode, either proper or improper, on the duct in the "transposed" medium.

Now let $\mathbf{E}_{\mathrm{II}}^{(\mathrm{T})} = \mathbf{E}_{-\hat{m},-\hat{n}}^{(\mathrm{T})}(\mathbf{r})$, $\mathcal{H}_{\mathrm{II}}^{(\mathrm{T})} = \mathcal{H}_{-\hat{m},-\hat{n}}^{(\mathrm{T})}(\mathbf{r})$. Then

$$
\begin{aligned}
J_\Sigma &= \oint_\Sigma \left(\mathbf{E}(\mathbf{r}) \times \mathcal{H}_{-\hat{m},-\hat{n}}^{(\mathrm{T})}(\mathbf{r}) - \mathbf{E}_{-\hat{m},-\hat{n}}^{(\mathrm{T})}(\mathbf{r}) \times \mathcal{H}(\mathbf{r}) \right) \cdot \hat{\mathbf{n}}\, d\sigma \\
&= \int_V \left(\mathcal{J}^e(\mathbf{r}) \cdot \mathbf{E}_{-\hat{m},-\hat{n}}^{(\mathrm{T})}(\mathbf{r}) - \mathbf{J}^m(\mathbf{r}) \cdot \mathcal{H}_{-\hat{m},-\hat{n}}^{(\mathrm{T})}(\mathbf{r}) \right) d\mathbf{r},
\end{aligned}
$$

$$(6.91)$$

in which integration on the right side is performed over a volume V occupied by currents. Using the orthogonality conditions given by (6.66) and (6.67) and noting that the result of integration over the distant surface S_R vanishes as $R \to \infty$, we obtain

$$J_\Sigma = \int_{S_+} \left(\mathbf{E}(\mathbf{r}) \times \mathcal{H}^{(T)}_{-\tilde{m},-\tilde{n}}(\mathbf{r}) - \mathbf{E}^{(T)}_{-\tilde{m},-\tilde{n}}(\mathbf{r}) \times \mathcal{H}(\mathbf{r}) \right) \cdot \hat{\mathbf{z}}_0 \, \mathrm{d}\sigma = a_{\tilde{m},\tilde{n}} \, N_{\tilde{m},\tilde{n}}.$$

By following the same procedure, but replacing the field $\mathbf{E}^{(T)}_{-\tilde{m},-\tilde{n}}(\mathbf{r})$, $\mathcal{H}^{(T)}_{-\tilde{m},-\tilde{n}}(\mathbf{r})$ in (6.91) by $\mathbf{E}^{(T)}_{-\tilde{m},\tilde{n}}(\mathbf{r})$, $\mathcal{H}^{(T)}_{-\tilde{m},\tilde{n}}(\mathbf{r})$, it is deduced that

$$\begin{aligned}
J_\Sigma &= \int_{S_-} \left(\mathbf{E}(\mathbf{r}) \times \mathcal{H}^{(T)}_{-\tilde{m},\tilde{n}}(\mathbf{r}) - \mathbf{E}^{(T)}_{-\tilde{m},\tilde{n}}(\mathbf{r}) \times \mathcal{H}(\mathbf{r}) \right) \cdot (-\hat{\mathbf{z}}_0) \, \mathrm{d}\sigma \\
&= -a_{\tilde{m},-\tilde{n}} \, N_{\tilde{m},-\tilde{n}} = a_{\tilde{m},-\tilde{n}} \, N_{\tilde{m},\tilde{n}}.
\end{aligned}$$

Finally, one can write the proper-mode excitation coefficients in the form

$$a_{m,\pm n} = \frac{1}{N_{m,n}} \int_V \left(\boldsymbol{\mathcal{J}}^e(\mathbf{r}) \cdot \mathbf{E}^{(T)}_{-m,\mp n}(\mathbf{r}) - \mathbf{J}^m(\mathbf{r}) \cdot \mathcal{H}^{(T)}_{-m,\mp n}(\mathbf{r}) \right) \mathrm{d}\mathbf{r}. \tag{6.92}$$

Similarly, by substituting the appropriate fields of improper modes, belonging to the continuous spectrum, for $\mathbf{E}^{(T)}_{\mathrm{II}}$, $\mathcal{H}^{(T)}_{\mathrm{II}}$ and using the orthogonality conditions given by (6.67) and (6.68), we deduce that

$$a_{m,\pm,\alpha}(q) = \frac{1}{N_{m,\alpha}(q)} \int_V \left(\boldsymbol{\mathcal{J}}^e(\mathbf{r}) \cdot \mathbf{E}^{(T)}_{-m,\mp,\alpha}(\mathbf{r},q) - \mathbf{J}^m(\mathbf{r}) \cdot \mathcal{H}^{(T)}_{-m,\mp,\alpha}(\mathbf{r},q) \right) \mathrm{d}\mathbf{r}. \tag{6.93}$$

The proof is made by analogy with that for (6.92). One should only take into account that the terms of the form

$$\rho \left(\mathbf{E}_{m,s,\alpha}(\mathbf{r},q) \times \mathcal{H}^{(T)}_{\tilde{m},\tilde{s},\tilde{\alpha}}(\mathbf{r},\tilde{q}) - \mathbf{E}^{(T)}_{\tilde{m},\tilde{s},\tilde{\alpha}}(\mathbf{r},\tilde{q}) \times \mathcal{H}_{m,s,\alpha}(\mathbf{r},q) \right) \cdot \hat{\boldsymbol{\rho}}_0,$$

which appear under the sign of the integral over S_R, do not contribute to the result of the integration in the limit $R \to \infty$. Here, this limit is understood in a sense of the theory of distributions. Thus, we have obtained the excitation coefficients in the expansions (6.87) and (6.88) giving the field everywhere outside the interval $z_1 < z < z_2$.

To determine the field between the planes $z = z_1$ and $z = z_2$, the following, somewhat roundabout, procedure is applied (see, for example, Vaynshteyn, 1988, § 76). We remove the sources from an infinitesimally narrow gap located between two planes with the coordinates $z - \Delta$ and $z + \Delta$, where $z \in (z_1, z_2)$ and $\Delta \to 0$. Then the field within the gap, $\mathbf{E}^{(I)}(\mathbf{r})$, $\mathcal{H}^{(I)}(\mathbf{r})$, is formally given by (6.89), regardless of whether $\rho > b$ or $\rho < b$, with the excitation coefficients

$$a_{m,n}(z) = \frac{1}{N_{m,n}} \int_{(z_1,z)} \left(\boldsymbol{\mathcal{J}}^e(\mathbf{r}) \cdot \mathbf{E}^{(\mathrm{T})}_{-m,-n}(\mathbf{r}) - \mathbf{J}^m(\mathbf{r}) \cdot \boldsymbol{\mathcal{H}}^{(\mathrm{T})}_{-m,-n}(\mathbf{r}) \right) d\mathbf{r},$$

$$a_{m,-n}(z) = \frac{1}{N_{m,n}} \int_{(z,z_2)} \left(\boldsymbol{\mathcal{J}}^e(\mathbf{r}) \cdot \mathbf{E}^{(\mathrm{T})}_{-m,n}(\mathbf{r}) - \mathbf{J}^m(\mathbf{r}) \cdot \boldsymbol{\mathcal{H}}^{(\mathrm{T})}_{-m,n}(\mathbf{r}) \right) d\mathbf{r}, \qquad (6.94)$$

$$a_{m,+,\alpha}(z,q) = \frac{1}{N_{m,\alpha}(q)} \int_{(z_1,z)} \left(\boldsymbol{\mathcal{J}}^e(\mathbf{r}) \cdot \mathbf{E}^{(\mathrm{T})}_{-m,-,\alpha}(\mathbf{r}) - \mathbf{J}^m(\mathbf{r}) \cdot \boldsymbol{\mathcal{H}}^{(\mathrm{T})}_{-m,-,\alpha}(\mathbf{r}) \right) d\mathbf{r},$$

$$a_{m,-,\alpha}(z,q) = \frac{1}{N_{m,\alpha}(q)} \int_{(z,z_2)} \left(\boldsymbol{\mathcal{J}}^e(\mathbf{r}) \cdot \mathbf{E}^{(\mathrm{T})}_{-m,+,\alpha}(\mathbf{r}) - \mathbf{J}^m(\mathbf{r}) \cdot \boldsymbol{\mathcal{H}}^{(\mathrm{T})}_{-m,+,\alpha}(\mathbf{r}) \right) d\mathbf{r}.$$

Here the notations (z_1, z) and (z, z_2) stand for the intervals of integration with respect to z. Now the only difference between the real current distribution and that without sources in the gap consists in the presence of current discontinuities across the boundaries of the gap in the latter case. These discontinuities lead to the appearance of surface charges with the densities $\sigma^{e,m} = (\mathrm{i}\omega)^{-1} j_z^{e,m}$ and $\sigma^{e,m} = -(\mathrm{i}\omega)^{-1} j_z^{e,m}$ on the planes at $z - \Delta$ and $z + \Delta$, respectively, thereby producing the field between these planes, which, for $\Delta \to 0$, is written as

$$\mathbf{E}^{(\mathrm{II})}(\mathbf{r}) = -\mathrm{i}\,(k_0\eta)^{-1} \mathcal{J}_z^e(\mathbf{r})\,\hat{\mathbf{z}}_0, \quad \boldsymbol{\mathcal{H}}^{(\mathrm{II})}(\mathbf{r}) = -\mathrm{i}\,k_0^{-1} J_z^m(\mathbf{r})\,\hat{\mathbf{z}}_0. \qquad (6.95)$$

Since the contribution to the total field due to terms (6.95) arises from the current discontinuities which appear by removing the currents from the gap, the actual field follows by subtracting (6.95) from (6.89) to give

$$\mathbf{E}(\mathbf{r}) = \mathbf{E}^{(\mathrm{I})} + \mathrm{i}\,(k_0\eta)^{-1} \mathcal{J}_z^e(\mathbf{r})\,\hat{\mathbf{z}}_0, \quad \boldsymbol{\mathcal{H}}(\mathbf{r}) = \boldsymbol{\mathcal{H}}^{(\mathrm{I})} + \mathrm{i}\,k_0^{-1} J_z^m(\mathbf{r})\,\hat{\mathbf{z}}_0. \qquad (6.96)$$

Thus, substituting for $\mathbf{E}^{(\mathrm{I})}(\mathbf{r})$, $\boldsymbol{\mathcal{H}}^{(\mathrm{I})}(\mathbf{r})$ from (6.89), along with expressions (6.94), yields the total field between the planes $z = z_1$ and $z = z_2$. Clearly expressions (6.96) reduce to (6.89) for $\rho > b$. Notice that the terms (6.95) are entirely analogous to those appearing in the expressions for source-excited fields in a *homogeneous* magnetoplasma, as seen from (3.23) and (3.25).

The formulas derived above are exact and allow us to determine the total field for arbitrary sources occupying a bounded volume, and for arbitrary values of driving frequency, static magnetic-field strength, collisional losses, plasma density and composition. Although the duct was assumed to be formed by variation of the density, the foregoing analysis remains applicable when dealing with any azimuthally symmetric modifications of the plasma parameters involved in the dielectric tensor $\hat{\varepsilon} = \hat{\varepsilon}(\mathbf{r})$. The above-stated method of finding the excitation coefficients is, in essence, analogous to that developed for closed waveguides (e.g., Vaynshteyn, 1988, Chapter 14) and open dielectric waveguides (e.g., Shevchenko, 1971 a, Chapter 2; Marcuse, 1972, § 8.5; Snyder and Love, 1983, Chapter 31).

As we now have expressions for modal excitation coefficients, the total field launched by a prescribed distribution of electric and magnetic currents can be determined without difficulty. As an example, consider ring sources with the current densities (5.14) and (5.15). Those currents can only excite the azimuthally symmetric fields so that the sum over m in (6.58) would contain only one term with $m = 0$. Consequently, in what follows, the azimuthal index may be omitted. Here we restrict ourselves to presenting the expression for the E_ϕ component in the source-free regions. That is

$$E_\phi(\mathbf{r}) = \sum_n a_n\, E_{\phi;\,n}(\rho) \exp(-\mathrm{i}k_0 p_n\,|z|)$$

$$+ \sum_{\alpha=0}^{\mathrm{x}} \int_\Gamma J_\alpha(q)\, \frac{(-1)\, k_0^2\, E_{\phi;\,s,\alpha}(\rho,q)}{16\pi\left(1 + \eta_a^{-1}\, n_{s,\alpha}^{(1)^2}\right) C_{s,\alpha}^{(1)}(q)\, \Delta C_{s,\alpha}(q)}\, \frac{\mathrm{d}p_\alpha}{\mathrm{d}q}\, \exp[-\mathrm{i}k_0 p_\alpha(q)|z|]\, \mathrm{d}q \tag{6.97}$$

where

$$J_\alpha(q) = 2\pi b \left[Z_0 I_0^e\, E_{\phi;\,+,\alpha}(b,q) - I_0^m\, \mathcal{H}_{\phi;\,+,\alpha}(b,q)\, \frac{\sin(k_0\, p_\alpha(q)\, d)}{k_0\, p_\alpha(q)\, d}\, \mathrm{sgn}\, z \right], \tag{6.98}$$

$$a_n = J_{\hat{\alpha}}(q_n)\, N_n^{-1}, \tag{6.99}$$

and $n = 1, 2, \ldots$. In deducing these results, we made use of the relationships (6.46) and (6.50), together with the evident formula $p_\alpha'(-q) = -p_\alpha'(q)$, to extend the q integration along the path Γ, which is the same as the path Γ_2 shown in Figure 3.2. Also, the relationships (6.7) and (6.75) between the forward-and backward-propagating waves were taken into account. Note that the use of the path Γ, instead of the previous integration path along the positive real half axis of q in (6.58), facilitates evaluation of the integrals in (6.97).

In the outer region of the duct, where the plasma is uniform, the E_ϕ component becomes

$$E_\phi(\mathbf{r}) = \sum_n \mathrm{i}a_n \left[C_{s,\hat{\alpha}}^{(2)}(q_n) H_1^{(2)}(k_0 q_n \rho) + C_{s,\hat{\alpha}}(q_n) H_1^{(2)}(k_0\, q_{\hat{\alpha}}(q_n)\rho) \right] \exp(-\mathrm{i}k_0 p_n |z|)$$

$$+ \sum_{\alpha=0}^{\mathrm{x}} \int_\Gamma J_\alpha(q)\, \frac{(-\mathrm{i})\, k_0^2 \left[\Delta C_{s,\alpha}(q)\, H_1^{(2)}(k_0 q \rho) + C_{s,\alpha}(q)\, H_1^{(2)}(k_0 q_\alpha(q)\,\rho) \right]}{16\pi\left(1 + \eta_a^{-1} n_{s,\alpha}^{(1)^2}\right) C_{s,\alpha}^{(1)}(q)\, \Delta C_{s,\alpha}(q)}$$

$$\times\ \frac{\mathrm{d}p_\alpha}{\mathrm{d}q}\, \exp[-\mathrm{i}k_0\, p_\alpha(q)\, |z|]\, \mathrm{d}q. \tag{6.100}$$

The other field components can be represented in a similar way, and formulas for them need not be given.

We shall return to analyzing field expressions such as (6.97) and (6.100) in §6.7.

6.6. Analytic properties of the functions $p_\alpha(q)$ and $q_\alpha(q)$

As shown earlier, the total field contains the discrete contributions from the eigenmodes and the contribution due to modes of the continuous spectrum. It is recognized that the latter contribution is still in the form of integrals which must be evaluated. For this purpose, special techniques of the contour integration in the complex q plane can be used, and their applications will be discussed in the forthcoming sections. Since the integrands in the field representation (6.58) (see also (6.97)) are multi-valued functions of q, we must determine the position and shape of the branch cuts in the complex q plane required to make the integrands single-valued. These integrands include the Hankel functions $H_m^{(1,2)}(k_0 q\rho)$, whose branch cut passes along the negative real q axis from the branch singularity at $q = 0$, and the functions $p_\alpha(q)$ and $q_\alpha(q)$ which also possess the branch points. To render the consideration more concrete, attention will be restricted to the whistler frequencies lying in the ranges (2.70) and (2.71).

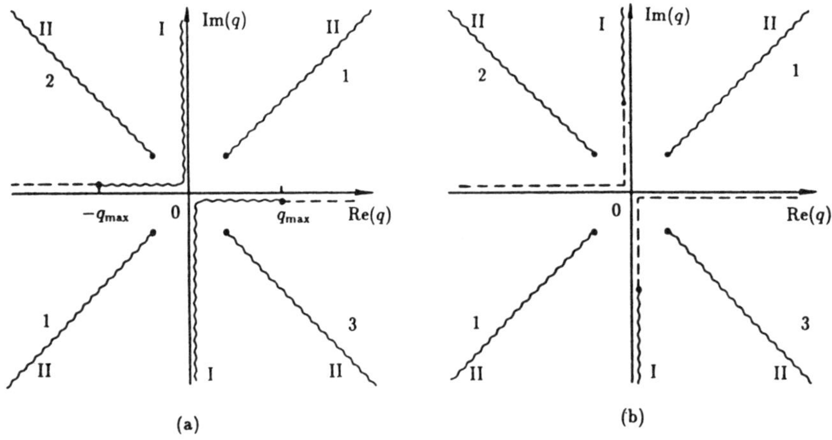

Figure 6.1. Location of branch singularities of functions $p_\alpha(q)$ in the complex q plane when $\Omega_H \ll \omega < \omega_{LH}$ (not to scale); $\mathrm{Im}\,(p_\alpha) < 0$ and $\mathrm{Re}\,(R_p) > 0$. (a) $\alpha = $ 'x', (b) $\alpha = $ 'o'. $\mathrm{Re}\,(p_\alpha) > 0$ in domain 1 and $\mathrm{Re}\,(p_\alpha) < 0$ in domains 2 and 3.

Consider first the branch singularities of the functions $p_\alpha(q)$ and $q_\alpha(q)$ for the *frequency interval* (2.70). Figure 6.1 shows the sheets of the multi-sheeted Riemann q plane of the functions $p_x(q)$ and $p_o(q)$; the prescribed conditions $\mathrm{Im}\,(p_{x,o}) < 0$ and $\mathrm{Re}\,(R_p) > 0$, which serve to define the requisite branch for each of the above two functions, are fulfilled in these sheets. In Figure 6.1 the branch cuts I given by $\mathrm{Im}\,(p_{x,o}) = 0$, and the dashed lines given by $\mathrm{Re}\,(p_{x,o}) = 0$ pass from the branch points $q = \pm\left[(\varepsilon_a^2 - g_a^2)/\varepsilon_a\right]^{1/2} = \pm q_{\max}$ for $\alpha = \mathrm{x}$, and $q = \pm\eta_a^{1/2}$ for $\alpha = \mathrm{o}$, which are zeros of $p_{\mathrm{x,o}}(q)$. The branch cuts II extend along the lines $\mathrm{Re}\,(R_p) = 0$ from the branch points

$$q = \pm\,\frac{g_a}{\eta_a - \varepsilon_a}\left\{2\eta_a\left[1 \pm \left(1 - \frac{(\eta_a - \varepsilon_a)^2}{g_a^2}\right)^{1/2}\right]\right\}^{1/2}, \qquad (6.101)$$

which are zeros of $R_p(q)$.

The analytic properties of the functions $q_{\mathrm{x}}(q)$ and $q_{\mathrm{o}}(q)$ in the frequency interval (2.70) are represented by their Riemann q planes shown in Figure 6.2, where $\mathrm{Im}\,(q_{\mathrm{x,o}}) < 0$ and $\mathrm{Re}\,(R_p) > 0$. The branch cuts I and the dashed lines defined by $\mathrm{Im}\,(q_{\mathrm{x,o}}) = 0$ and $\mathrm{Re}\,(q_{\mathrm{x,o}}) = 0$, respectively, pass here from the branch points $q = \pm\left[-(\eta_a + \varepsilon_a + g_a)\,g_a/\varepsilon_a\right]^{1/2}$ for $\alpha = \mathrm{x}$, and $q = \pm\left[(\eta_a + \varepsilon_a - g_a)\,g_a/\varepsilon_a\right]^{1/2}$ for $\alpha = \mathrm{o}$, which are zeros of $q_{\mathrm{x,o}}(q)$. (Recall that for the time dependence assumed, $\mathrm{Re}\,(g_a) < 0$ at the whistler frequencies.) The branch cuts II in the q planes exhibited in Figure 6.2 are like those in Figure 6.1.

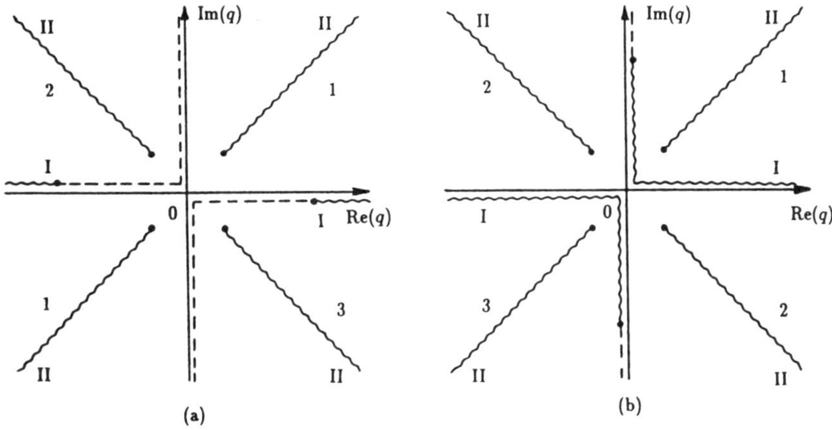

Figure 6.2. Location of branch singularities of functions $q_\alpha(q)$ in the complex q plane when $\Omega_{\mathrm{H}} \ll \omega < \omega_{\mathrm{LH}}$ (not to scale); $\mathrm{Im}\,(q_\alpha) < 0$ and $\mathrm{Re}\,(R_p) > 0$. (a) $\alpha = $ 'x', $\mathrm{Re}\,(q_{\mathrm{x}}) > 0$ in domains 2 and 3, and $\mathrm{Re}\,(q_{\mathrm{x}}) < 0$ in domain 1. (b) $\alpha = $ 'o', $\mathrm{Re}\,(q_{\mathrm{o}}) > 0$ in domain 2 and $\mathrm{Re}\,(q_{\mathrm{o}}) < 0$ in domains 1 and 3.

Now let us discuss the analytic properties of the functions $p_\alpha(q)$ and $q_\alpha(q)$ in the *frequency interval* (2.71). We begin again with the functions $p_{\mathrm{x}}(q)$ and $p_{\mathrm{o}}(q)$ whose Riemann sheets are shown in Figure 6.3. As before, the sheets are chosen by stipulating that the inequalities $\mathrm{Im}\,(p_{\mathrm{x,o}}) < 0$ and $\mathrm{Re}\,(R_p) > 0$ are valid in them.

Here branch cuts I, dashed curves, and branch cuts II pass along the lines defined by $\mathrm{Im}\,(p_{x,o}) = 0$, $\mathrm{Re}\,(p_{x,o}) = 0$, and $\mathrm{Re}\,(R_p) = 0$, respectively. Unlike the previous frequency interval (2.70), now there are no branch points where $p_\alpha(q) = 0$ in part (a) of Figure 6.3, relating to the case $\alpha = \mathrm{x}$, while part (b) relating to the case $\alpha = \mathrm{o}$ shows the presence of four branch points $q = \pm\left[(\varepsilon_a^2 - g_a^2)/\varepsilon_a\right]^{1/2}$ and $q = \pm\eta_a^{1/2}$, which are zeros of the function $p_o(q)$. Such a difference in behavior of $p_{x,o}(q)$ in the ranges (2.70) and (2.71) is associated only with the nomenclature of the terms "ordinary" and "extraordinary" accepted here. In the frequency interval (2.71), the function $p_o(q)$ vanishes indeed at the above four points, so that

$$p_o(\pm\eta_a^{1/2}) = p_o\left(\pm\sqrt{(\varepsilon_a^2 - g_a^2)/\varepsilon_a}\right) = 0, \qquad (6.102)$$

whereas the function $p_x(q)$ vanishes nowhere while we remain in the Riemann sheet where $\mathrm{Re}\,(R_p) > 0$.

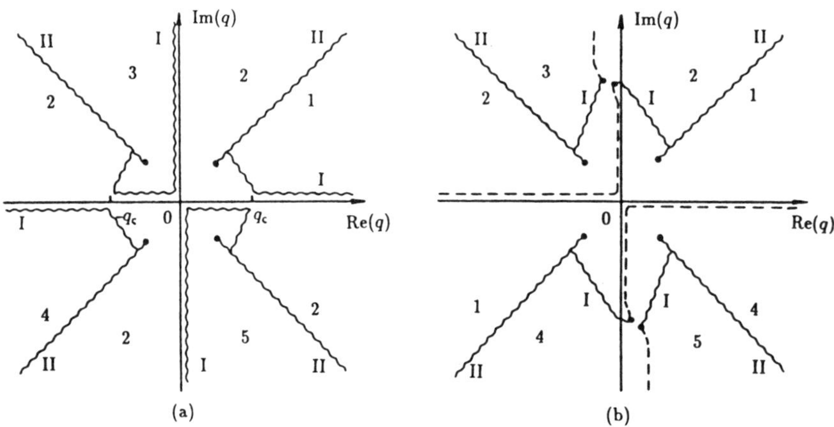

Figure 6.3. Location of branch singularities of functions $p_\alpha(q)$ in the complex q plane when $\omega_{\mathrm{LH}} < \omega < \omega_{\mathrm{H}}/2$ (not to scale); $\mathrm{Im}\,(p_\alpha) < 0$ and $\mathrm{Re}\,(R_p) > 0$. (a) $\alpha = \text{'x'}$, $\mathrm{Re}\,(p_x) > 0$ in domain 2 and $\mathrm{Re}\,(p_x) < 0$ in domains 1 and 3–5. (b) $\alpha = \text{'o'}$, $\mathrm{Re}\,(p_o) > 0$ in domains 1, 3 and 5, and $\mathrm{Re}\,(p_o) < 0$ in domains 2 and 4.

As to functions $q_{x,o}(q)$, their Riemann sheets where $\mathrm{Im}\,(q_{x,o}) < 0$ and $\mathrm{Re}\,(R_p) > 0$ are shown, for frequencies (2.71), in Figure 6.4. Here the branch cuts I and dashed lines defined by $\mathrm{Im}\,(q_{x,o}) = 0$ and $\mathrm{Re}\,(q_{x,o}) = 0$, respectively, extend from the branch points $q = \pm\left[(\eta_a + \varepsilon_a - g_a)\,g_a/\varepsilon_a\right]^{1/2} = \pm q_p$ if $\alpha = \mathrm{x}$, and $q = \pm\left[-(\eta_a + \varepsilon_a + g_a)\,g_a/\varepsilon_a\right]^{1/2}$ if $\alpha = \mathrm{o}$. These points are zeros of $q_\alpha(q)$. The cuts II are like those shown in the preceding figures. To avoid ambiguity in position of branch singularities, small collisional losses were assumed when preparing Riemann sheets presented in Figures 6.1 to 6.4.

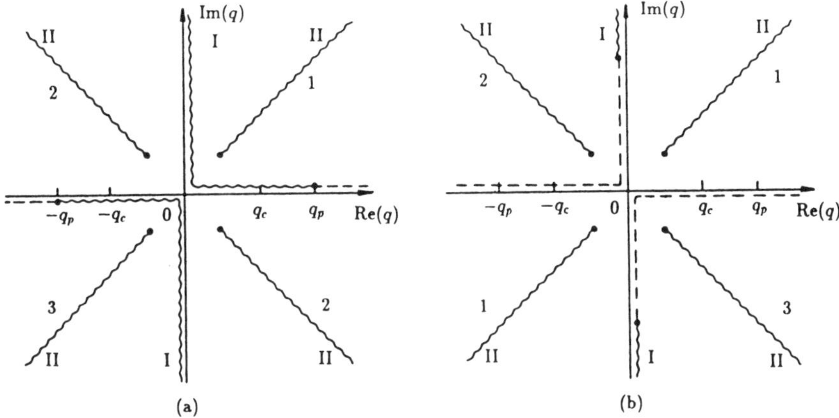

Figure 6.4. Location of branch singularities of functions $q_\alpha(q)$ in the complex q plane when $\omega_{\mathrm{LH}} < \omega < \omega_{\mathrm{H}}/2$ (not to scale); $\mathrm{Im}\,(q_\alpha) < 0$ and $\mathrm{Re}\,(R_p) > 0$. (a) $\alpha = $ 'x', $\mathrm{Re}\,(q_{\mathrm{x}}) > 0$ in domains 1 and 3, and $\mathrm{Re}\,(q_{\mathrm{x}}) < 0$ in domain 2. (b) $\alpha = $ 'o', $\mathrm{Re}\,(q_{\mathrm{o}}) > 0$ in domain 1 and $\mathrm{Re}\,(q_{\mathrm{o}}) < 0$ in domains 2 and 3.

The branch singularities listed above must be taken into account when deforming the path of integration in (6.58).

Before doing this, we make some general remarks concerning the location of the points $q_{m,n}$ in the complex q plane for the chosen frequency intervals. These points corresponding to proper modes are poles of the integrand with $\alpha = \hat{\alpha}$. Referring to the theory of § 4.4, it is evident that, for density enhancements, at frequencies (2.70), there may exist proper modes with their eigenvalues $q_{m,n}$. These are imaginary in the absence of losses, and lie on the left side of the branch cut I in the lower q half-plane, shown in part (a) of Figure 6.1. It is thus found for these eigenmodes that $\hat{\alpha} = $ x so that $p_{m,n} = p_{\mathrm{x}}(q_{m,n})$. The auxiliary transverse wavenumbers $q_{\mathrm{x}}(q_{m,n})$ lie on the negative imaginary q axis below the branch points of $p_{\mathrm{o}}(q)$ (see part (b) of Figure 6.1), ensuring the relation $p_{\mathrm{o}}\Big(q_{\mathrm{x}}(q_{m,n})\Big) = p_{m,n}$.

At frequencies from the interval (2.71), there may exist the interface-type eigenmode (§ 4.3) and the eigenmodes of a density trough (§ 4.4). For these types of modes, the eigenvalues are found to be complex, but the propagation constants remain *real* if the duct is non-absorbing. The auxiliary transverse wavenumbers of these modes obey the relation $q_{\hat{\alpha}}(q_{m,n}) = -q_{m,n}^{*}$. It is seen from Figure 6.3 that this is due to the fact that, for frequencies (2.71), the branch cuts I defined by $\mathrm{Im}\,(p_\alpha) = 0$ are symmetric about the imaginary q axis everywhere in the complex q plane[*], except close to the branch points of the function $p_o(q)$.

For the *complex* eigenmodes mentioned in §4.9, the eigenvalues $q_{m,n}$, the auxiliary transverse wavenumbers $q_\check{\alpha}(q_{m,n})$, and the propagation constants $p_{m,n} = p_\check{\alpha}(q_{m,n})$ are all complex. It was already explained that these modes appears in pairs, so that if $q_{m,n}$ is the eigenvalue, then $\tilde{q}_{m,n} = -q_{m,n}^*$ is also the eigenvalue, and, accordingly, $q_\check{\alpha}(q_{m,n}) = -q_\check{\alpha}^*(\tilde{q}_{m,n})$ and $p_\check{\alpha}(q_{m,n}) = -p_\check{\alpha}^*(\tilde{q}_{m,n})$ for a non-absorbing duct.

6.7. Separation of leaky modes from the continuous spectrum

In Chapter 4 we saw that the duct with enhanced density can support, at frequencies belonging to the interval (2.71), the waveguide-like propagation of slightly leaky waves called the *leaky modes*. Although they are not members of the eigenfunction expansion, the leaky modes may be separated from the continuous spectrum to obtain an alternative field representation rapidly convergent in certain spatial regions. Also, it was mentioned that the true eigenmodes cannot exist in ducts with enhanced density, except that the interface-type eigenmode may be found under some conditions if the duct is sharply bounded (Adachi, 1966). Since this mode is not encountered in actual practice we shall assume that only improper modes of the continuous spectrum contribute to the total field.

Starting from the general eigenfunction expansion, the radiation field can be expressed exactly as a sum of leaky modes together with a *space-wave* contribution which accounts for the remaining part of the field. It is not difficult to verify that the leaky modes are associated with nonspectral (improper) poles of the integrands of (6.58). For the frequency interval (2.71), there may be poles $q = q_{m,\nu}$ and $q = (e^{i\pi} q_{m,\nu})^*$ which satisfy Equation (6.52), and poles $q = \tilde{q}_{m,\nu} \equiv e^{i\pi} q_{m,\nu}$ and $q = e^{i\pi} \tilde{q}_{m,\nu}^*$ which satisfy Equation (6.53), provided that $\alpha = $ x. Also, it is found that $-3\pi/2 < \arg q_{m,\nu} < -\pi$. In what follows, the poles $q = \tilde{q}_{m,\nu}$ and $q = e^{i\pi} \tilde{q}_{m,\nu}^*$ will be discarded as they do not give new modes compared with the poles $q = q_{m,\nu}$ and $q = (e^{i\pi} q_{m,\nu})^*$. We emphasize that only the poles which are roots of (6.52) are important when passing from integration over the positive real q axis to integration along the path Γ, because only the quantity $C_{m,s,\alpha}^{(1)}$ is involved in the denominator of the integrand after changing the integration path (see, for example, (6.97)). The remaining poles are divided into two sets. The poles of the first set, $q = q_{m,\nu}$, correspond to leaky modes whose Poynting vectors are directed outwards at infinity. The complex propagation constants of these modes are evidently given

---

by $p_{m,\nu} = p_{\hat{\alpha}}(q_{m,\nu})$ with $\hat{\alpha} = \mathrm{x}$, the condition $|\mathrm{Im}\,(q_{m,\nu})| < |\mathrm{Im}\,(q_{\hat{\alpha}}(q_{m,\nu}))|$ again being valid (see, for comparison, (6.55)). The poles of the second set, $q = (e^{i\pi}\,q_{m,\nu})^{*}$, correspond to modes whose Poynting vectors at infinity are directed towards the duct. With further analysis, the poles of the second set will be naturally rejected, and they will not be shown in the Riemann q planes to be appeared later on.

Next, taking into account the analytic properties of all the functions involved in the integrands, we present, for the frequency range (2.71), the Riemann sheets defined by stipulating that $\mathrm{Im}\,(p_{\mathrm{x},o}) < 0$, $\mathrm{Im}\,(q_{\mathrm{x},o}) < 0$, $\mathrm{Re}\,(R_p) > 0$, and $-\pi < \arg q < \pi$. These sheets given in Figure 6.5 are obtained by superposing Figures 6.3 and 6.4, and adding the branch cut along the negative real q axis. In Figure 6.5 we also show the integration path Γ appropriate to the expression (6.97). The poles $q = q_{m,\nu}$ are located on the other Riemann sheet defined by the inequalities $\mathrm{Im}\,(p_{\mathrm{x}}) < 0$, $\mathrm{Im}\,(q_{\mathrm{x}}) < 0$, $\mathrm{Re}\,(R_p) > 0$, and $-3\pi < \arg q < -\pi$ (for simplicity, only poles $q_\nu = q_{0,\nu}$ are presented). It was shown in § 4.4 that the leakage is, as a rule, rather small in the frequency range (2.71), provided that the wave frequency is well below half the electron gyrofrequency and the propagation constants lie in the interval (4.75). The above conditions will hereinafter be assumed throughout. Then the poles $q = q_{m,\nu}$ are located in the vicinity of the negative real q half axis, and the following inequality can be shown to be valid:

$$\bar{\gamma} = \max_{m,\nu}\left\{\arctan\left[-\mathrm{Im}\,(q_{m,\nu})/\mathrm{Re}\,(q_{m,\nu})\right]\right\} \ll \pi. \tag{6.103}$$

For $m = 0$, the inequality (6.103) reduces to (5.49), if one makes the replacement $q_{m,\nu} \to q_2(p_{m,\nu})$ in order to return to the notations of Chapter 5. The leaky-mode poles are seen to lie to the left of the branch point $q = -q_p$ where $q_p = [(\eta_a + \varepsilon_a - g_a)\,g_a/\varepsilon_a]^{1/2}$. It is curious that q_p exactly satisfies $p_{\mathrm{x}}(q_p) = p_{\mathrm{x}}(0)$ when no collisional losses are included.

Now we deform the contour Γ so as to coincide with a straight line defined by $\mathrm{Im}\,(q) = -\mathrm{Re}\,(q)\tan\gamma$ where $\bar{\gamma} < \gamma < \pi/4$. This line is supplemented at infinity by two arcs $\tilde{\Gamma}_\infty$ and Γ_∞, as shown in Figure 6.5. Figure 6.5 consists of two parts, (a) and (b), relating to the cases $\alpha = \mathrm{x}$ and $\alpha = \mathrm{o}$, respectively. In both parts, the segments of the distorted contour which are shown as chain (dot-and-dash) lines pass on the Riemann sheet where $\mathrm{Im}\,(p_{\mathrm{x},o}) < 0$ and $\mathrm{Im}\,(q_{\mathrm{x},o}) < 0$ but $-3\pi < \arg q < -\pi$. The dotted line, in part (a) of Figure 6.5, passes on the sheet defined by $\mathrm{Im}\,(p_{\mathrm{x}}) > 0$, $\mathrm{Im}\,(q_{\mathrm{x}}) < 0$, and $-3\pi < \arg q < -\pi$, while the dashed line is on the sheet defined by $\mathrm{Im}\,(p_{\mathrm{x}}) > 0$, $\mathrm{Im}\,(q_{\mathrm{x}}) < 0$, and $-\pi < \arg q < \pi$ (note that $\mathrm{Re}\,(R_q) > 0$ in all the sheets mentioned). It is readily found that the arc $\tilde{\Gamma}_\infty$ in the lower half of the q plane does not contribute anything to the integrals since both integrands, with $\alpha = \mathrm{x}$ and $\alpha = \mathrm{o}$, become infinitely attenuated on $\tilde{\Gamma}_\infty$. Furthermore, if the current sources are confined to the bounded region defined by $\rho < b$ and $|z| < d$, then, for the domain

$$|z| > \max\left\{ d + (\varrho - a)(-\eta_a/\varepsilon_a)^{1/2}, \, d + (\varrho - a)\tan\gamma \right\}, \qquad (6.104)$$

where ϱ is given by (5.51), the arc Γ_∞ in the upper half of the q plane contributes nothing to the integrals. Note that one condition in (6.104), namely $|z| > d + (\varrho - a)(-\eta_a/\varepsilon_a)^{1/2}$, already arose in §5.5 (see (5.50) and (5.51)). Now it provides the zero contribution when integrating the terms with $\alpha = \mathrm{x}$ along Γ_∞. The other condition built in (6.104), $|z| > d + (\varrho - a)\tan\gamma$, plays the analogous role, but for the integral with $\alpha = \mathrm{o}$.

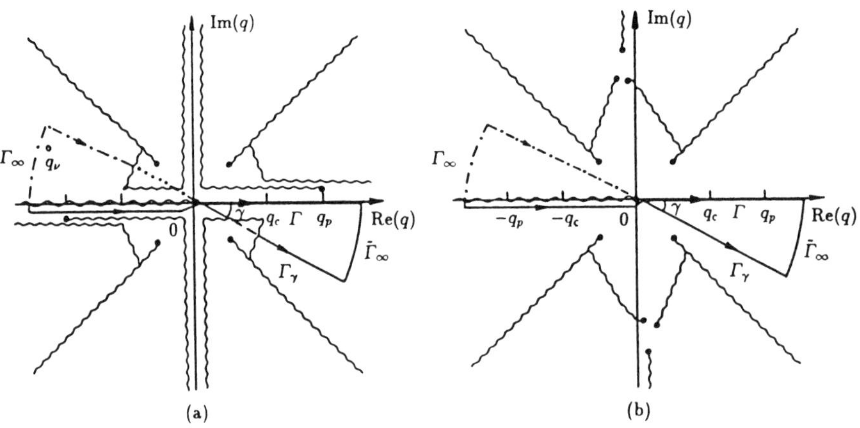

Figure 6.5. Location of leaky-mode poles q_ν and contours Γ and Γ_γ in the complex q plane when $\omega_{\mathrm{LH}} < \omega < \omega_{\mathrm{H}}/2$ (not to scale). (a) $\alpha = \text{'x'}$. (b) $\alpha = \text{'o'}$.

With the above points in mind, we calculate residues of the leaky-wave poles intercepted during the path deformation and then pass from integration along the entire contour given by $\mathrm{Im}\,(q) = -\mathrm{Re}\,(q)\tan\gamma$ to integration along its part Γ_γ lying in the fourth quadrant of the complex q plane. Remembering that no proper modes are assumed to exist, expression (6.97) for the E_ϕ component in the case of ring sources (5.14) and (5.15) is then written in domain (6.104), for positive z, as follows:

$$E_\phi(\mathbf{r}) = \sum_\nu a_\nu E_{\phi;\,\nu}(\rho)\exp(-\mathrm{i}k_0 p_\nu z) \qquad (6.105)$$

$$+ \sum_{\alpha=\mathrm{o}}^{\mathrm{x}} \int_{\Gamma_\gamma} J_\alpha(q)\,\frac{(-1)\,k_0^2\,E_{\phi;\,+,\alpha}(\rho,q)}{16\pi\left(1 + \eta_a^{-1}\,n_{+,\alpha}^{(1)^2}\right) C_{+,\alpha}^{(1)}(q)\,C_{+,\alpha}^{(2)}(q)}\,\frac{\mathrm{d}p_\alpha}{\mathrm{d}q}\,\exp[-\mathrm{i}k_0 p_\alpha(q)z]\,\mathrm{d}q,$$

where

$$E_{\phi;\,\nu}(\rho) = E_{\phi;\,+,\mathrm{x}}(\rho, q_\nu), \quad a_\nu = 2\pi \mathrm{i} J_\mathrm{x}(q)\left(\frac{\mathrm{d}N_\mathrm{x}}{\mathrm{d}q}\right)^{-1}\Bigg|_{q=q_\nu}, \tag{6.106}$$

with $J_\mathrm{x}(q)$ being obtainable from (6.98).

It is now clear that the leaky-modes fields appear as residue contributions arising from the presence of the nonspectral poles $q = q_{0,\nu} \equiv q_\nu$. Interestingly, the representation (6.105) is functionally identical with the previous eigenfunction expansion (6.97) involving a sum of the proper modes together with the continuous spectrum contribution.

The representation of the form (6.105) can be generalized to the case when the field is excited by arbitrary sources (but confined to some region $\rho < b$, $|z| < d$). The generalized representation, valid for domain (6.104), is written thus:

$$\mathbf{E}(\mathbf{r}) = \sum_{m=-\infty}^{+\infty}\left[\sum_\nu a_{m,\nu}\,\mathbf{E}_{m,\nu}(\mathbf{r}) + \sum_{\alpha=0}^{\mathrm{x}}\int_{\Gamma_\gamma} a_{m,s,\alpha}(q)\,\mathbf{E}_{m,s,\alpha}(\mathbf{r},q)\,\mathrm{d}q\right],$$

$$\mathcal{H}(\mathbf{r}) = \sum_{m=-\infty}^{+\infty}\left[\sum_\nu a_{m,\nu}\,\mathcal{H}_{m,\nu}(\mathbf{r}) + \sum_{\alpha=0}^{\mathrm{x}}\int_{\Gamma_\gamma} a_{m,s,\alpha}(q)\,\mathcal{H}_{m,s,\alpha}(\mathbf{r},q)\,\mathrm{d}q\right], \tag{6.107}$$

where $\nu = 1, 2, \ldots$ and $s = (+)$ for positive z, and are replaced by $(-\nu)$ and $(-s)$ otherwise. The terms

$$\mathbf{E}_{m,\pm\nu}(\mathbf{r}) = \mathbf{E}_{m,\pm\nu}(\rho)\exp[-\mathrm{i}m\phi \mp \mathrm{i}k_0\,p_{m,\nu}\,z],$$

$$\mathcal{H}_{m,\pm\nu}(\mathbf{r}) = \mathcal{H}_{m,\pm\nu}(\rho)\exp[-\mathrm{i}m\phi \mp \mathrm{i}k_0\,p_{m,\nu}\,z], \tag{6.108}$$

corresponding to the nonspectral poles, represent the leaky-mode fields. Thus, the representation (6.107) expresses the field in the duct in the form of an expansion in the leaky modes plus the integrals describing the remaining portion of the radiation field. The quantities $a_{m,\nu}$ may be called the excitation coefficients of leaky modes.

It is important to note that the orthogonality relations of §6.4 are invalid for the leaky modes. This is evident from the fact that the field of each leaky mode tend to infinity as $\rho \to \infty$. However, the modal fields $\mathbf{E}_{m,\nu}$, $\mathcal{H}_{m,\nu}$ and $\mathbf{E}_{m,s,\alpha}$, $\mathcal{H}_{m,s,\alpha}$ occurring in the expansion (6.107) satisfy some different orthogonality relations which may be written

$$\begin{aligned}
J_{\tilde{m},\tilde{\nu}}^{m,\nu} &= \int_0^{2\pi}\mathrm{d}\phi\int_{L_\gamma}\left(\mathbf{E}_{m,\nu}(\mathbf{r}) \times \mathcal{H}_{\tilde{m},\tilde{\nu}}^{(\mathrm{T})}(\mathbf{r}) - \mathbf{E}_{\tilde{m},\tilde{\nu}}^{(\mathrm{T})}(\mathbf{r}) \times \mathcal{H}_{m,\nu}(\mathbf{r})\right)\cdot \hat{\mathbf{z}}_0\rho\,\mathrm{d}\rho \\
&= N_{m,\nu}\,\delta_{m,-\tilde{m}}\,\delta_{\nu,-\tilde{\nu}},
\end{aligned} \tag{6.109}$$

$$J^{m,s,\alpha}_{\tilde{m},\tilde{\nu}} = \int_0^{2\pi} \mathrm{d}\phi \int_{L_\gamma} \Big(\mathbf{E}_{m,s,\alpha}(\mathbf{r},q) \times \mathcal{H}^{(\mathrm{T})}_{\tilde{m},\tilde{\nu}}(\mathbf{r})$$

$$- \mathbf{E}^{(\mathrm{T})}_{\tilde{m},\tilde{\nu}}(\mathbf{r}) \times \mathcal{H}_{m,s,\alpha}(\mathbf{r},q) \Big) \cdot \hat{\mathbf{z}}_0 \rho \, \mathrm{d}\rho = 0, \qquad (6.110)$$

$$J^{m,s,\alpha}_{\tilde{m},\tilde{s},\tilde{\alpha}} = \int_0^{2\pi} \mathrm{d}\phi \int_{L_\gamma} \Big(\mathbf{E}_{m,s,\alpha}(\mathbf{r},q) \times \mathcal{H}^{(\mathrm{T})}_{\tilde{m},\tilde{s},\tilde{\alpha}}(\mathbf{r},\tilde{q})$$

$$- \mathbf{E}^{(\mathrm{T})}_{\tilde{m},\tilde{s},\tilde{\alpha}}(\mathbf{r},\tilde{q}) \times \mathcal{H}_{m,s,\alpha}(\mathbf{r},q) \Big) \cdot \hat{\mathbf{z}}_0 \rho \, \mathrm{d}\rho$$

$$= N_{m,s,\alpha}(q)\, \delta(q-\tilde{q})\, \delta_{m,-\tilde{m}}\, \delta_{s,-\tilde{s}}\, \delta_{\alpha,\tilde{\alpha}}, \qquad (6.111)$$

where the normalization $N_{m,\nu}$ is obtainable from (6.70) by making the replacement $n \to \nu$ and setting, for the frequency range (2.71), $\hat{\alpha} = \mathrm{x}$. The contour L_γ is shown in Figure 6.6, and it consists of two straight lines. For the first line given by $0 < \rho < \varrho_0$ where $\varrho_0 = \max\{a,b\}$, the variable ρ is purely real. On the other part of the contour where $\mathrm{Re}\,(\rho) > \varrho_0$, $\rho = \mathrm{Re}\,(\rho) + \mathrm{i}[\mathrm{Re}\,(\rho) - \varrho_0]\tan\gamma$, so that ρ is complex. Other notations in the relations (6.109) to (6.111) are the same as in (6.66) to (6.68). A procedure of establishing (6.109) to (6.111) is analogous with that used in obtaining the orthogonality relations of §6.4 if one takes into account that, for $\rho \to \infty$ and $\rho \in L_\gamma$, the fields $\mathbf{E}_{m,\nu}$, $\mathcal{H}_{m,\nu}$ of leaky modes vanish, whereas the fields $\mathbf{E}_{m,s,\alpha}$, $\mathcal{H}_{m,s,\alpha}$ with $q \in \Gamma_\gamma$ remain to be oscillating quantities with finite variation (Kondrat'ev et al., 1994, 1996). Consequently, the orthogonality conditions for leaky modes and improper modes with $q \in \Gamma_\gamma$ are functionally identical with the previous conditions (6.66) to (6.68), provided that the infinite real cross-section is replaced by the cross-section which is real for $\rho < \varrho_0$ and becomes complex for $\rho > \varrho_0$.

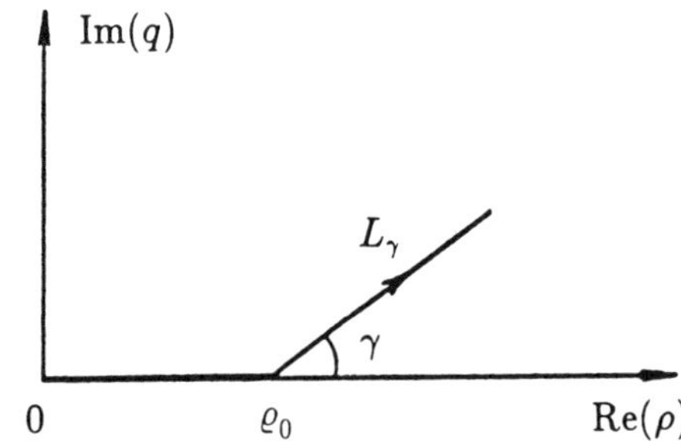

Figure 6.6. The L_γ contour in the complex ρ plane.

Now discuss briefly what happens when the proper modes exist. For example, let the eigenmode guided by the duct interface contribute to the total field (see Adachi, 1966). Then the proper-mode pole in the lower half of q plane appears. If this pole is not intercepted when passing to the contour Γ_γ, then the field of the proper mode is added to (6.107) and thereby allowed within the summation sign. Otherwise, the residue, reversed in sign, of the proper-mode pole appears, and the residue contribution together with the proper-mode contribution will cancel in pair. This implies, in effect, that the field of the proper mode turns out to be allowed within the integration sign. Regardless, the forms of the expansion (6.107) and the orthogonality relations (6.109) to (6.111) preserve, so that allowance for the proper modes will not change the above formulation.

As we now have orthogonality relations and normalization expressions for leaky modes and improper continuous modes whose transverse wavenumbers q belong to Γ_γ, results which were derived for proper modes in earlier sections can simply be extended to apply to leaky modes. In particular, with the help of the orthogonality relations (6.109) to (6.111), the excitation coefficient standing in the expansion (6.107) can be calculated by using the method of § 6.6, based on Lorentz's lemma, if instead of integration over a real cross-section of the duct, as made in § 6.6, one performs integration over a complex cross-section with $\rho \in L_\gamma$. The resulting expressions for coefficients $a_{m,\pm\nu}$ and $a_{m,s,\alpha}(q)$ will then coincide in form with (6.92) and (6.93) if we make the replacement $n \rightarrow \nu$. This allows us to operate further with leaky modes, in a sense, as with proper modes. We shall see that this is much more reasonable when the source most-effectively excites leaky modes, and the field close to and within the duct core is approximately given by the leaky-mode contributions. In particular, we shall later show that a ring source with the uniform electric-current distribution, being immersed in a high-density duct, is efficient in exciting the leaky modes.

It is important that, for practical implementation of the orthogonality relations (6.109) to (6.111), it is not necessary to know all the details of L_γ. More importantly, the complex cross-sectional area can always be chosen so as to ensure leaky-mode fields to be vanishing when $\rho \rightarrow \infty$. We emphasize that although our consideration was applied to leaky modes in the whistler band, it can be extended to other bands permitting the existence of leaky modes.

We also note that the theory of this section is a generalization of the approach previously suggested for leaky modes on open dielectric waveguides (Shevchenko, 1971 b; Marcuse, 1974; Sammut and Snyder, 1975).

6.8. On the relation between alternative field representations

We have seen that the total field of a duct can be determined either by Green's function methods, as discussed in Chapter 5, or by modal methods. The total field, of course, is independent of what method is used, and the possibility therefore exists to pass from one representation to another. As an example, show how this can be made, starting from the integral representation (5.11), for the simplest case of a uniform duct excited by ring currents. It was shown in §5.5 that (5.11) reduces to the form (5.47), with integration contours Γ_x and Γ_o exhibited in Figure 5.2. Now let q_1 be replaced by q on the sides of the branch cut defined by $\mathrm{Im}\,(q_1) = 0$, and let q_2 be replaced by q on the sides of the branch cut defined by $\mathrm{Im}\,(q_2) = 0$, when performing integration in (5.47) along the path Γ_x. Similarly, let q_2 be replaced by q on the sides of the branch cut given by $\mathrm{Im}\,(q_2) = 0$ when integrating along the path Γ_o. The integration is now along the real q axis. Further, it is noted that, for real q, each point p of the path Γ_x is determined as $p = p_x(q)$, while that of the path Γ_o as $p = p_o(q)$. Next, upon integrating along the entire real q axis, it is found that the terms $\tilde{f}_k\,\tilde{G}_k^{(1)}$ and $f_k\,G_k^{(1)}$ (see expressions (5.27) to (5.30)) involved in the integrands yield a zero result, and a nonzero contribution arises from the remainders of the integrands. Passing then to the integration along the *positive* real q axis, one arrives at the representation of the form (6.58).

Thus, it is clear that one representation of the total field may be obtained from another by suitably deforming the path of integration and changing the variables under the integration signs. Since both methods of representing the total electromagnetic field are quite equivalent, the advantages in using either Green's functions or eigenfunction expansions depend on the particular problem in question.

6.9. Note on the limiting transition to the case of a uniform plasma

It is useful to examine the transition to the limit of a uniform magnetoplasma in the eigenfunction expansion of §6.2. We consider the expression (6.100) for the E_ϕ component in the outer region of a uniform duct excited by a ring electric current. With the help of formulas (6.46) and (6.59), the expression (6.100) can be written as

$$
\begin{aligned}
E_\phi(\mathbf{r}) \;=\; & \sum_{\alpha=o}^{x} \int_0^\infty I_0^e Z_0\, \frac{k_0^2 b}{8}\, \frac{B_{s,\alpha}^{(1)}(q)\,J_1(k_0\tilde{q}_\alpha^{(1)}b) + B_{s,\alpha}^{(2)}(q)\,J_1(k_0\tilde{q}_\alpha^{(2)}b)}{(1 + \eta_a^{-1}n_{s,\alpha}^{(1)^2})\,C_{s,\alpha}^{(1)}(q)\,C_{s,\alpha}^{(2)}(q)} \\[2mm]
& \times\; \left[C_{s,\alpha}^{(1)}(q)\,H_1^{(1)}(k_0 q\rho) + C_{s,\alpha}^{(2)}(q)\,H_1^{(2)}(k_0 q\rho) + C_{s,\alpha}(q)\,H_1^{(2)}(k_0\,q_\alpha(q)\,\rho) \right] \\[2mm]
& \times\; \frac{\mathrm{d}p_\alpha}{\mathrm{d}q}\, \exp\left[-ik_0\,p_\alpha(q)\,|z| \right]\, \mathrm{d}q.
\end{aligned}
\tag{6.112}
$$

The contributions of possible proper modes are omitted here, as they disappear in the limiting case of a uniform plasma. Referring to formulas (6.62) to (6.64) for the coefficients $B_{s,\alpha}^{(1,2)}$, $C_{s,\alpha}^{(1,2)}$, $C_{s,\alpha}$, it is now a straightforward matter to show that, for $\tilde{N} \to N_a$,

$$
B_{s,\alpha}^{(1)}(q)\,J_1(k_0\tilde{q}_\alpha^{(1)}b) + B_{s,\alpha}^{(2)}(q)\,J_1(k_0\tilde{q}_\alpha^{(2)}b) = 2C_{s,\alpha}^{(1)}(q)\,J_1(k_0 q b),
$$

$$
C_{s,\alpha}^{(1)}(q) = C_{s,\alpha}^{(2)}(q), \qquad C_{s,\alpha}(q) = 0,
$$

whence

$$E_\phi(\mathbf{r}) = \sum_{\alpha=o}^{x} \int_0^\infty I_0^e Z_0 \, \frac{k_0^2 b}{2} \, \frac{J_1(k_0 q b) \, J_1(k_0 q \rho)}{1 + \eta_a^{-1} \, n_{s,\alpha}^{(1)^2}} \, \frac{\mathrm{d}p_\alpha}{\mathrm{d}q} \, \exp\left[-\mathrm{i}k_0 \, p_\alpha(q)\,|z|\right] \mathrm{d}q. \qquad (6.113)$$

By making use of the identity

$$\left(1 + \eta_a^{-1} n_{s,\alpha}^{(1)^2}\right)^{-1} \frac{\mathrm{d}p_\alpha}{\mathrm{d}q} = \chi_\alpha \, \frac{g_a^2}{\eta_a} \, \frac{q^2 - \eta_a}{q^2 + p_\alpha^2(q) - \varepsilon_a} \, \frac{q}{p_\alpha(q)} \, \frac{1}{\left[q^4\left(1 - \dfrac{\varepsilon_a}{\eta_a}\right)^2 - 4q^2 \dfrac{g_a^2}{\eta_a} + 4g_a^2\right]^{1/2}},$$

where $\chi_o = -\chi_x = -1$, the formula (6.113) reduces to

$$E_\phi(\mathbf{r}) = \sum_{\alpha=o}^{x} \int_0^\infty I_0^e Z_0 \, \frac{k_0^2 b}{2} \, \chi_\alpha \, \frac{g_a^2}{\eta_a} \, \frac{q^2 - \eta_a}{q^2 + p_\alpha^2(q) - \varepsilon_a} \, \frac{q}{p_\alpha(q)} \, \frac{J_1(k_0 q b) \, J_1(k_0 q \rho)}{\left[q^4\left(1 - \dfrac{\varepsilon_a}{\eta_a}\right)^2 - 4q^2 \dfrac{g_a^2}{\eta_a} + 4g_a^2\right]^{1/2}}$$

$$\times \quad \exp\left[-\mathrm{i}k_0 \, p_\alpha(q)\,|z|\right] \mathrm{d}q. \qquad (6.114)$$

But this is exactly the expression for the E_ϕ component in the case of a uniform magnetoplasma with the density $N = N_a$ (see (3.96) and the following expressions of § 3.5). Although we started from the field expression in the outer region of the duct in order to obtain the result (6.114), it is evident that the same formula can also be deduced when starting from the field expression in the duct core. Similarly, when passing to the limit $\tilde{N} \to N_a$, all the remaining components of the field on a duct transform into those relating to the case of a uniform magnetoplasma.

It is clear that if the uniform magnetoplasma were viewed as a "waveguide" with axis along z, the formal solution of the eigenvalue problem formulated in § 6.1 would give the continuous eigenvalue spectrum $q \in [0, \infty)$ and the corresponding vector eigenfunctions, and the total source-excited field could be found by application of guided wave techniques discussed above.

6.10. The radiation field

Now we consider the field behavior far from sources in the outer region of the duct. This is the radiation field, and it is given by the integrals over the continuous eigenvalue spectrum. For definiteness, we limit ourselves to frequency intervals (2.70) and (2.71) and take either a ring electric current or ring magnetic current with the densities (5.14) and (5.15) as the sources that excite the field in the duct. The field will evidently be azimuthally symmetric. The extension to nonsymmetric field distributions is not difficult, and will not be given here.

At the chosen frequencies, only the "extraordinary" mode for which $p_x(q)$ may be real is important far from the source, as the "ordinary" mode for which $p_o(q)$ is imaginary will decay exponentially with distance from the source. Bearing this in mind, the radiation field in the outer region can be written, by analogy with (6.100), as

$$\mathcal{F}_\ell = \sum_{j=1}^{2} \frac{k_0}{4} \int_\Gamma p_\mathrm{x}'(q)\, f_\ell^{(j)}(q)\, V^{(j)}(q)\, H_n^{(2)}\!\left(Q^{(j)}\frac{\rho}{a}\right) \exp\left[-ik_0\, p_\mathrm{x}(q)\, |z|\right] \mathrm{d}q, \quad (6.115)$$

where the quantities $\mathcal{F}_\ell$ ($\ell = 1, 2, \ldots, 6$) stand for E_ρ, E_ϕ, E_z, $\mathcal{H}_\rho$, $\mathcal{H}_\phi$, $\mathcal{H}_z$, respectively; and $n = 0$ for $\ell = 3, 6$, while $n = 1$ for $\ell = 1, 2, 4, 5$. The other notations in (6.115) are as follows:

$$\begin{aligned}
&f_1^{(j)} = -(p_{s,\mathrm{x}}\, n_{s,\mathrm{x}}^{(j)} + g_a)/\varepsilon_a, \quad f_2^{(j)} = \mathrm{i}, \quad f_3^{(j)} = \mathrm{i} n_{s,\mathrm{x}}^{(j)} Q^{(j)}/(k_0 a \eta_a), \\
&f_4^{(j)} = -\mathrm{i} p_{s,\mathrm{x}}, \quad f_5^{(j)} = -n_{s,\mathrm{x}}^{(j)}, \quad f_6^{(j)} = -Q^{(j)}/k_0 a, \\
&Q^{(1)} = k_0 a q, \quad Q^{(2)} = k_0 a q_\mathrm{x}, \quad V^{(1)} = \Delta C_{s,\mathrm{x}} V_0, \quad V^{(2)} = C_{s,\mathrm{x}} V_0, \\
&V_0 = -k_0\, J_\mathrm{x} \left[4\pi\left(1 + \eta_a^{-1} n_{s,\mathrm{x}}^{(1)^2}\right) C_{s,\mathrm{x}}^{(1)}\, \Delta C_{s,\mathrm{x}}\right]^{-1},
\end{aligned} \quad (6.116)$$

where J_x is given by the expression (6.98) with $\alpha = \mathrm{x}$.

A close examination of (6.115) for the *frequency interval* (2.70) shows that only the first integral corresponding to the superscript $j = 1$ in (6.115) contributes to the radiation field in the far zone where $r = \left(\rho^2 + z^2\right)^{1/2}$ is large enough, while the second integral corresponding to the superscript $j = 2$ describes the field exponentially decaying with the radial coordinate. If we expand the Hankel function $H_n^{(2)}\left(Q^{(1)}\frac{\rho}{a}\right)$ into its large-argument approximation, the final integration can be done by the method of stationary phase, as explained in section 3.4.1 of § 3.4. Sometimes it is more convenient to evaluate integrals of the type (6.115) by the close method of steepest descents (see, for example, Felsen and Marcuvitz, 1973, Chapter 4). It consists first in determining saddle points of the integrands, deforming the integration contour so that it coincides with the lines of steepest descent through the saddle points, and then evaluating the contributions to the integral from the vicinity of each saddle point and adding the results. In the case under consideration, the saddle points $q = q_{\varsigma i}$ are roots of Equation (3.64). Here the subscript "ς" is used to distinguish the saddle points from the poles $q = q_1, q_2, \ldots$ of the integrand. When evaluating the integral with $j = 1$ in (6.115) by using the method of steepest descents, it is not difficult to verify that the contribution from the neighborhood of each saddle point coincides with that obtained by means of the method of stationary phase. In general, both methods give the same result if the integration path used in one of them can be transformed into that used in the other, without intercepting the singularities of the integrand. Since some poles of the integrand may exist, the method of steepest descents seems more preferable and therefore it will be used in what follows.

An examination of the saddle-point equation reveals three saddle points when the observation point is between the two conical caustic surfaces defined by $\theta = \theta_{S1}$ and $\theta = \theta_{S2}$ (i.e., for $\theta_{S2} < \theta < \theta_{S1}$), and only one saddle point for $\theta \leq \theta_{S2}$ and $\theta_{S1} \leq \theta \leq \pi/2$ (see, for details, §2.6 and the refractive index surface of the extraordinary wave in Figure 2.1). Note that because of symmetry about the plane $\theta = \pi/2$, we consider the field only in the half-space $0 \leq \theta \leq \pi/2$. We emphasize that for practical implementation of the method of steepest descents, it is sufficient to know the exact behavior of the lines of steepest descent only in the close vicinity of each saddle point. As an example, for the special case when $\theta_{S2} < \theta < \theta_{S1}$, Figure 6.7 shows the integration contour distorted so as to coincide with the steepest-descent lines in the neighborhoods of the saddle points. Note that the part of distorted contour which is shown dashed passes on the Riemann sheet where $\mathrm{Im}(p_x) > 0$. We stress that Figure 6.7 exhibits the situation when the proper modes are tightly bound, so that the poles $q = q_n$ of the integrand are well below the real q axis and not intercepted during the path deformation. In passing, it may be interesting to mention that a transform of the integration contour to that enclosing the point $q = q_n$ leads to the appearance, with the sign reversed, of the residue of this pole, which annuls the nth item involved in the sum in the eigenfunction expansion of the total field.

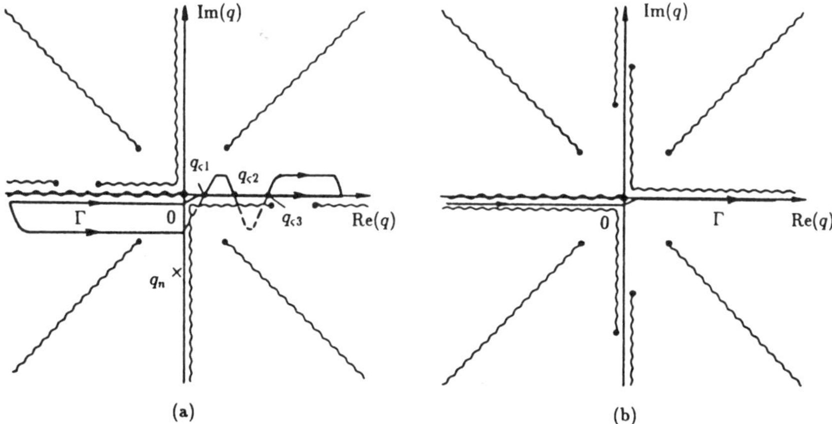

Figure 6.7. Branch singularities, saddle points and integration contours in the complex q plane for the frequency interval $\Omega_{\mathrm{H}} \ll \omega < \omega_{\mathrm{LH}}$. (a) $\alpha = $ 'x', location of the pole appropriate to the nth axisymmetric proper mode. (b) $\alpha = $ 'o'.

The asymptotic formulas for the field components in the far zone may now be written down directly. By the ordinary saddle-point procedure, one obtains

$$\mathcal{F}_\ell(\mathbf{r}) \;=\; \sum_i \frac{p'_{\mathrm{x}}(q_{\varsigma i})\, W_\ell(q_{\varsigma i})}{2r\left[q_{\varsigma i}\, p''_{\mathrm{x}}(q_{\varsigma i})\sin\theta\cos\theta\right]^{1/2}}$$

$$\times\quad \exp\left[-ik_0 q_{\varsigma i}\rho - ik_0 p_{\mathrm{x}}(q_{\varsigma i})\,|z| + i\,\frac{\pi n}{2}\right], \tag{6.117}$$

where

$$W_\ell(q) = f_\ell^{(1)}(q)\, V^{(1)}(q), \tag{6.118}$$

while the other notations are like those in (3.62). By analogy with (3.62), the 'i' summation in (6.117) covers all the saddle points $q_{\varsigma i}$. One sees from (6.117) that the field does satisfy the radiation condition at infinity, as predicted. We mentioned earlier that the improper modes of the continuous spectrum are "unphysical" since, individually, they are not absolutely square integrable (with finite power) nor do they obey the radiation condition at infinity. The result (6.117) shows that only by superposing a continuum of these modes can the physical radiation field be correctly represented.

Let us continue to find the radiation field for the *frequency interval* (2.71), assuming, in addition, that $\omega \ll \omega_H$ (the latter condition was explained in § 4.4). Unlike the preceding case, now the radiation is observed only within the Storey cone defined by $\theta = \theta_S$ (see § 2.6). The saddle points $q_{\varsigma i}$ for the first integral ($j = 1$) in (6.115) are determined again from Equation (3.64). Recall that in the frequency interval considered, Equation (3.64) gives three saddle points for $\theta < \theta_r$, and two saddle points for $\theta_r \le \theta < \theta_S$, where θ_r is the resonance-cone half angle in the observation space (see §§ 2.6 and 3.4). The second integral ($j = 2$) in (6.115) now contributes to the radiation field if $\theta < \theta_p$ where $\theta_p = \arctan p'_{\mathrm{x}}(q_p) < \theta_r$. Three saddle points $\tilde{q}_{\varsigma i}$ may be found for this integral. They are roots of the equation

$$p'_{\mathrm{x}}(q) + q'_{\mathrm{x}}(q)\tan\theta = 0. \tag{6.119}$$

As an example, for the case $\theta < \theta_p$, parts (a) and (b) of Figure 6.8, referring respectively to integrals with $j = 1$ and $j = 2$ in (6.115), show the location of the saddle points and the lines of steepest descent in the neighborhoods of the saddle points. In Figure 6.8 the dot-and-dashed lines with double and triple dots pass on the Riemann sheets defined by $\mathrm{Im}\,(p_\alpha) < 0$, $\mathrm{Im}\,(q_\alpha) > 0$ and $\mathrm{Im}\,(p_\alpha) > 0$, $\mathrm{Im}\,(q_\alpha) > 0$, respectively, the conditions $\mathrm{Re}\,(R_p) > 0$ and $-\pi < \arg q < \pi$ for the sheets mentioned being understood. Other lines in Figure 6.8 are marked by the rules accepted for Figure 6.5.

It can be shown that, for a *non-absorbing duct*, a summary contribution from the saddle points $q = q_{\varsigma 3}$ and $q = \tilde{q}_{\varsigma 1}$ lying on the negative real q axis is equal to

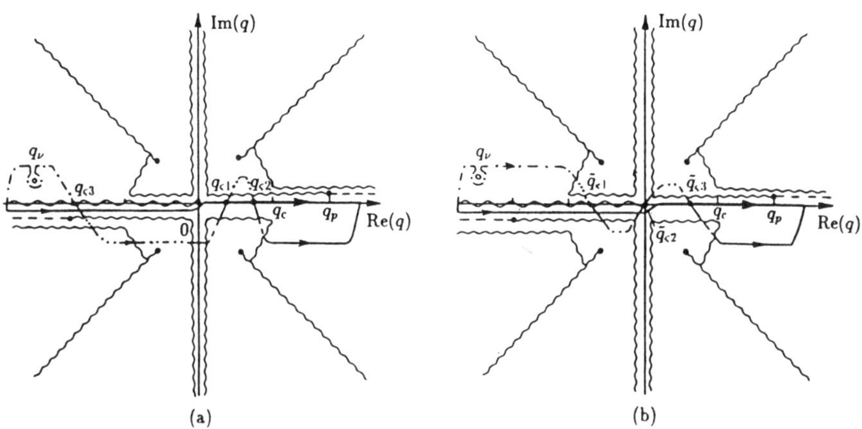

Figure 6.8. Saddle points and integration contours in the complex q plane for the frequency interval $\omega_{\mathrm{LH}} < \omega < \omega_{\mathrm{H}}/2$. (a) α = 'x', location of the pole appropriate to the νth axisymmetric leaky mode. (b) α = 'o'.

zero[*], the relations $q_{\varsigma 3} = q_{\mathrm{x}}(\tilde{q}_{\varsigma 3})$ and $\tilde{q}_{\varsigma 1} = -\tilde{q}_{\varsigma 3} = q_{\mathrm{x}}(q_{\varsigma 3})$ being valid. Moreover, the point $\tilde{q}_{\varsigma 2} = 0$ does not contribute to the radiation field. Hence the result for the range $\theta < \theta_p$ will ultimately be determined by contributions from the saddle points $q = q_{\varsigma 1,2}$ and $q = \tilde{q}_{\varsigma 3}$. For the ranges $\theta_p < \theta < \theta_r$ and $\theta_r \leq \theta < \theta_S$, only the first integral in (6.115) contributes to the radiation field. Finally, bearing in mind the location of all the saddle points involved, it can be shown that the radiation field within the Storey cone is formally represented by (6.117), with $q_{\varsigma i}$ being defined by Equation (3.64), if we adopt the following definition for $W_\ell(q)$:

$$W_\ell(q) = \begin{cases} f_\ell^{(1)}(q)\, V^{(1)}(q) & \text{if } q \in (-\infty, -q_p] \text{ and } q \in [0, q_c], \\ f_\ell^{(2)}(\hat{q})\, V^{(2)}(\hat{q}) & \text{if } q \in [-q_p, -q_c], \end{cases} \tag{6.120}$$

where $\hat{q} = -q_{\mathrm{x}}(q)$. Other intervals of q are unessential since they do not contain saddle points.

It may be emphasized that magnitudes of the field components have sharp maxima at $\theta = \theta_\nu = -\arctan\left[\mathrm{Re}\left(p'_{\mathrm{x}}(q_\nu)\right)\right]$. These may be attributed to leaky modes with $\mathrm{Im}(q_\nu) \ll |\mathrm{Re}(q_\nu)|$. The last relations implies that if $\theta = \theta_\nu$ and hence

---

[*] This fact becomes evident from Figure 5.2 by noting that the result of p integration between P_{x} and P_c along the *lower* side of the branch cut $\mathrm{Im}(q_1) = 0$, where q varies from 0 to $-q_c$, together with the result of integration between P_c and P_{x} along the *upper* side of the branch cut $\mathrm{Im}(q_2) = 0$, where q varies from $-q_c$ to $-q_p$, cancel in the case of a lossless plasma because the above two segments of the entire integration path Γ_{x} coalesce in the lossless limit, but have the opposite directions of integration.

$q_{<3} = \mathrm{Re}\,(q_\nu)$, then the coefficient $C^{(1)}_{s,x}$ is close to zero, and the quantity $V^{(1)}(q)$ in (6.120) reaches the maximum value, thereby leading to the appearance of the maximum in the field pattern.

It was shown in §§ 4.3 and 4.4 that, unlike bound (proper) modes, the field of each leaky mode increases as $\rho \to \infty$ although it is still decreasing with z. There exists, however, an angle interval, individual for each leaky mode, within which its field decays exponentially with r and where the leaky mode has a physical meaning. For a given leaky mode, with indices m, ν say, this angle is specified as $\theta < \theta^{(r)}_{m,\nu}$ for the half-space $\theta < \pi/2$ (respectively, $\pi - \theta^{(r)}_{m,\nu} < \theta < \pi$ for the half-space $\pi/2 < \theta < \pi$), where $\theta^{(r)}_{m,\nu}$ is the angle such that the line of steepest descent through the saddle point appropriate to the direction $\theta = \theta^{(r)}_{m,\nu}$ passes also through the pole $q = q_{m,\nu}$. Further, we shall deal with axisymmetric modes, and the index m will be omitted. It is clear from Figure 6.8 (a) that $\theta^{(r)}_\nu < \theta_\nu < \theta_r$ for each leaky mode. The situation is depicted schematically in Figure 6.9. From Figure 6.8 (a), it is evident that the leaky-mode poles may contribute in the steepest-descent formulation. Thus, one sees that for the region $\theta < \theta^{(r)}_\nu$ the νth leaky-mode pole may be intercepted during the path deformation into the line of steepest descent. As a result, the geometric-optical field (6.117) will be augmented by the leaky-mode contribution arising from the residue of the pole q_ν. However, since this pole lies in the "valley" region in the vicinity of the appropriate saddle point, the leaky-mode field decays exponentially away in all radial directions within the angular region $\theta < \theta^{(r)}_\nu$, and does not contribute to the radiation field. Outside the region $\theta < \theta^{(r)}_\nu$, the νth leaky mode cannot appear in the steepest-descent formulation because the pole q_ν is not intercepted. We note that in the region $\theta > \theta^{(r)}_\nu$ the power leakage of the νth leaky mode does not occur directly into the field but is carried into the radiation field via the space wave contribution (the contribution of the continuous spectrum).

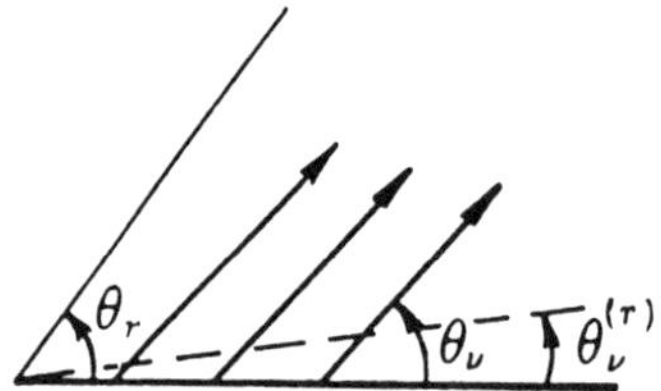

Figure 6.9. Diagram to show power flow for a leaky mode; see text for discussion.

We emphasize that the above expressions for the radiation field obtained within the usual geometrical optics approximation must be refined in the vicinity of the caustic directions. This can be made by analogy with that in a homogeneous magnetoplasma. Also, we note that in the preceding analysis, we omitted the contribution from a lateral wave (e. g., Felsen and Marcuvitz, 1973, Chapter 5, § 3). This contribution may become apparent when the saddle point occurs in the vicinity of the branch point, either $q = q_{max}$ for the frequency interval (2.70) or $q = -q_p$ for the frequency interval (2.71). However, the lateral-wave contribution drops with r more rapidly than the geometric-optical radiation field (6.117), and is therefore unimportant for our future discussion of the radiation pattern in the outer region of the duct.

6.11. The radiation pattern and the total radiated power

Once the radiation fields are known, the radiation pattern in an ambient medium may be found by using the general formulation of § 3.7. Let us discuss the main features of the radiation pattern in the whistler band, again using the ring sources with the current densities (5.14) and (5.15). Starting from the formula (6.117) for the radiation field which is emitted from the chosen sources into the outer region of the duct, and employing the procedure presented in § 3.7, the radiation pattern can be shown to be given by

$$\mathcal{D}(\theta, \phi) \equiv \mathcal{D}(\theta) = \sum_i \frac{p_x'^2(q_{\varsigma i})\, P(q_{\varsigma i})}{8 \sin \theta \cos^2 \theta \, |q_{\varsigma i}\, p_x''(q_{\varsigma i})|}, \tag{6.121}$$

where

$$P(q) = Z_0^{-1} \operatorname{Re}\left[W_1(q)\, W_5^*(q) - W_2(q)\, W_4^*(q) \right]. \tag{6.122}$$

The notations used in the preceding expressions are defined in § 6.10. The radiation pattern (6.121) can be examined by analogy with § 3.7. Observe that in the frequency interval (2.70), the radiation pattern has integrable singularities $\mathcal{D}(\theta) \sim O\left(\sqrt{\theta_{S1} - \theta}\right)$ and $\mathcal{D}(\theta) \sim O\left(\sqrt{\theta - \theta_{S2}}\right)$ in the vicinity of the caustic directions when $\theta \to \theta_{S1} - 0$ and $\theta \to \theta_{S2} + 0$, respectively. In the frequency interval (2.71), the radiation pattern $\mathcal{D}(\theta)$ has an integrable singularity given by $\mathcal{D}(\theta) \sim O\left(\sqrt{\theta_S - \theta}\right)$ near the caustic direction $\theta = \theta_S$. Unlike the case of a homogeneous magnetoplasma (see § 3.7), $\mathcal{D}(\theta)$ has no singularity at $\theta \to 0$ since conical-refraction whistler-mode waves cannot be excited in the outer region when we are dealing with an infinitely long duct. This fact follows (theoretically) from our expressions and is intuitive from structures of refractive index surface and ray surface for the extraordinary (whistler) mode. Note that $\mathcal{D}(\theta)$ has at $\theta = \theta_\nu$ ($\nu = 1, 2, \ldots$) sharp peaks corresponding to leaky modes, all the peaks lying within the resonance cone: $\theta_\nu < \theta_r$.

The power P_r radiated outwards is found by integrating the radiation pattern over the total solid angle:

$$P_r = \int_0^{2\pi} d\phi \int_0^{\pi} \mathcal{D}(\theta, \phi) \sin \theta d\theta. \tag{6.123}$$

The integration in (6.123) is more easily handled if one employs the following relations which are readily verified:

$$p_x'(q) \operatorname{sgn} q = \frac{n_x'(\vartheta) \cos \vartheta - n_x(\vartheta) \sin \vartheta}{n_x'(\vartheta) \sin \vartheta + n_x(\vartheta) \cos \vartheta} \,,$$

$$p_x''(q) = -\frac{n_x^2(\vartheta) + 2n_x'^2(\vartheta) - n_x''(\vartheta)\, n_x(\vartheta)}{(n_x'(\vartheta) \sin \vartheta + n_x(\vartheta) \cos \vartheta)^3} \,, \tag{6.124}$$

$$\frac{1}{\cos^2 \theta} = 1 + \tan^2\theta = 1 + p_x'^2(q_{\varsigma i}) = \frac{n_x'^2(\vartheta_i) + n_x^2(\vartheta_i)}{(n_x'(\vartheta_i) \sin \vartheta_i + n_x(\vartheta_i) \cos \vartheta_i)^2} \,,$$

$$\vartheta_i = \arctan\left[|q_{\varsigma i}|/p_x(q_{\varsigma i})\right].$$

In the above formulas, the prime indicates a derivative with respect to an argument, and $n_x(\vartheta)$ is obtained from (2.58) where we put $\varepsilon = \varepsilon_a$, $g = g_a$, and $\eta = \eta_a$. Substituting from (6.124) into (6.121), we get

$$\mathcal{D}(\theta) = \frac{\sin \theta}{8 \cos^2 \theta} \sum_i P(n_x(\vartheta_i) \sin \vartheta_i) \frac{n_x'(\vartheta_i) \sin \vartheta_i + n_x(\vartheta_i) \cos \vartheta_i}{n_x(\vartheta_i) \sin \vartheta_i}$$

$$\times \frac{n_x'^2(\vartheta_i) + n_x^2(\vartheta_i)}{|n_x^2(\vartheta_i) - 2n_x'^2(\vartheta_i) - n_x''(\vartheta_i)\, n_x(\vartheta_i)|} \,. \tag{6.125}$$

This form is suitable for readily performing the integration of $\mathcal{D}(\theta)$ over the total solid angle. To do this in the whistler band (2.69), the saddle-point condition (3.64) can be replaced by the equivalent relation

$$\pm \theta = \vartheta - \arctan\left[n_x'(\vartheta)/n_x(\vartheta)\right], \tag{6.126}$$

which allows determining the quantities $\vartheta_i(\theta)$, whence

$$\left|\frac{\partial \vartheta}{\partial \theta}\right| = \frac{n_x'^2(\vartheta) + n_x^2(\vartheta)}{\left|n_x^2(\vartheta) + 2n_x'^2(\vartheta) - n_x''(\vartheta)\, n_x(\vartheta)\right|} \tag{6.127}$$

(Wang and Bell, 1972 a). If (6.125) is inserted in (6.123), and use is made of the relation (6.127), then a change of variables $q = n_x(\vartheta) \sin \vartheta$ yields the convenient form

$$P_r = \int_0^{\hat{q}} \frac{\pi}{2} q^{-1} p_x'^2(q)\, P(q)\, dq, \tag{6.128}$$

where $\hat{q} = q_{\max} = [(\varepsilon_a^2 - g_a^2)/\varepsilon_a]^{1/2}$ for the frequency interval (2.70), and $\hat{q} = \infty$ for the frequency interval (2.71). Note that the latter choice of $\hat{q}$ is also valid for interval (2.72) if the quantity $P(q)$ is suitably calculated.

This covers the procedure for finding the radiation pattern and the power radiated into the ambient medium. The theory has been given for the simplest azimuthally symmetric ring currents and the frequencies belonging to the whistler band. The extension to arbitrary sources and other frequency bands is not difficult.

To determine the *total radiated power*, that is the total radiation losses of a source, one needs to know the power P_r launched into the outer region of the duct and the "partial" powers $P_{m,n}$ radiated in the individual eigenmodes specified by indices m, n. For the general case, the total radiated power P_Σ is then

$$P_\Sigma = \sum_{m,n} P_{m,n} + P_r, \tag{6.129}$$

in which the "partial" powers $P_{m,n}$ are summed up. Adding these quantities is possible due to the power orthogonality of the eigenmodes, as explained in § 6.4. One can calculate the values of $P_{m,n}$ by integrating the Poynting flux of each eigenmode through two distant cross-sections of the duct, between which the sources are located.

The total radiated power can also be computed by using the alternative power-integral method described in § 3.3 (section 3.3.1). Although it gives the same result as the approach discussed above, its practical implementation seems more complicated in the frequency intervals where the whistler mode possesses an open refractive index surface.

6.12. Radiated power distribution over the spatial spectrum of excited waves

When studying the effect of a duct on the radiation characteristics of given sources, the following question is sometimes asked: What fraction of the total radiated power goes to guided modes of the duct? For a waveguide capable of sustaining proper modes, the answer is clear from the expression (6.129). In many other cases, when, for example, a considerable amount of power goes to leaky modes, the answer is not so evident because leaky modes do not obey the power orthogonality and hence the fractions of power associated with them cannot, in general, be merely added. To clarify this question, consider leaky modes of a density enhancement, at frequencies of the interval (2.71). As mentioned earlier, even if such a duct is capable of guiding the interface-type eigenmode (Adachi, 1966), its field is tightly bound to the interface and, practically, possesses a very small amplitude when the source immersed in the duct. Hence, the power going to this mode is small, and one may write $P_\Sigma \approx P_r$.

An examination shows that if the radiation losses of leaky modes are rather small (i.e., $|\mathrm{Im}\,(p_{m,\nu})| \ll \mathrm{Re}\,(p_{m,\nu})$ and consequently $\mathrm{Im}\,(q_{m,\nu}) \ll |\mathrm{Re}\,(q_{m,\nu})|$), the integrand in the integral power representation such as (6.128) has sharp peaks at $q = |\mathrm{Re}(q_{m,\nu})| \simeq |q_{m,\nu}|$. The result of integration over each peak, that is over a certain interval $|q_{m,\nu}| \pm \Delta q$, ceases to depend on the concrete magnitude of Δq, beginning with the definite value Δq_{max} which is much less than a distance between the neighboring peaks. The result of integration over the corresponding interval can be identified as the "partial" power $P_{m,\nu}$ going to the leaky mode with indices m, ν. The summary power going to leaky modes is then calculated to a reasonable approximation by adding the powers $P_{m,\nu}$.

Apart from the above method, there is an alternative way of computing the powers $P_{m,\nu}$. One should integrate the Poynting flux of each leaky mode through two distant cross-sectional areas given by $z = \pm z_0$ and $0 \leq \rho \leq \rho^{(r)}$, where z_0 obeys the inequalities $\varrho_0 \, |\eta_a/\varepsilon_a|^{1/2} \ll |z_0| \ll |k_0 \, \mathrm{Im}\,(p_{m,\nu})|^{-1}$ and $\rho^{(r)} = |z_0| \, \min_{m,\nu} (\tan \theta_{m,\nu}^{(r)})$; here $\varrho_0 = \max\{a, b\}$ and the quantities $\theta_{m,\nu}^{(r)}$ are defined in § 6.10. As calculations show, a reasonable agreement is observed between the results obtained by this and the above methods. We emphasize that the possibility of adding the partial powers $P_{m,\nu}$ attests to the fact that the power orthogonality takes place approximately for these modes when $|\mathrm{Im}\,(p_{m,\nu})| \ll \mathrm{Re}\,(p_{m,\nu})$.

In Chapter 4 we saw that the leakage out of a sharp-walled duct may be rather small if the real propagation constants lie in the range (4.75) and in the range $\mathrm{Re}\,(p) > \tilde{\mathcal{P}}$ (see §§ 4.3 and 4.4). For the case of a smooth-walled duct, when the interface-type mode disappears, the leaky modes of the range $p > \tilde{\mathcal{P}}$ become rapidly radiating, and only the modes of range (4.75) remain slightly leaky. Therefore, in order to determine the amount of energy going to guided waves in this case, it is reasonable to find only the power going to leaky waves of range (4.75).

When dealing with calculation of the radiated power, sources must be specified so as to secure convergence of the integral that gives the power P_r. For a uniform duct at frequencies (2.71), it is instructive to verify whether the distributions (5.14) and (5.15) of ring electric and magnetic currents secure boundedness of the power P_r given by (6.128). Let $\tilde{P}^e(q)$ and $\tilde{P}^m(q)$ stand for the integrands of (6.128) when respectively either the electric current (5.14) or the magnetic current (5.15) is specified. Let, for definiteness, $b < a$. Then it can be shown that in the limiting case $q^2 \gg q_p^2$, the integrands $\tilde{P}^e(q)$ and $\tilde{P}^m(q)$ become:

$$\tilde{P}^e(q) = |I_0^e|^2 Z_0 \, \frac{\pi}{2} \, (k_0 b)^2 \, \frac{g_a^2}{(\varepsilon_a \, |\eta_a|)^{1/2}} \, q^{-2} \, J_1^2(k_0 q b) \, \tilde{W}(q), \qquad (6.130)$$

$$\tilde{P}^m(q) = |I_0^m|^2 Z_0^{-1} \, \frac{\pi}{2} \, (k_0 b)^2 \, (\varepsilon_a \, |\eta_a|)^{1/2} \, J_1^2(k_0 q b) \, \frac{\sin^2\left(k_0 q d \, \sqrt{\varepsilon_a/|\eta_a|}\right)}{(k_0 q d)^2 \, (\varepsilon_a/|\eta_a|)} \, \tilde{W}(q), \qquad (6.131)$$

where

$$\tilde{W}(q) = \frac{2}{\pi k_0 q a} \left[J_1^2(k_0 q a) + \frac{\eta_a^2}{\tilde{\eta}^2} J_0^2(k_0 q a) \right]^{-1} \tag{6.132}$$

(Zaboronkova *et al.*, 1993 c). It is clear from (6.130) and (6.131) that the integral in (6.128) converges for the ring electric current, but will diverge for the ring magnetic current if $d \to 0$. The situation is thus entirely analogous to that observed for the corresponding sources in a homogeneous magnetoplasma (see §§ 3.5 and 3.6). It is curious that the behavior of the integrands (6.130) and (6.131) at large values of q is nearly similar to that of the corresponding integrands in the power integrals for the same ring currents immersed in a homogeneous medium, except that now there appears the additional term $\tilde{W}(q)$ which accounts for the effect of the duct.

6.13. Some numerical results for ring currents

We can now proceed to some numerical applications of the above theory. These were made for ring currents (5.14) and (5.15) immersed in the duct with enhanced density. For these sources, the total power radiated from them and the power going to guided modes were found numerically. The ambient plasma parameters were chosen to be typical of the earth's ionosphere: $N_a = 10^6 \, \text{cm}^{-3}$ and $B_0 = 0.5 \, \text{Gs}$. With these values the plasma had $\omega_\text{p} = 5.64 \times 10^7 \, \text{s}^{-1}$ and $\omega_\text{H} = 8.78 \times 10^6 \, \text{s}^{-1}$. The density distribution of the duct was approximated by the step profile (4.39). The density $\tilde{N}$ within the duct and its radius a were taken to be $\tilde{N} = (3 \times 10^7 \div 3 \times 10^8) \, \text{cm}^{-3}$ and $a = 5 \, \text{m}$. These parameters are typical of the active experiments referred to in § 1.3. The frequency ω was chosen to be in the interval (2.71) and equal to $\omega = 1.88 \times 10^5 \, \text{s}^{-1}$ ($\omega/2\pi = 30 \, \text{kHz}$). The source dimensions were $b = 2.5 \, \text{m}$ and $d = 0.01 \, \text{m}$. Table 6.1 gives the total radiation resistance $R_\Sigma^e = 2P_\Sigma^e/|I_0^e|^2$ and the radiation resistance $R_W^e = 2P_W^e/|I_0^e|^2$ in the whistler leaky modes belonging to range (4.75), for the ring electric current (5.14), and the analogous quantities $G_\Sigma^m = 2P_\Sigma^m/|I_0^m|^2$ and $G_W^m = 2P_W^m/|I_0^m|^2$ for the ring magnetic current (5.15). The quantities G_Σ^m and G_W^m may be called the total "radiation conductivity" and the "radiation conductivity" in whistler leaky modes, respectively, of the magnetic current. In the preceding, P_Σ^e and P_Σ^m are the total radiated powers of the electric and magnetic currents, respectively, and P_W^e and P_W^m are the powers radiated from these sources in the whistler leaky modes of range (4.75); m_W is the number of the corresponding modes. For comparison, it is worth mentioning that when the same sources are immersed in a homogeneous plasma with the density $N = N_a$, the quantities R_Σ^e and G_Σ^m are equal to $R_{\Sigma,a}^e = 3.53 \times 10^{-3} \, \Omega$ and $G_{\Sigma,a}^m = 1.23 \times 10^{-20} \, \Omega^{-1}$.

We emphasize that leaky modes regarded in Tables 6.1 can be distinguished from each other by the number of large-scale and fine-scale variations over the radius (see §4.4). Thus, for $\tilde{N} = 3 \times 10^7\,\text{cm}^{-3}$ all the leaky modes of range (4.75) have one large-scale variation, while for $\tilde{N} = 3 \times 10^8\,\text{cm}^{-3}$ the number of large-scale variations in the modes varies from one to three. As an example, for $\tilde{N} = 10^8\,\text{cm}^{-3}$, Figure 6.10 shows the radial distribution of the E_ϕ component in the least-attenuated lower-order mode and the radiation resistance of the ring electric current in this mode versus the source radius (the complex propagation constant of the mode is $p_\nu = 2.482 \times 10^2 - \mathrm{i}\,3.762 \times 10^{-4}$).

Table 6.1. R_W^e, R_Σ^e, G_W^m, G_Σ^m, and m_W versus $\tilde{N}$

$\tilde{N}, \text{cm}^{-3}$	m_W	R_W^e, Ω	R_Σ^e, Ω	G_W^m, Ω^{-1}	G_Σ^m, Ω^{-1}
3×10^7	10	7.31×10^{-2}	7.98×10^{-2}	1.42×10^{-19}	3.66×10^{-19}
5×10^7	13	1.09×10^{-1}	1.19×10^{-1}	2.56×10^{-19}	6.08×10^{-19}
8×10^7	17	1.54×10^{-1}	1.67×10^{-1}	4.38×10^{-19}	9.67×10^{-19}
1×10^8	19	1.79×10^{-1}	1.94×10^{-1}	5.49×10^{-19}	1.20×10^{-18}
3×10^8	34	3.30×10^{-1}	3.40×10^{-1}	2.10×10^{-18}	3.47×10^{-18}

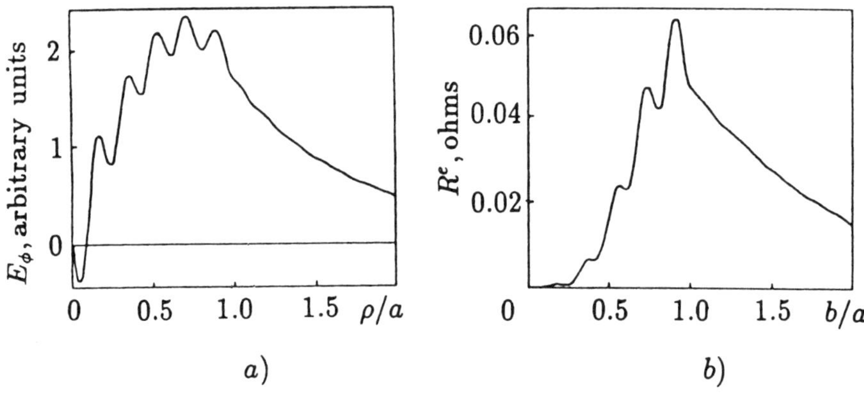

Figure 6.10. (a) E_ϕ component in the least attenuated lower-order axisymmetric leaky mode and (b) radiation resistance in it, of a ring electric current; see text for discussion.

From the results given in Table 6.1 we can draw two important conclusions. First, the presence of a duct with enhanced density leads to a noticeable increase in the total power radiated from ring sources in the VLF (whistler) band. Second, the major portion of the radiated power of a ring electric current is spent for modes conveyed by the duct if the density in it is high enough. The physical reason for this becomes fairly clear if we return to the theory of §3.5 where it has been shown that, for a ring electric current, an increase in either the plasma density or the ring radius leads to an increase in the total radiated power and to a redistribution of the power to the long-wavelength part of the spatial spectrum of emitted whistler-mode waves. The waves of this part (of ranges I and III, as defined by (3.44)) are most effectively excited if $k_0 b\,|\tilde{\eta}|^{1/2} \gg 1$, and can be trapped in the duct, as explained in §4.6 by applying Brillouin's concept.

In contrast to ring electric currents, the ring magnetic current usually launches the major portion of energy into quasi-electrostatic whistler-mode waves, so that a lesser part of the total power radiated from this current goes to guided modes. From the theory given in §§ 3.4 and 3.5, it is also clear that a similar situation would occur for electric dipoles and for sufficiently nonuniform ring electric currents immersed in the duct.* Thus we have seen that the theory of Chapter 3 together with Brillouin's concept can qualitatively explain the character of a distribution of radiated power over different types of modes in the duct.

In the above, the duct was assumed to be of infinite extent. Therefore, all leaky modes will ultimately radiate in quasi-electrostatic whistler-mode waves of the ambient magnetoplasma. The same situation will occur for the case of a duct with a finite extent $L_\parallel$ if $L_\parallel \gg |k_0 \operatorname{Im}(p_\nu)|^{-1}$. In the other possible case, when the inequalities $[k_0 \operatorname{Re}(p_\nu)]^{-1} \ll L_\parallel \ll |k_0 \operatorname{Im}(p_\nu)|^{-1}$ are satisfied, the power distribution between the guided and radiating waves would be approximately the same as for an infinitely long duct, but the radiation characteristics in the ambient plasma would change since the guided waves would re-emit mainly from the duct end. This topic will be discussed in Chapter 8.

The preceding results refer to ducts with enhanced density. The presence of a duct within which the density is diminished normally leads to a decrease in the total radiated power. Some results of the pertinent calculations for this case can be found in the work of Zaboronkova *et al.* (1993 c).

Summarizing, we now dwell on the implications of the results of this chapter. The object was to set out the modal methods for examination of the problem of radiation from given sources in the presence of a duct, and to show what is involved in the mathematically simplest special cases. We have demonstrated that there exists a convenient and simple way of obtaining the eigenfunction expansion for a source-excited field on a duct and that the eigenfunctions of a duct can be determined by solving the source-free Maxwell equations and stipulating that the solutions satisfy the boundedness conditions (6.8) when $\rho \to \infty$. An intriguing

---

* As the density increases, a nonuniformity of the electric current in a real loop antenna becomes noticeable, and the loop will develop an electric dipole moment which strongly affects the radiation characteristics (see §3.5). The required current uniformity can apparently be ensured by covering the antenna wire by a special dielectric shell.

feature of this procedure is that although the inequalities (6.8) essentially differ from the conventional radiation condition, the resultant expression, furnished via the eigenfunction expansion, yields the field satisfying that condition at infinity. We can conclude that the modal approach for representing fields on ducts has important mathematical and physical advantages. Among them are unified forms of orthogonality relations for proper and improper modes as well as convenient and physically clear expressions for mode excitation coefficients. Also, we emphasize that it is the modal representation that readily allows a subsequent generalization to the more realistic case of axially nonuniform ducts. This topic is discussed in the following chapter.

Problems

6.1. Derive expressions for proper and improper eigenfunctions of the uniform cylindrical duct for the general nonsymmetric case ($\partial/\partial\phi = 0$).

6.2. Starting from expression (6.128), derive formulas (6.130) to (6.132).

6.3. Derive an expression for the radiation power of the ring current (5.14) in the presence of a uniform duct, for the frequency interval (2.70), by the use of a power-integral method of section 3.3.1. Prove that the result coincides with that obtained in § 6.11 by integrating the radiation pattern over a total solid angle.

6.4. Consider the nonuniform ring currents specified in problem 5.1 to Chapter 5. Find the eigenfunction expansion for the field excited by such currents in the presence of a uniform duct and determine the radiation power in the whistler band (2.69) when the density in the duct is enhanced. Discuss the radiation power distribution over the spatial spectrum of excited waves for different intervals of the whistler band.

Chapter 7

Wave Propagation Along Axially Nonuniform Ducts

7.1. Introduction

In this chapter, we describe wave propagation along ducts with parameters that vary with distance z along the duct axis. Such guides, which are no longer translationally invariant, are called the *axially nonuniform*, or *z-dependent*, ducts. The description employs the eigenfunction set of the axially uniform, translationally invariant duct and leads to coupled equations for the modal amplitudes. We shall specialize our consideration to ducts whose parameters vary rather slowly with z. As a rule, such a character of variation is a common feature of both artificial and natural (e. g., magnetospheric) ducts (see Chapter 1).

We begin by writing the field equations in the form convenient for later work in this chapter. In § 6.1 we decomposed the electric and magnetic fields of the duct into longitudinal and transverse components, parallel to and perpendicular to the duct axis, respectively, and denoted by subscripts z and $\perp$. This decomposition is expressed by (6.5). If we substitute the field representation of (6.5) into the source-free Maxwell equations, i. e. Equations (2.31) with $\boldsymbol{J}^e = 0$, $\mathbf{J}^m = 0$, and compare transverse and longitudinal components, we deduce

$$\hat{\mathbf{z}}_0 \times \left(\frac{\partial \mathbf{E}_\perp}{\partial z} - \nabla_\perp E_z \right) = -\mathrm{i}k_0 \boldsymbol{\mathcal{H}}_\perp, \tag{7.1}$$

$$\hat{\mathbf{z}}_0 \times \left(\frac{\partial \boldsymbol{\mathcal{H}}_\perp}{\partial z} - \nabla_\perp \mathcal{H}_z \right) = \mathrm{i}k_0 \hat{\varepsilon}_\perp \, \mathbf{E}_\perp, \tag{7.2}$$

$$\hat{\mathbf{z}}_0 \cdot (\nabla_\perp \times \mathbf{E}_\perp) = -\mathrm{i}k_0 \mathcal{H}_z, \tag{7.3}$$

$$\hat{\mathbf{z}}_0 \cdot (\nabla_\perp \times \boldsymbol{\mathcal{H}}_\perp) = \mathrm{i}k_0 \eta E_z, \tag{7.4}$$

where $\nabla_\perp = (\nabla - \hat{\mathbf{z}}_0\, \partial/\partial z)$ represents a transverse (to z) gradient operator and $\hat{\varepsilon}_\perp$ is a transverse dyadic whose representation is as follows:

$$\hat{\varepsilon}_\perp = \begin{pmatrix} \varepsilon & -ig \\ ig & \varepsilon \end{pmatrix}. \tag{7.5}$$

Recall that the electric and magnetic fields of the *axially uniform* (z-*independent*) duct are expressible as a superposition of fields with the separable form

$$\mathbf{E}(\mathbf{r}) = \mathbf{E}(\rho, \phi)\exp(-ik_0 p z), \quad \mathcal{H}(\mathbf{r}) = \mathcal{H}(\rho, \phi)\exp(-ik_0 p z). \tag{7.6}$$

Substituting these forms into Equations (7.1) and (7.2) yields

$$\hat{\mathbf{z}}_0 \times (ik_0 p\mathbf{E}_\perp + \nabla_\perp E_z) = ik_0\mathcal{H}_\perp, \tag{7.7}$$

$$\hat{\mathbf{z}}_0 \times (ik_0 p\mathcal{H}_\perp + \nabla_\perp \mathcal{H}_z) = -ik_0\hat{\varepsilon}_\perp \mathbf{E}_\perp. \tag{7.8}$$

Clearly the fields of modes of the z-independent duct satisfy Equations (7.7) and (7.8), in which one should only take the appropriate propagation constants and modal fields.

In the forthcoming analysis, the plasma density will be assumed to be a smooth function $N(\rho, z)$, of both radial and longitudinal coordinates. It is implied that, beginning with a certain value $\rho = a(z)$ which, in general, is individual for each position z along the duct axis, that is for $\rho > a(z)$, the function $N(\rho, z)$ becomes equal to the constant density N_a of an ambient uniform plasma. For a discontinuous (step) profile of the density, we can also obtain the correct results, provided we interpret the fields as the limit of those for a continuous, or smoothed-out, profile.

7.2. The local field-structure. The method of local modes

On a waveguide that is not translationally invariant, we can no longer express the fields $\mathbf{E}$ and $\mathcal{H}$ in the separable form of representation (7.6). To describe the wave propagation along such ducts, we apply the *method of local modes*, known also as the *method of cross-sections* (Katzenelenbaum, 1961). This method is well developed for closed and open dielectric z-dependent waveguides (e. g., Katzenelenbaum, 1961; Shevchenko, 1971 a; Snyder and Love, 1983), and is extended here to the case of an open magnetized plasma waveguide. According to this method, the field of an axially nonuniform duct in each cross-section $z = $ const is expanded over the vector eigenfunctions, $\mathbf{E}_{m,n}$, $\mathcal{H}_{m,n}$ and $\mathbf{E}_{m,s,\alpha}$, $\mathcal{H}_{m,s,\alpha}$, of some other (auxiliary) axially uniform duct, provided that all the parameters of this auxiliary duct vary with ρ in the same manner as the corresponding parameters of the z-dependent duct at the given position z. We therefore express the total transverse field of the z-dependent duct by

$$
\mathbf{E}_\perp(\rho, \phi, z) = \sum_{m=-\infty}^{+\infty} e^{-im\phi} \left[\sum_{n=1}^{n_m} b_{m,n}\, \mathbf{E}_{\perp;m,n} \right.
$$
$$
\left. + \sum_{\alpha=o}^{x} \int_0^\infty b_{m,s,\alpha}(q)\, \mathbf{E}_{\perp;m,s,\alpha}(q)\, \mathrm{d}q \right],
\tag{7.9}
$$

$$
\mathcal{H}_\perp(\rho, \phi, z) = \sum_{m=-\infty}^{+\infty} e^{-im\phi} \left[\sum_{n=1}^{n_m} c_{m,n}\, \mathcal{H}_{\perp;m,n} \right.
$$
$$
\left. + \sum_{\alpha=o}^{x} \int_0^\infty c_{m,s,\alpha}(q)\, \mathcal{H}_{\perp;m,s,\alpha}(q)\, \mathrm{d}q \right],
\tag{7.10}
$$

where n_m is the total number of proper modes with the azimuthal index m. It is important to note that for the *z-independent duct*, the proper eigenfunctions $\mathbf{E}_{m,n}$, $\mathcal{H}_{m,n}$ depend only on ρ, while the improper eigenfunctions $\mathbf{E}_{m,s,\alpha}$, $\mathcal{H}_{m,s,\alpha}$ depend on ρ and q, and dependence on z is involved only in amplitudes $b_{m,n}$, $c_{m,n}$ and $b_{m,s,\alpha}(q)$, $c_{m,s,\alpha}(q)$ through factors $\exp(-ik_0\, p_{m,n}\, z)$ and $\exp[-ik_0\, p_{s,\alpha}(q)\, z]$, respectively. For the *z-dependent duct*, the vector functions on the right sides of (7.9) and (7.10) depend *implicitly* on z because they are found individually for each position z along the duct, being the eigenfunctions of the auxiliary z-independent duct appropriate to this position. The dependence on z of amplitudes $b_{m,n}$, $c_{m,n}$ and $b_{m,s,\alpha}(q)$, $c_{m,s,\alpha}(q)$ is yet to be determined. Thus, all the quantities involved in the right-hand sides of expressions (7.9) and (7.10), except for $p_{s,\alpha}(q)$, depend on z. Note that functions $p_{s,\alpha}(q)$ cannot depend on z since they relate to the ambient magnetoplasma which is uniform. For notational simplicity we drop the variables ρ and z in the corresponding expressions of the field.

We emphasize that the transverse field representations (7.9) and (7.10) contain the waves propagating only in one direction along the duct. Observe that amplitudes for $\mathbf{E}_\perp$ and $\mathcal{H}_\perp$ are different. By including the functions appropriate to forward- and backward-propagating waves and making use of the relationships of (6.7) between their fields, we can introduce new modal amplitudes

$$
\begin{aligned}
d_{m,n} &= (b_{m,n} + c_{m,n})/2, & d_{m,-n} &= (b_{m,n} - c_{m,n})/2, \\
d_{m,s,\alpha} &= (b_{m,s,\alpha} + c_{m,s,\alpha})/2, & d_{m,-s,\alpha} &= (b_{m,s,\alpha} - c_{m,s,\alpha})/2,
\end{aligned}
\tag{7.11}
$$

to obtain, instead of (7.9) and (7.10),

$$\mathbf{E}_{\perp}(\rho,\phi,z) = \sum_{m=-\infty}^{+\infty} e^{-im\phi} \left[\sum_{\pm n} d_{m,n}\, \mathbf{E}_{\perp;m,n} \right.$$

$$\left. + \sum_{s,\alpha} \int_0^{\infty} d_{m,s,\alpha}(q)\, \mathbf{E}_{\perp;m,s,\alpha}(q)\, \mathrm{d}q \right], \qquad (7.12)$$

$$\boldsymbol{\mathcal{H}}_{\perp}(\rho,\phi,z) = \sum_{m=-\infty}^{+\infty} e^{-im\phi} \left[\sum_{\pm n} d_{m,n}\, \boldsymbol{\mathcal{H}}_{\perp;m,n} \right.$$

$$\left. + \sum_{s,\alpha} \int_0^{\infty} d_{m,s,\alpha}(q)\, \boldsymbol{\mathcal{H}}_{\perp;m,s,\alpha}(q)\, \mathrm{d}q \right], \qquad (7.13)$$

where the symbols $\sum\limits_{\pm n}$ and $\sum\limits_{s,\alpha}$ denote the following operations:

$$\sum_{\pm n}(\cdots) = \sum_{n=-n_m}^{-1}(\cdots) + \sum_{n=1}^{n_m}(\cdots),$$

$$\sum_{s,\alpha}(\cdots) = \sum_{s=(-)}^{(+)} \sum_{\alpha=0}^{x}(\cdots). \qquad\qquad (7.14)$$

The longitudinal fields $E_z(\rho,\phi,z)$ and $\mathcal{H}_z(\rho,\phi,z)$ are found by substituting representations (7.12) and (7.13) into Equations (7.3) and (7.4), respectively, and noting that the longitudinal and transverse components of the fields $\mathbf{E}_{m,n}$, $\boldsymbol{\mathcal{H}}_{m,n}$ and $\mathbf{E}_{m,s,\alpha}$, $\boldsymbol{\mathcal{H}}_{m,s,\alpha}$ are also related by Equations (7.3) and (7.4). Hence

$$E_z(\rho,\phi,z) = \sum_{m=-\infty}^{+\infty} e^{-im\phi} \left[\sum_{\pm n} d_{m,n}\, E_{z;m,n} \right.$$

$$\left. + \sum_{s,\alpha} \int_0^{\infty} d_{m,s,\alpha}(q)\, E_{z;m,s,\alpha}(q)\, \mathrm{d}q \right], \qquad (7.15)$$

$$\mathcal{H}_z(\rho,\phi,z) = \sum_{m=-\infty}^{+\infty} e^{-im\phi} \left[\sum_{\pm n} d_{m,n}\, \mathcal{H}_{z;m,n} \right.$$

$$\left. + \sum_{s,\alpha} \int_0^{\infty} d_{m,s,\alpha}(q)\, \mathcal{H}_{z;m,s,\alpha}(q)\, \mathrm{d}q \right]. \qquad (7.16)$$

We have thus obtained the representations for transverse and longitudinal components with the common amplitudes $d_{m,\pm n}$ and $d_{m,\pm,\alpha}$. Note that in the derivation of E_z and $\mathcal{H}_z$, we have carried the transverse gradient operator $\nabla_{\perp}$ through the integration over the improper modes. This is permissible if the first derivatives of

$\mathbf{E}_{\perp;m,s,\alpha}$ and $\mathcal{H}_{\perp;m,s,\alpha}$ are continuous. The modal fields of ducts with continuous density profiles satisfy this condition.

Combining expressions (7.12), (7.13), (7.15) and (7.16), we can write

$$
\begin{aligned}
\mathbf{E}(\rho,\phi,z) &= \sum_{m=-\infty}^{+\infty} e^{-im\phi}\left[\sum_{\pm n} d_{m,n}\,\mathbf{E}_{m,n} + \sum_{s,\alpha}\int_0^\infty d_{m,s,\alpha}(q)\,\mathbf{E}_{m,s,\alpha}(q)\,\mathrm{d}q\right], \\
\mathcal{H}(\rho,\phi,z) &= \sum_{m=-\infty}^{+\infty} e^{-im\phi}\left[\sum_{\pm n} d_{m,n}\,\mathcal{H}_{m,n} + \sum_{s,\alpha}\int_0^\infty d_{m,s,\alpha}(q)\,\mathcal{H}_{m,s,\alpha}(q)\,\mathrm{d}q\right].
\end{aligned}
\tag{7.17}
$$

The representation of the field on a z-dependent duct in the form (7.17) is commonly called the expansion over the set of forward- and backward-propagating local modes. The quantities $d_{m,\pm n}$ and $d_{m,\pm,\alpha}$ may then be called the *local-mode amplitudes*. Our further purpose is to derive the set of coupled equations for these amplitudes.

7.3. Coupled local-mode equations

To obtain the required equations for the modal amplitudes in the expansion (7.17), we substitute from (7.17) into Equations (7.1) and (7.2) and note that the local-mode fields satisfy Equations (7.7) and (7.8), where p is to be replaced by the local propagation constants $p_{m,n}(z)$ for proper modes and by the propagation constants $p_{s,\alpha}(q)$ for improper modes (note: the quantities $p_{s,\alpha}$ remain independent of z for the z-dependent duct). We thus deduce

$$
\begin{aligned}
\sum_{\tilde{m}=-\infty}^{+\infty} e^{-i\tilde{m}\phi}&\left\{\sum_{\pm\tilde{n}}(d'_{\tilde{m},\tilde{n}} + ik_0 p_{\tilde{m},\tilde{n}}\,d_{\tilde{m},\tilde{n}})(\hat{\mathbf{z}}_0\times\mathbf{E}_{\tilde{m},\tilde{n}})\right. \\
&\left.+ \sum_{\tilde{s},\tilde{\alpha}}\int_0^\infty\left[d'_{\tilde{m},\tilde{s},\tilde{\alpha}}(\tilde{q}) + ik_0 p_{\tilde{s},\tilde{\alpha}}(\tilde{q})\,d_{\tilde{m},\tilde{s},\tilde{\alpha}}(\tilde{q})\right]\left(\hat{\mathbf{z}}_0\times\mathbf{E}_{\tilde{m},\tilde{s},\tilde{\alpha}}(\tilde{q})\right)\mathrm{d}\tilde{q}\right\} \\
= -\sum_{\tilde{m}=-\infty}^{+\infty} e^{-i\tilde{m}\phi}&\left[\sum_{\pm\tilde{n}} d_{\tilde{m},\tilde{n}}\,(\hat{\mathbf{z}}_0\times\mathbf{E}'_{\tilde{m},\tilde{n}}) + \sum_{\tilde{s},\tilde{\alpha}}\int_0^\infty d_{\tilde{m},\tilde{s},\tilde{\alpha}}(\tilde{q})\left(\hat{\mathbf{z}}_0\times\mathbf{E}'_{\tilde{m},\tilde{s},\tilde{\alpha}}(\tilde{q})\right)\mathrm{d}\tilde{q}\right],
\end{aligned}
\tag{7.18}
$$

$$
\begin{aligned}
\sum_{\tilde{m}=-\infty}^{+\infty} e^{-i\tilde{m}\phi}&\left\{\sum_{\pm\tilde{n}}(d'_{\tilde{m},\tilde{n}} + ik_0 p_{\tilde{m},\tilde{n}}\,d_{\tilde{m},\tilde{n}})(\hat{\mathbf{z}}_0\times\mathcal{H}_{\tilde{m},\tilde{n}})\right. \\
&\left.+ \sum_{\tilde{s},\tilde{\alpha}}\int_0^\infty\left[d'_{\tilde{m},\tilde{s},\tilde{\alpha}}(\tilde{q}) + ik_0 p_{\tilde{s},\tilde{\alpha}}(\tilde{q})\,d_{\tilde{m},\tilde{s},\tilde{\alpha}}(\tilde{q})\right]\left(\hat{\mathbf{z}}_0\times\mathcal{H}_{\tilde{m},\tilde{s},\tilde{\alpha}}(\tilde{q})\right)\mathrm{d}\tilde{q}\right\} \\
= -\sum_{\tilde{m}=-\infty}^{+\infty} e^{-i\tilde{m}\phi}&\left[\sum_{\pm\tilde{n}} d_{\tilde{m},\tilde{n}}\,(\hat{\mathbf{z}}_0\times\mathcal{H}'_{\tilde{m},\tilde{n}}) + \sum_{\tilde{s},\tilde{\alpha}}\int_0^\infty d_{\tilde{m},\tilde{s},\tilde{\alpha}}(\tilde{q})\left(\hat{\mathbf{z}}_0\times\mathcal{H}'_{\tilde{m},\tilde{s},\tilde{\alpha}}(\tilde{q})\right)\mathrm{d}\tilde{q}\right],
\end{aligned}
\tag{7.19}
$$

where the prime denotes the operation $\partial/\partial z$, and the notations $\tilde{m}$, $\tilde{n}$, $\tilde{s}$, $\tilde{\alpha}$, $\tilde{q}$ stand for the quantities m, n, s, α, q, respectively, which are summed (integrated) over. We dot-multiply Equation (7.18) with $e^{im\phi}\mathcal{H}^{(\mathrm{T})}_{-m,-n}$, Equation (7.19) with $e^{im\phi}\mathbf{E}^{(\mathrm{T})}_{-m,-n}$ and add the results. Then we integrate over the infinite cross-section, using the orthogonality relations of §6.4. Next, we dot-multiply Equations (7.18) and (7.19) with $e^{im\phi}\mathcal{H}^{(\mathrm{T})}_{-m,-s,\alpha}(q)$ and $e^{im\phi}\mathbf{E}^{(\mathrm{T})}_{-m,-s,\alpha}(q)$, respectively, and perform similar operations, with the subsequent integration with respect to $\tilde{q}$ on the left side of resultant equations. We then obtain

$$d'_{m,n} + \mathrm{i}k_0 p_{m,n} d_{m,n} = \sum_{\pm\tilde{n}} S^m_{n;\tilde{n}}\, d_{m,\tilde{n}} + \sum_{\tilde{s},\tilde{\alpha}} \int_0^\infty S^m_{n;\tilde{s},\tilde{\alpha}}(\tilde{q})\, d_{m,\tilde{s},\tilde{\alpha}}(\tilde{q})\, \mathrm{d}\tilde{q}, \qquad (7.20)$$

$$\begin{aligned} d'_{m,s,\alpha}(q) + \mathrm{i}k_0 p_{s,\alpha}(q)\, d_{m,s,\alpha}(q) \;=\;& \sum_{\pm\tilde{n}} S^m_{s,\alpha;\tilde{n}}(q)\, d_{m,\tilde{n}} \\ &+ \sum_{\tilde{s},\tilde{\alpha}} \int_0^\infty S^m_{s,\alpha;\tilde{s},\tilde{\alpha}}(q,\tilde{q})\, d_{m,\tilde{s},\tilde{\alpha}}(\tilde{q})\, \mathrm{d}\tilde{q}, \end{aligned} \qquad (7.21)$$

where $S^m_{n;\tilde{n}}$, $S^m_{n;\tilde{s},\tilde{\alpha}}$, $S^m_{s,\alpha;\tilde{n}}$, $S^m_{s,\alpha;\tilde{s},\tilde{\alpha}}$ are *coupling coefficients* defined by

$$S^m_{n;n} = -\frac{N'_{m,n}}{2N_{m,n}}, \qquad (7.22)$$

$$S^m_{n;\tilde{n}} = \frac{2\pi}{N_{m,n}} \int_0^\infty \left(\mathbf{E}'_{m,\tilde{n}} \times \mathcal{H}^{(\mathrm{T})}_{-m,n} + \mathbf{E}^{(\mathrm{T})}_{-m,n} \times \mathcal{H}'_{m,\tilde{n}} \right) \cdot \hat{\mathbf{z}}_0 \rho\, \mathrm{d}\rho, \quad n \neq \tilde{n}, \quad (7.23)$$

$$S^m_{n;\tilde{s},\tilde{\alpha}}(\tilde{q}) = \frac{2\pi}{N_{m,n}} \int_0^\infty \left(\mathbf{E}'_{m,\tilde{s},\tilde{\alpha}}(\tilde{q}) \times \mathcal{H}^{(\mathrm{T})}_{-m,n} + \mathbf{E}^{(\mathrm{T})}_{-m,n} \times \mathcal{H}'_{m,\tilde{s},\tilde{\alpha}}(\tilde{q}) \right) \cdot \hat{\mathbf{z}}_0 \rho\, \mathrm{d}\rho, \quad (7.24)$$

$$\begin{aligned} S^m_{s,\alpha;\tilde{n}}(q) \;=\;& \frac{2\pi}{N_{m,s,\alpha}(q)} \int_0^\infty \left(\mathbf{E}'_{m,\tilde{n}} \times \mathcal{H}^{(\mathrm{T})}_{-m,s,\alpha}(q) \right. \\ &\left. + \; \mathbf{E}^{(\mathrm{T})}_{-m,s,\alpha}(q) \times \mathcal{H}'_{m,\tilde{n}} \right) \cdot \hat{\mathbf{z}}_0 \rho\, \mathrm{d}\rho, \end{aligned} \qquad (7.25)$$

$$\begin{aligned} S^m_{s,\alpha;\tilde{s},\tilde{\alpha}}(q,\tilde{q}) \;=\;& \frac{2\pi}{N_{m,s,\alpha}(q)} \int_0^\infty \left(E'_{m,\tilde{s},\tilde{\alpha}}(\tilde{q}) \times \mathcal{H}^{(\mathrm{T})}_{-m,s,\alpha}(q) \right. \\ &\left. + \; \mathbf{E}^{(\mathrm{T})}_{-m,s,\alpha}(q) \times \mathcal{H}'_{m,\tilde{s},\tilde{\alpha}}(\tilde{q}) \right) \cdot \hat{\mathbf{z}}_0 \rho\, \mathrm{d}\rho. \end{aligned} \qquad (7.26)$$

In obtaining the above expressions, we made use of formulas (6.7) by which the backward-propagating waves are related to the forward-propagating waves. The derivation of (7.22) is made by using the relation (6.66), and by noting that without loss of generality we may adopt the following convention:

$$E^{(\mathrm{T})}_{\rho;\,-m,-n} = -E_{\rho;\,m,n}, \quad E^{(\mathrm{T})}_{\phi;\,-m,-n} = E_{\phi;\,m,n}, \quad E^{(\mathrm{T})}_{z;\,-m,-n} = E_{z;\,m,n},$$

$$\mathcal{H}^{(\mathrm{T})}_{\rho;\,-m,-n} = -\mathcal{H}_{\rho;\,m,n}, \quad \mathcal{H}^{(\mathrm{T})}_{\phi;\,-m,-n} = \mathcal{H}_{\phi;\,m,n}, \quad \mathcal{H}^{(\mathrm{T})}_{z;\,-m,-n} = \mathcal{H}_{z;\,m,n}, \tag{7.27}$$

the proof of which is evident from Equations (4.10) to (4.15) if one takes into account that $p^{(\mathrm{T})}_{-m,n} = p_{m,n}$.

Equations (7.20) and (7.21) are the coupled local-mode equations which define the modal amplitudes $d_{m,n}$ and $d_{m,s,\alpha}(q)$. One may observe that the derivation of these equations is nearly similar to that for closed waveguides and open dielectric waveguides (see, for example, Katzenelenbaum, 1961; Shevchenko, 1971 a).

7.4. Alternative form of the coupling coefficients

The coupling coefficients given by (7.22) to (7.26) may be written in a more concise and convenient form. To do this, we adopt the following normalization for proper and improper modes:

$$N_{m,n} = p_{m,n}, \quad N_{m,s,\alpha} = p_{s,\alpha}(q). \tag{7.28}$$

This can always be provided by suitably renormalizing the modal fields. Then by applying the special procedure suggested in the book by Katzenelenbaum (1961), we bring the coupling coefficients to the following form:

$$S^m_{n;\,n} = -\frac{p'_{m,n}}{2p_{m,n}} = \frac{\pi}{p^2_{m,n}} \int_0^\infty \mathbf{E}^{(\mathrm{T})}_{-m,-n} \cdot \hat{\varepsilon}' \cdot \mathbf{E}_{m,n}\, \rho\, \mathrm{d}\rho, \tag{7.29}$$

$$S^m_{n;\,\tilde{n}} = \frac{2\pi}{p_{m,n}\,(p_{m,\tilde{n}} - p_{m,n})} \int_0^\infty \mathbf{E}^{(\mathrm{T})}_{-m,-n} \cdot \hat{\varepsilon}' \cdot \mathbf{E}_{m,\tilde{n}}\, \rho\, \mathrm{d}\rho, \quad n \neq \tilde{n}, \tag{7.30}$$

$$S^m_{n;\,\tilde{s},\tilde{\alpha}}(\tilde{q}) = \frac{2\pi}{p_{m,n}\,[p_{\tilde{s},\tilde{\alpha}}(\tilde{q}) - p_{m,n}]} \int_0^\infty \mathbf{E}^{(\mathrm{T})}_{-m,-n} \cdot \hat{\varepsilon}' \cdot \mathbf{E}_{m,\tilde{s},\tilde{\alpha}}(\tilde{q})\, \rho\, \mathrm{d}\rho, \tag{7.31}$$

$$S^m_{s,\alpha;\,\tilde{n}}(q) = \frac{2\pi}{p_{s,\alpha}(q)\,[p_{m,\tilde{n}} - p_{s,\alpha}(q)]} \int_0^\infty \mathbf{E}^{(\mathrm{T})}_{-m,-s,\alpha}(q) \cdot \hat{\varepsilon}' \cdot \mathbf{E}_{m,\tilde{n}}\, \rho\, \mathrm{d}\rho, \tag{7.32}$$

$$S^m_{s,\alpha;\,\tilde{s},\tilde{\alpha}}(q,\tilde{q}) = \mathscr{P}\,\frac{2\pi}{p_{s,\alpha}(q)\,[p_{\tilde{s},\tilde{\alpha}}(\tilde{q}) - p_{s,\alpha}(q)]} \int_0^\infty \mathbf{E}^{(\mathrm{T})}_{-m,-s,\alpha}(q) \cdot \hat{\varepsilon}' \cdot \mathbf{E}_{m,\tilde{s},\tilde{\alpha}}(\tilde{q})\, \rho\, \mathrm{d}\rho, \tag{7.33}$$

where

$$\hat{\varepsilon}' = \begin{pmatrix} \varepsilon' & -\mathrm{i}g' & 0 \\ \mathrm{i}g' & \varepsilon' & 0 \\ 0 & 0 & \eta' \end{pmatrix} \tag{7.34}$$

The symbol $\mathscr{P}$ in (7.33) shows that the Cauchy principal value of the integral with respect to $\tilde{q}$ is to be taken on the right side of Equation (7.21) at the point $\tilde{q} = q$ when $\tilde{s} = s$ and $\tilde{\alpha} = \alpha$.

Let us present the salient steps of the derivation of expressions (7.29) to (7.33). Consider, for some position z, the appropriate auxiliary z-independent duct, as introduced in the above. Let ζ be a longitudinal coordinate along the axis of this duct. The fields of proper modes in it can then be written as

$$
\overset{o}{\mathbf{E}}_{m,n} = \mathbf{E}_{m,n}(\rho, z) \exp[-\mathrm{i}m\phi - \mathrm{i}k_0 p_{m,n}(z)\zeta],
$$

$$
\overset{o}{\mathcal{H}}_{m,n} = \mathcal{H}_{m,n}(\rho, z) \exp[-\mathrm{i}m\phi - \mathrm{i}k_0 p_{m,n}(z)\zeta].
$$

$$(7.35)$$

Here the vector functions $\mathbf{E}_{m,n}(\rho, z)$ and $\mathcal{H}_{m,n}(\rho, z)$ are the same as those that stand under the summation sign $\sum_{\pm n}$ in (7.17). Denoting the gradient operator in (x, y, ζ)-space by $\overset{o}{\nabla}$ and differentiating the both sides of Maxwell's equations with respect to z, we can write

$$
\overset{o}{\nabla} \times \overset{o}{\mathbf{E}}{}'_{m,\tilde{n}} = -\mathrm{i}k_0 \overset{o}{\mathcal{H}}{}'_{m,\tilde{n}},
$$

$$(7.36)$$

$$
\overset{o}{\nabla} \times \overset{o}{\mathcal{H}}{}'_{m,\tilde{n}} = \mathrm{i}k_0 \hat{\varepsilon} \overset{o}{\mathbf{E}}{}'_{m,\tilde{n}} + \mathrm{i}k_0 \hat{\varepsilon}' \overset{o}{\mathbf{E}}_{m,\tilde{n}},
$$

$$(7.37)$$

$$
\overset{o}{\nabla} \times \overset{o}{\mathbf{E}}{}^{(\mathrm{T})}_{-m,-n} = -\mathrm{i}k_0 \overset{o}{\mathcal{H}}{}^{(\mathrm{T})}_{-m,-n},
$$

$$(7.38)$$

$$
\overset{o}{\nabla} \times \overset{o}{\mathcal{H}}{}^{(\mathrm{T})}_{-m,-n} = \mathrm{i}k_0 \hat{\varepsilon}^{(\mathrm{T})} \overset{o}{\mathbf{E}}{}^{(\mathrm{T})}_{-m,-n},
$$

$$(7.39)$$

where the primes stand, as in the above, for $\partial/\partial z$. By dot-multiplying Equations (7.36), (7.37), (7.38), (7.39) with $\overset{o}{\mathcal{H}}{}^{(\mathrm{T})}_{-m,-n}$, $\overset{o}{\mathbf{E}}{}^{(\mathrm{T})}_{-m,-n}$, $-\overset{o}{\mathcal{H}}{}'_{m,\tilde{n}}$, $-\overset{o}{\mathbf{E}}{}'_{m,\tilde{n}}$, respectively, and adding the results we get

$$
\overset{o}{\nabla} \cdot (\overset{o}{\mathbf{E}}{}'_{m,\tilde{n}} \times \overset{o}{\mathcal{H}}{}^{(\mathrm{T})}_{-m,-n} - \overset{o}{\mathbf{E}}{}^{(\mathrm{T})}_{-m,-n} \times \overset{o}{\mathcal{H}}{}'_{m,\tilde{n}}) = \mathrm{i}k_0 \overset{o}{\mathbf{E}}{}^{(\mathrm{T})}_{-m,-n} \cdot \hat{\varepsilon}' \cdot \overset{o}{\mathbf{E}}_{m,\tilde{n}}.
$$

$$(7.40)$$

Integrating (7.40) over an infinite cross-sectional area defined by $\zeta = \mathrm{const}$ using the two-dimensional form (6.19) of the divergence theorem then yields

$$
(p_{m,\tilde{n}} - p_{m,n}) \int_0^\infty (\mathbf{E}'_{m,\tilde{n}} \times \mathcal{H}^{(\mathrm{T})}_{-m,n} + \mathbf{E}^{(\mathrm{T})}_{-m,n} \times \mathcal{H}'_{m,\tilde{n}}) \cdot \hat{\mathbf{z}}_0 \rho \, \mathrm{d}\rho
$$

$$
- p'_{m,\tilde{n}} [1 - \mathrm{i}k_0\zeta(p_{m,\tilde{n}} - p_{m,n})] \int_0^\infty (\mathbf{E}_{m,\tilde{n}} \times \mathcal{H}^{(\mathrm{T})}_{-m,-n} - \mathbf{E}^{(\mathrm{T})}_{-m,-n} \times \mathcal{H}_{m,\tilde{n}}) \cdot \hat{\mathbf{z}}_0 \rho \, \mathrm{d}\rho
$$

$$
= \int_0^\infty \mathbf{E}^{(\mathrm{T})}_{-m,-n} \cdot \hat{\varepsilon}' \cdot \mathbf{E}_{m,\tilde{n}} \rho \, \mathrm{d}\rho + \lim_{\rho \to \infty} I^m_{n;\tilde{n}},
$$

$$(7.41)$$

where

$$
\begin{aligned}
I_{n;\tilde{n}}^{m} &= ik_0^{-1}\rho \left[\mathbf{E}'_{m,\tilde{n}} \times \mathcal{H}^{(T)}_{-m,-n} - \mathbf{E}^{(T)}_{-m,-n} \times \mathcal{H}'_{m,\tilde{n}} \right. \\
&\quad - \left. ik_0 p'_{m,\tilde{n}}\, \zeta \left(\mathbf{E}_{m,\tilde{n}} \times \mathcal{H}^{(T)}_{-m,-n} - \mathbf{E}^{(T)}_{-m,-n} \times \mathcal{H}_{m,\tilde{n}} \right) \right] \cdot \hat{\boldsymbol{\rho}}_0.
\end{aligned}
\tag{7.42}
$$

In obtaining the above expressions, we also made use of the relations (6.7) and of the fact that $p^{(T)}_{-m,n} = p_{m,n}$. For proper modes the last item on the right side of (7.41) goes to zero. From (7.41), using the orthogonality relation (6.66), we immediately arrive at (7.29) if $n = \tilde{n}$. Similarly, in the case when $n \neq \tilde{n}$, we come to (7.30).

Replacing the fields of proper modes of the discrete spectrum by the required fields of improper modes of the continuous spectrum and repeating the above procedure with the use of appropriate orthogonality relations of §6.4, we can deduce the expressions (7.31), (7.32), and the expression (7.33), except for the case when $s = \tilde{s}$ and $\alpha = \tilde{\alpha}$. We emphasize that in these situations, the terms which appear instead of $I_{n;\tilde{n}}^{m}$ give a zero contribution. It is important to note that when deriving the expression (7.33) for $S^m_{s,\alpha;\tilde{s},\tilde{\alpha}}(q,\tilde{q})$, we deal with weak convergence of the term

$$
I^m_{s,\alpha;\tilde{s},\tilde{\alpha}} = ik_0^{-1}\rho \left(\mathbf{E}'_{m,\tilde{s},\tilde{\alpha}}(\tilde{q}) \times \mathcal{H}^{(T)}_{-m,-s,\alpha}(q) - \mathbf{E}^{(T)}_{-m,-s,\alpha}(q) \times \mathcal{H}'_{m,\tilde{s},\tilde{\alpha}}(\tilde{q}) \right) \cdot \hat{\boldsymbol{\rho}}_0
\tag{7.43}
$$

for $\rho \to \infty$, as made in the theory of distributions. If $s = \tilde{s}$ and $\alpha = \tilde{\alpha}$, the term (7.43) does not converge weakly to zero, as $\rho \to \infty$, and we deduce

$$
[p_{s,\alpha}(\tilde{q}) - p_{s,\alpha}(q)] \int_0^\infty \left(\mathbf{E}'_{m,s,\alpha}(\tilde{q}) \times \mathcal{H}^{(T)}_{-m,s,\alpha}(q) + \mathbf{E}^{(T)}_{-m,s,\alpha}(q) \times \mathcal{H}'_{m,s,\alpha}(\tilde{q}) \right) \cdot \hat{\mathbf{z}}_0 \rho\, \mathrm{d}\rho
$$

$$
= \lim_{R\to\infty} \left[1 - \cos\left(k_0 R(q - \tilde{q}) \right) \right] \int_0^R \mathbf{E}^{(T)}_{-m,-s,\alpha}(q) \cdot \bar{\varepsilon}' \cdot \mathbf{E}_{m,s,\alpha}(\tilde{q})\, \rho\, \mathrm{d}\rho.
\tag{7.44}
$$

The proof may be left as an exercise for the reader. Taking into account that, weakly,

$$
\lim_{R\to\infty} \frac{1 - \cos Rz}{z} = \pi i \left[\delta_-(z) - \delta_+(z) \right],
\tag{7.45}
$$

where

$$
\delta_\pm(z) = \int_0^\infty e^{\pm i 2\pi \xi z}\, \mathrm{d}\xi = \frac{1}{2}\left(\delta(z) \pm \frac{i}{\pi}\, \mathscr{P}\, \frac{1}{z} \right)
\tag{7.46}
$$

(Madelung, 1957), we deduce from (7.44) the final expression (7.33).

7.5. The WKB solutions for guided modes

The system of singular integro-differential coupled equations (7.20) and (7.21) is too complicated to be of much use for straightforward computation, and there is no general method of seeking their exact solutions, as far as the authors are aware. The commonest way of using the coupled mode equations is in an asymptotic method of successive approximations. There is one important case where successive approximations can easily be applied, that is, when the duct parameters vary slowly enough with z, and no two of the $p_{m,n}$ nor $p_{m,n}$ and $p_{s,\alpha}$ become nearly equal. Then the coupling coefficients are small quantities, and there is a zero-order approximation solution in which coupling may be ignored. This means that all the coupling coefficients, except one, $S^m_{n;n}$, are set equal to zero, and only the term involving $S^m_{n;n}$ is retained as its contribution to the resultant value of $d_{m,n}$ is accumulating with z. Bearing this in mind, the solutions are written

$$d_{m,n}(z) = d_{m,n}(0) \left(\frac{p_{m,n}(0)}{p_{m,n}(z)}\right)^{1/2} \exp[-\mathrm{i}k_0\gamma_{m,n}(z)], \qquad (7.47)$$

where

$$\gamma_{m,n}(z) = \int_0^z p_{m,n}(\zeta)\,\mathrm{d}\zeta, \qquad (7.48)$$

and

$$d_{m,s,\alpha}(z,q) = d_{m,s,\alpha}(0,q)\,\frac{1}{p_{s,\alpha}(q)}\,\exp[-\mathrm{i}k_0 p_{s,\alpha}(q)\,z]. \qquad (7.49)$$

It is evident that (7.47) is a customary WKB solution for proper modes. It may be rewritten:

$$d_{m,n}(z) = d_{m,n}(0) \left(\frac{N_{m,n}(0)}{N_{m,n}(z)}\right)^{1/2} \exp[-\mathrm{i}k_0\gamma_{m,n}(z)]. \qquad (7.50)$$

By comparing the expressions (7.22) and (7.29) for $S^m_{n;n}$ it is not difficult to verify that the WKB solution in the form (7.50) is valid regardless of whether the normalization (7.28) for proper modes is used or not.

The above expressions give the simplest approximate solutions, and to test whether the approximation is good, we should find the next (first-order) approximation for the solutions, taking into account the weak mode coupling, and then compare the results obtained within both approximations.

7.6. The use of successive approximations. Conditions for the validity of the WKB solutions

Before examining accuracy of zero-order solutions to the coupled mode equations, it is convenient to introduce new notations $\tilde{d}_{m,n}$ and $\tilde{d}_{m,s,\alpha}$ (instead of $d_{m,n}$ and $d_{m,s,\alpha}$) via

$$d_{m,n}(z) = \tilde{d}_{m,n}(z) \left(\frac{p_{m,n}(0)}{p_{m,n}(z)}\right)^{1/2} \exp[-\mathrm{i}k_0\gamma_{m,n}(z)],$$

$$(7.51)$$

$$d_{m,s,\alpha}(z,q) = \tilde{d}_{m,s,\alpha}(z,q)\frac{1}{p_{s,\alpha}(q)}\exp[-\mathrm{i}k_0 p_{s,\alpha}(q)\,z],$$

On substituting from (7.51) into (7.20) and (7.21), we get

$$
\begin{aligned}
\tilde{d}'_{m,n}(z) &= \sum_{\pm\tilde{n}} \left(\frac{p_{m,\tilde{n}}(0)}{p_{m,n}(0)}\right)^{1/2} S^m_{n;\tilde{n}} \left(\frac{p_{m,n}(z)}{p_{m,\tilde{n}}(z)}\right)^{1/2} (1 - \delta_{n,\tilde{n}})\, \tilde{d}_{m,\tilde{n}}(z) \\
&\quad \times\ \exp\left\{ - ik_0[\gamma_{m,\tilde{n}}(z) - \gamma_{m,n}(z)] \right\} \\
&\quad + \sum_{\tilde{s},\tilde{\alpha}} \left(\frac{p_{m,n}(z)}{p_{m,n}(0)}\right)^{1/2} \int_0^\infty S^m_{n;\tilde{s},\tilde{\alpha}}(\tilde{q})\, \frac{1}{p_{\tilde{s},\tilde{\alpha}}(\tilde{q})}\, \tilde{d}_{m,s,\alpha}(z,\tilde{q}) \\
&\quad \times\ \exp\left\{ - ik_0[p_{\tilde{s},\tilde{\alpha}}(\tilde{q})\, z - \gamma_{m,n}(z)] \right\}d\tilde{q},
\end{aligned}
\tag{7.52}
$$

$$
\begin{aligned}
\tilde{d}'_{m,s,\alpha}(z,q) &= \sum_{\pm\tilde{n}} p_{s,\alpha}(q)\, S^m_{s,\alpha;\tilde{n}}(q) \left(\frac{p_{m,\tilde{n}}(0)}{p_{m,\tilde{n}}(z)}\right)^{1/2} \tilde{d}_{m,\tilde{n}}(z) \\
&\quad \times\ \exp\left\{ - ik_0[\gamma_{m,\tilde{n}}(z) - p_{s,\alpha}(q)\, z] \right\} \\
&\quad + \sum_{\tilde{s},\tilde{\alpha}} \int_0^\infty S^m_{s,\alpha;\tilde{s},\tilde{\alpha}}(q,\tilde{q})\, \frac{p_{s,\alpha}(q)}{p_{\tilde{s},\tilde{\alpha}}(\tilde{q})}\, \tilde{d}_{m,\tilde{s},\tilde{\alpha}}(z,\tilde{q}) \\
&\quad \times\ \exp\left\{ - ik_0[p_{\tilde{s},\tilde{\alpha}}(\tilde{q}) - p_{s,\alpha}(q)]\, z \right\}d\tilde{q},
\end{aligned}
\tag{7.53}
$$

where the bar on the integral sign indicates a principal value. With new notations introduced via (7.51), the zero-order solutions of § 7.5 may be written

$$
\tilde{d}_{m,n}(z) = d_{m,n}(0), \quad \tilde{d}_{m,s,\alpha}(z,q) = d_{m,s,\alpha}(0,q).
\tag{7.54}
$$

These can be inserted in the right sides of Equations (7.52) and (7.53), and the equations can be solved to give a better approximation. To simplify matters, we shall discuss this topic for the simplest case of azimuthally symmetric fields.

In this instance, $m = 0$ and the azimuthal index will be omitted in the following analysis in accordance with the convention adopted in Chapter 6. The mode coupling will now be illustrated by considering a specific problem in which the coupling coefficients are assumed to be distinct from zero only in the interval $0 \leq z \leq L$. Based on the results of solving this problem, we shall be able to formulate some general conclusions for the cases of interest.

Suppose that only one proper mode of number k, with its amplitude $\tilde{d}_k$ equal to unity, is incident from the half-space $z < 0$ at the endface $z = 0$ of the chosen interval, so that the zero-order solutions are

$$
\begin{aligned}
\tilde{d}_k(z) &= 1, \\
\tilde{d}_n(z) &= 0, \quad \text{for}\ \ n \neq k, \\
\tilde{d}_{\pm,\alpha}(z,q) &= 0.
\end{aligned}
\tag{7.55}
$$

The required solutions must satisfy the following terminal conditions at $z = 0$ and $z = L$:

$$\tilde{d}_k(0) = 1,$$
$$\tilde{d}_n(0) = 0, \quad \text{for} \quad n > 0, \quad n \neq k,$$
$$\tilde{d}_n(L) = 0, \quad \text{for} \quad n < 0, \tag{7.56}$$
$$\tilde{d}_{+,\alpha}(0, q) = \tilde{d}_{-,\alpha}(L, q) = 0.$$

The zero-order approximation (7.55) is now substituted in the right sides of Equations (7.52) and (7.53), and the equations are solved with the use of the conditions (7.56). This leaves $\tilde{d}_k(z)$ unaffected so that the zero-order solution for the kth mode is unchanged. Solutions for the other quantities $\tilde{d}_n(z)$ and $\tilde{d}_{s,\alpha}(z, q)$, however, become:

$$\tilde{d}_n(z) = \left(\frac{p_k(0)}{p_n(0)}\right)^{1/2} \int_{z_0}^z S_{n;\,k} \left(\frac{p_n(\zeta)}{p_k(\zeta)}\right)^{1/2} \exp\left\{-ik_0[\gamma_k(\zeta) - \gamma_n(\zeta)]\right\}d\zeta, \tag{7.57}$$
$$n \neq k,$$

$$\tilde{d}_{s,\alpha}(z, q) = p_{s,\alpha}(q) \int_{z_0}^z S_{s,\alpha;\,k} \left(\frac{p_k(0)}{p_k(\zeta)}\right)^{1/2} \exp\left\{-ik_0[\gamma_k(\zeta) - p_{s,\alpha}(q)\,z]\right\}d\zeta, \tag{7.58}$$

where the lower limit z_0 is defined by

$$z_0 = \begin{cases} 0, & \text{for} \quad n > 0 \quad \text{and} \quad s = (+), \\ L, & \text{for} \quad n < 0 \quad \text{and} \quad s = (-). \end{cases} \tag{7.59}$$

The above formulas represent the first-order approximation for the solutions of the coupled mode equations. If one wishes, further iterations could be effected.

The first-order approximation shows that one WKB solution for a guided mode can generate some of the others, and this constitutes the process of a distributed transformation of one mode into backward- and forward-propagating waves that forms reflection from the layer between $z = 0$ and $z = L$ and transmission through this layer.

Under the conditions

$$|d_n(z)| \ll |d_k(z)|, \quad n \neq k, \tag{7.60}$$

$$|d_{s,\alpha}(z, q)| \ll |d_k(z)|, \tag{7.61}$$

we are restricted to considering the zero-order approximation. To illustrate an application of conditions (7.60), (7.61) in more detail, we turn to the whistler frequencies belonging to the interval (2.70). Among the proper modes propagating along a density enhancement in that frequency range, in which case

$$\mathcal{P} < \ldots < p_k < \ldots < \tilde{\mathcal{P}}$$

(compare (4.74)), the greatest coupling with the kth mode will be observed for modes of numbers $n = k \pm 1$. Taking $(p_k - p_n)$ outside the integral sign in

$$\gamma_k(\zeta) - \gamma_n(\zeta) = \int_0^\zeta \left[p_k(\zeta') - p_n(\zeta') \right] \mathrm{d}\zeta'$$

at some middle point of the integration interval and neglecting variation in $(p_k - p_n)$ over the interval $0 \leq z \leq L$, the inequality (7.60) reduces to

$$\frac{|S_{k\pm1;\,k}|}{k_0\,|p_{k\pm1} - p_k|} \ll 1. \tag{7.62}$$

From (7.29) and (7.30) it is seen that the coupling coefficients $S_{k\pm1;\,k}$ may be estimated roughly by

$$|S_{k\pm1;\,k}| \sim \frac{|p_k|}{|p_{k\pm1} - p_k|\,\Lambda}, \tag{7.63}$$

with Λ being the scale of variation of the propagation constant p_k with z. In fact, for a rather gradual variation of duct parameters with z, one may set $\Lambda \sim L$. Using (7.63), the inequality (7.62) can be rewritten as the more definitive condition

$$\left(\frac{p_k}{p_{k\pm1} - p_k} \right)^2 \ll k_0\,p_k\,\Lambda. \tag{7.64}$$

If the propagation constant of the kth mode is the nearest to $\mathcal{P}$, the quantity p_{k-1} in the last inequality needs to be replaced by $\mathcal{P}$. This gives

$$\left(\frac{p_k}{p_k - \mathcal{P}} \right)^2 \ll k_0\,p_k\,\Lambda. \tag{7.65}$$

The last condition follows from (7.61) if one takes into account that it is necessary to provide the weak coupling between the kth proper mode and the improper modes of the continuous spectrum. Clearly the inequalities (7.64) and (7.65) are the conditions for the validity of the WKB solutions found in §7.5 for proper modes. Although these conditions are given for the azimuthally symmetric fields, generalization to the case of nonsymmetric field distributions can be made easily—one should only replace in (7.64) and (7.65) the propagation constants of axisymmetric modes by those of nonsymmetric modes to obtain the required conditions.

Thus, while the conditions (7.64) and (7.65) hold, the proper mode propagate adiabatically along the z-dependent duct. Conversely, when these inequalities are violated, some coupling terms become large, and therefore the WKB solutions fail. This may occur for many typical ducts with enhanced density where the plasma density decreases with distance along the duct axis. Then, as z increases, each propagation constant reaches $\mathcal{P}$ in some cross-section, commonly known as the *critical cross-section*. In its vicinity the coupling between a proper mode and improper continuous modes becomes noticeable, giving rise to radiation from the duct. To the authors' knowledge, no detailed study of coupled local-mode equations in the vicinity of the critical cross-sections has yet been made for a duct in a magnetoplasma.[*]

It is important to note that our reasoning may be extended to include leaky modes in the frequency interval (2.71), which were discussed in §§ 4.4, 4.8, and 6.7. For the frequency region (2.71), using the representation (6.107) and orthogonality relations (6.109) to (6.111), we can follow steps analogous to those which led to formulas of §§ 7.2–7.4. As a final result, we come to expressions which may be obtained from Equations (7.20) and (7.21), if we make the replacements $n \to \nu$ and $\bar{n} \to \bar{\nu}$ (note: ν and $\bar{\nu}$ are indices for leaky modes) and take Γ_γ as an integration path in the integrals with respect to q and $\bar{q}$, and L_γ as an integration path in the integrals with respect to ρ. The contours Γ_γ and L_γ were defined in § 6.7. The procedure of solving the equations for $d_{m,\nu}(z)$ and $d_{m,s,\alpha}(z,q)$ is identical with that used above for proper and improper modes, and will not be repeated here. We note that when a typical attenuation length of leaky modes considerably exceeds the longitudinal extent of the duct, the conditions for the validity of the WKB solutions for leaky modes are obtained from (7.64) and (7.65) by replacing $p_k \to \mathrm{Re}(p_\nu)$. Propagation of leaky modes along a z-dependent duct and their re-emission from the duct endface into the surrounding magnetoplasma are discussed in Chapter 8.

In conclusion, the theory of this chapter is really an extension of a similar approach described in many texts on isotropic guides (see, for example, Katzenelenbaum, 1961; Shevchenko, 1971a; Snyder and Love, 1983) to the case when a magnetoplasma is used as a filling medium. Much is known about the methods of treating coupled local-mode equations for isotropic z-dependent guides. Some of these methods have been extended here to axially-magnetized z-dependent ducts. Still, some relevant problems remain, such as the need for a rigorous description of the field behavior in the vicinity of the critical cross-section. A satisfactory solution of this problem would be of great practical and academic interest.

Problem

7.1. Derive the relation given by Equation (7.44).

---

[*] For account of whistler-mode propagation investigated by ray-tracing for ducts whose parameters vary with distance along the axis, see, for example, Strangeways (1981).

Chapter 8

Wave Re-emission from a Density Duct

8.1. Introduction

In earlier chapters, we considered possible modes and their excitation and propagation for a duct whose longitudinal extent was assumed to be infinite. It is natural to make an attempt to apply the results of previous chapters to a problem of re-emission of guided waves when they meet, traveling along the z-dependent duct, the critical cross-sections. In fact, this implies that the duct is now assumed to possess a finite extent. This case may have an important bearing on understanding of the whistler wave propagation along artificial ducts observed in some active ionospheric and modeling laboratory experiments (§ 1.3), as well as along magnetospheric ducts near the polar regions.

To the authors' knowledge, no rigorous solution has been found for a terminated open guide, even if it is filled with an isotropic medium. The problem becomes much more difficult with a magnetoplasma as a filling medium. Therefore, we will employ some approximate heuristic method, in contrast to the rigorous theory of earlier chapters. The approximation to be used is justified, in fact, by the same general physical reasons that are usually used for open-ended metallic waveguides and abruptly terminated dielectric fibers (e. g., Brown and Spector, 1957; Angulo and Chang, 1959; Vaynshteyn, 1969, 1988; see also Tiberio *et al.* (1996) and references therein). These reasons are based on making use of known rigorous solutions found for a few canonical problems (Vaynshteyn, 1969).

Although some aspects of consideration in this chapter can be related to general quasi-cylindrical plasma nonuniformities, we shall focus on the case of an artificial duct having the form of a field-aligned cylindrical enhancement of plasma density. It is assumed that the duct is excited by an electromagnetic source inside of it, and the plasma density in the duct core decays gradually to the ambient (background) value with distance from a source. Such density perturbations, which can be formed in a self-consistent way in the magnetoplasma near radiators (see § 1.3), seem promising for the efficient control of the radiation characteristics of VLF/ELF sources under terrestrial ionospheric conditions (Zaboronkova *et al.*, 1996 b). As the radiator, we take a ring source with uniform electric current. Provided that the density in the duct core is high enough, the major portion of power radiated from such a source is spent for leaky modes with small attenuation (§ 6.13). Since an extent of a typical artificial duct is much less than the attenuation lengths of leaky modes, they would re-emit almost entirely from the duct ends to the ambient plasma.

The total radiated power and its distribution over excited modes in the duct are determined by the local plasma parameters in the immediate vicinity of the source. At the same time, the structure of the radiated field in the plasma environment is greatly dependent on the integral properties of the plasma distribution in the duct as a whole. Our goal in this chapter is to give some insight into the theory of mode re-emission that enables us to estimate how the power launched from the duct end will distribute over the spatial spectrum of waves excited in the ambient plasma.

8.2. Fields of local modes

Consider a system which consists of a cylindrical field-aligned duct and an electromagnetic source inside of it with a given ring electric current. Let the current density be defined by (5.14). The plasma density $N(\rho, z)$ in the duct is assumed to be decaying gradually to the ambient density $N = N_a$ with distance z along the duct axis. Let, for definiteness, the driving frequency ω obey the inequalities $\omega_{\mathrm{LH}} < \omega \ll \omega_{\mathrm{H}} \ll \omega_{\mathrm{p}}$. It is thus implied that it belongs to the interval (2.71) appropriate to whistler waves. As mentioned earlier, at the chosen frequencies such a duct can sustain the waveguide propagation only of leaky modes. In view of the discussion presented in §§ 6.7 and 7.6, the field on the duct is now expressible as

$$
\begin{aligned}
\mathbf{E}(\rho, z) &= \sum_{\pm\nu} d_\nu\, \mathbf{E}_\nu + \sum_{s,\alpha} \int_{\Gamma_\gamma} d_{s,\alpha}(q)\, \mathbf{E}_{s,\alpha}(q)\, \mathrm{d}q, \\
\mathcal{H}(\rho, z) &= \sum_{\pm\nu} d_\nu\, \mathcal{H}_\nu + \sum_{s,\alpha} \int_{\Gamma_\gamma} d_{s,\alpha}(q)\, \mathcal{H}_{s,\alpha}(q)\, \mathrm{d}q,
\end{aligned}
\tag{8.1}
$$

where $\mathbf{E}_\nu$, $\mathcal{H}_\nu$ and $\mathbf{E}_{s,\alpha}$, $\mathcal{H}_{s,\alpha}$ represent (for each position z) the fields of leaky modes and improper continuous modes, respectively, of the auxiliary z-independent duct, whose density profile coincides with that of our z-dependent duct for the chosen position z. Representation (8.1) thus shows how the field on the z-dependent duct is expanded in its local leaky and continuous modes. Note that the integration contour Γ_γ used in (8.1) was defined in § 6.7; the symbols $\sum_{\pm\nu}$ and $\sum_{s,\alpha}$ have the same meaning as the analogous symbols $\sum_{\pm n}$ and $\sum_{s,\alpha}$ in (7.14). We emphasize that in (8.1) use is made of the fact that the chosen source excites the azimuthally symmetric field, in which case only modes with the azimuthal index $m = 0$ contribute to the total field. The index m is therefore omitted in the field expressions and the representation (8.1) can be deduced by applying the procedure described in §7.2 to the field expansion (6.107). In what follows, we shall give field expressions related only to the half-space where $z > 0$, since the field pattern excited by the ring current (5.14) is symmetric about the ring plane $z = 0$.

Now let the density in the vicinity of the ring source within the duct core be high enough, so that the major portion of the radiated power goes to leaky modes whose local real propagation constants $p'_\nu = \mathrm{Re}\,(p_\nu)$ lie in the range

$$\max\left\{\mathcal{P}, \tilde{\mathcal{P}}_c(z)\right\} < p'_1 < \ldots < p'_\nu < \ldots < \tilde{\mathcal{P}}(z), \qquad (8.2)$$

as shown in §6.13. In the preceding, the quantities $\tilde{\mathcal{P}}_c(z)$ and $\tilde{\mathcal{P}}(z)$ relate, for each position z along the duct, to the local maximum value $\tilde{N}(z) = N(0, z)$ of the density $N(\rho, z)$. The quantity $\mathcal{P}$ relates, as earlier, to the ambient plasma. The sense of inequalities (8.2) was discussed in §4.3. Recall that attenuation constants $p''_\nu = -\mathrm{Im}\,(p_\nu)$ of leaky modes obeying (8.2) are sufficiently small in the chosen frequency interval, so that $p''_{\nu,\mathrm{max}} \ll p'_\nu$ (see §§4.4 and 4.8). Moreover, the following condition is usually met for the artificial ducts encountered in actual practice:

$$k_0 p''_\nu L_\parallel \ll 1, \qquad (8.3)$$

where $L_\parallel$ is the characteristic extent of the duct along the ambient static magnetic field. This condition enables one to neglect, in a first approximation, the leakage of guided modes through the duct walls, and is assumed throughout.

In what follows, we shall also adopt the conditions

$$\left(\frac{p'_\nu}{p'_{\nu\pm1} - p'_\nu}\right)^2 \ll k_0 p'_\nu L_\parallel, \qquad \left(\frac{p'_1}{p'_1 - \mathcal{P}}\right)^2 \ll k_0 p'_1 L_\parallel \qquad (8.4)$$

to be satisfied. As explained in §7.5, while these conditions hold, the propagation of leaky modes along the duct may be considered within the WKB (adiabatic) approximation. We note that inequalities (8.3) and (8.4) can be simultaneously satisfied for artificial ducts if $\tilde{N} \gg N_a$.

Given that the density on the duct axis is gradually reducing with distance from the source, the wavelengths of leaky modes and the widths of their radial distributions would increase with z. It is clear that beginning with a certain critical cross-sectional position $z = z_{c,\nu}$, where $p'_\nu(z_{c,\nu}) = \mathcal{P}$, the duct is no longer able to support the waveguide propagation of the νth mode and the mode re-radiates into the ambient magnetoplasma. Assuming that the mode separation in p value is not very large near the duct end, it is intuitive that all the quantities $z_{c,\nu}$ are close to each other, being of the order $L_\parallel$. We note that the WKB approximation fails even at somewhat smaller values of z than the critical positions $z_{c,\nu}$ because the inequalities (8.4) cease to be valid somewhere at $z \lesssim z_{c,\nu}$. It is important that if the plasma density in the vicinity of the duct end is close enough to the ambient value, then the amplitudes of reflected waves are obviously small, and the backward-traveling modes can merely be neglected.

Bearing all the foregoing assumptions in mind, the total field in the duct core and in some neighboring region of the outer zone can be approximated to a good accuracy by

$$\mathbf{E}(\rho, z) \approx \sum_\nu d_\nu(z)\,\mathbf{E}_\nu, \quad \mathcal{H}(\rho, z) \approx \sum_\nu d_\nu(z)\,\mathcal{H}_\nu, \qquad (8.5)$$

where

$$d_\nu(z) = d_\nu(0)\left(\frac{N_\nu(0)}{N_\nu(z)}\right)^{1/2} \exp\left(-ik_0 \int_0^z p_\nu(\zeta)\,\mathrm{d}\zeta\right) \qquad (8.6)$$

(compare (7.50)), N_ν is the normalization factor of the νth leaky mode, and the 'ν' summation covers all the leaky modes whose propagation constants belong to range (8.2). In turn, the quantities $d_\nu(0)$ can be found by an application of the Lorentz-lemma method, as discussed in §§ 6.5 and 6.7. This gives

$$d_\nu(0) = a_\nu = \left.\frac{1}{N_\nu}\, 2\pi b Z_0 I_0^e\, E_{\phi;\,-\nu}^{(\mathrm{T})}\right|_{\substack{\rho=b\\z=0}}. \qquad (8.7)$$

The above field description can be used for $z < z_s$ where z_s is a certain position such that $z_s \lesssim \min_\nu(z_{c,\nu})$. It is the cross-sectional area $z = z_s$ that will then be considered as an effective radiating aperture. The exact z-coordinate of the radiating aperture is not so important when an imprecision in specifying $z_{c,\nu}$ is small enough, compared to the local wavelengths of whistler modes in the near vicinity of the duct endface. Thus, the problem of finding the field in the surrounding plasma has become one of studying the re-emission of leaky modes from the duct end. In the next sections, we show how this re-emission can be treated by an application of the standard procedure using Huygens' principle.

8.3. The use of Huygens' principle and Kirchhoff's approximation

Recall that, according to the electrodynamical formulation of Huygens' principle, if a prescribed distribution of sources occupies a volume V bounded by a closed surface S, and produces a field $\mathbf{E}^{(s)}$, $\mathbf{H}^{(s)}$ on S, then in order to find the field exterior to V, we must introduce fictitious sources on S with surface electric and magnetic currents of densities

$$\mathbf{i}_s^e = \hat{\mathbf{n}} \times \mathbf{H}^{(s)}, \quad \mathbf{i}_s^m = -\hat{\mathbf{n}} \times \mathbf{E}^{(s)},$$

where $\hat{\mathbf{n}}$ is the unit outward normal on S, and determine the field excited by these fictitious sources. For an open-ended duct, Huygens' principle, in Kirchhoff's approximation, gives the following fictitious currents:

$$\boldsymbol{\mathcal{J}}_s^e = \left(\hat{\mathbf{z}}_0 \times \boldsymbol{\mathcal{H}}^{(s)}\right)\delta(z - z_s), \quad \mathbf{J}_s^m = -\left(\hat{\mathbf{z}}_0 \times \mathbf{E}^{(s)}\right)\delta(z - z_s), \qquad (8.8)$$

where the tangential components of the field $\mathbf{E}^{(s)}$, $\boldsymbol{\mathcal{H}}^{(s)}$ at the effective radiating aperture $z = z_s$ are set approximately equal to those in the forward-propagating field on a duct. To find the field emitted from the duct endface into the half-space $z > z_s$, it is necessary to determine the field excited by currents (8.8) as if they were immersed in the uniform unbounded magnetoplasma of density N_a.

Because the leaky-mode fields are used in the representation (8.5), it is difficult to specify the field $\mathbf{E}^{(s)}$, $\boldsymbol{\mathcal{H}}^{(s)}$ in formulas (8.8) for the fictitious sources at the effective radiating aperture. For example, consider the field (8.5) for the position $z = z_s$:

$$\mathbf{E}^{(s)} = \mathbf{E}(\rho, z_s) \approx \sum_\nu \mathbf{E}_\nu^{(s)}, \quad \boldsymbol{\mathcal{H}}^{(s)} = \boldsymbol{\mathcal{H}}(\rho, z_s) \approx \sum_\nu \boldsymbol{\mathcal{H}}_\nu^{(s)}, \qquad (8.9)$$

where the quantities $\mathbf{E}_\nu^{(s)} = d_\nu(z_s)\,\mathbf{E}_\nu(\rho, z_s)$, $\boldsymbol{\mathcal{H}}_\nu^{(s)} = d_\nu(z_s)\,\boldsymbol{\mathcal{H}}_\nu(\rho, z_s)$ represent the leaky-mode contributions to the total field at $z = z_s$. Let $a(z)$ be a local radius of the duct at some position z. The density $N(\rho, z)$ varies with ρ for $\rho < a(z)$, and becomes constant for $\rho > a(z)$. As follows from the theory of §6.10, the field expansion in leaky modes of the form (8.9) is permissible only inside the duct, i.e., for $\rho \leq a(z_s)$), and in its neighboring region given by $a(z_s) < \rho < \rho^{(r)}(z_s)$. The quantity $\rho^{(r)}(z_s)$ can be approximated as $\rho^{(r)}(z_s) \simeq a(0) + (-\varepsilon_a/\eta_a)^{1/2}z_s$. Note that for the chosen frequencies, $(-\varepsilon_a/\eta_a)^{1/2} \approx Y^{-1} \ll 1$. In the region $\rho > \rho^{(r)}(z_s)$, where the field is basically determined by improper continuous modes emitting from the duct walls, the representation (8.9) is not appropriate. This becomes evident by noting that the leaky-mode fields increase to infinity as $\rho \to \infty$, while the actual total field calculated accurately for this region tends to zero with ρ. Moreover, while the condition (8.3) holds, the leaked power from the duct walls is rather small, so that the total field would rapidly decay with ρ in the zone $\rho > \rho^{(r)}(z_s)$. Hence, this field may be approximated on that part of the aperture plane by any smooth functions vanishing rather rapidly with ρ. Another difficulty arises in the course of linking these fields at the boundary $\rho = \rho^{(r)}(z_s)$ when we are dealing with a resonant magnetoplasma where excitation of quasi-electrostatic waves is allowed. In such a medium, the field approximation at the radiating aperture must provide continuity of the fictitious surface currents (8.8). The current discontinuity is usually unimportant in isotropic and non-resonant gyrotropic media, but in a resonant magnetoplasma it can lead to a divergence of the integral giving the total power radiated from the duct end. The physical nature of this divergence, which is associated with the appearance of line charges due to the current discontinuity, was explained in §§ 3.4–3.6.

Specifying the requisite fictitious sources will be discussed in some detail in the following section.

8.4. The distribution of fictitious sources on the radiating aperture

The analytical representation of the field quantities $\mathbf{E}_\nu^{(s)}$, $\mathcal{H}_\nu^{(s)}$ on the aperture plane $z = z_s$ is permissible only when the plasma density is approximated by the step-like profile such that

$$N(\rho, z) = \begin{cases} \tilde{N}(z) & \text{if } \rho < a(z), \\ N_a & \text{if } \rho > a(z). \end{cases} \tag{8.10}$$

It is assumed that with distance along the z axis, the function $\tilde{N}(z)$ gradually reduces from its maximum value, N_m, at $z = 0$ to the ambient density N_a, whereas the local radius $a(z)$ increases from its minimum value $a(0)$. The density within the duct and the local radius at the position $z = z_s$ will be denoted as $\tilde{N}_s$ and a_s, respectively, that is $\tilde{N}_s = \tilde{N}(z_s)$ and $a_s = a(z_s)$.

To represent the field in the domain $\rho < a_s$ of the radiating aperture, we use the following rigorous expressions:

$$E_{\rho;\nu}^{(s)} = -\sum_{k=1}^{2} B_k^{(s)} \frac{\tilde{n}_k^{(s)} p + \tilde{g}_s}{\tilde{\varepsilon}_s} J_1\left(\tilde{Q}_k^{(s)} \frac{\rho}{a_s}\right),$$

$$E_{\phi;\nu}^{(s)} = \mathrm{i} \sum_{k=1}^{2} B_k^{(s)} J_1\left(\tilde{Q}_k^{(s)} \frac{\rho}{a_s}\right),$$

$$\mathcal{H}_{\rho;\nu}^{(s)} = -\mathrm{i}p \sum_{k=1}^{2} B_k^{(s)} J_1\left(\tilde{Q}_k^{(s)} \frac{\rho}{a_s}\right), \tag{8.11}$$

$$\mathcal{H}_{\phi;\nu}^{(s)} = -\sum_{k=1}^{2} B_k^{(s)} \tilde{n}_k^{(s)} J_1\left(\tilde{Q}_k^{(s)} \frac{\rho}{a_s}\right),$$

and in the domain $\rho > a_s$, the following approximate expressions:

$$E_{\rho;\nu}^{(s)} \approx -C^{(s)} \frac{n_1 p + g_a}{\varepsilon_a} K_1\left(S_1^{(s)} \frac{\rho}{a_s}\right),$$

$$E_{\phi;\nu}^{(s)} \approx iC^{(s)} K_1\left(S_1^{(s)} \frac{\rho}{a_s}\right),$$

$$\mathcal{H}_{\rho;\nu}^{(s)} \approx -ipC^{(s)} K_1\left(S_1^{(s)} \frac{\rho}{a_s}\right), \tag{8.12}$$

$$\mathcal{H}_{\phi;\nu}^{(s)} \approx -C^{(s)} n_1 K_1\left(S_1^{(s)} \frac{\rho}{a_s}\right),$$

where

$$\tilde{Q}_k^{(s)} = k_0 a_s\, q_k(p, \tilde{N}_s), \quad S_1^{(s)} = k_0 a_s\, s_1, \quad s_1^2 = -q_1^2(p, N_a),$$

$$\tilde{n}_k^{(s)} = -\frac{\tilde{\varepsilon}_s}{p \tilde{g}_s}\left(p^2 + q_k^2(p, \tilde{N}_s) + \frac{\tilde{g}_s^2}{\tilde{\varepsilon}_s} - \tilde{\varepsilon}_s\right), \tag{8.13}$$

$$n_1 = -\frac{\varepsilon_a}{p g_a}\left(p^2 - s_1^2 + \frac{g_a^2}{\varepsilon_a} - \varepsilon_a\right)$$

(compare (4.58) and (4.59)). In the above expressions, the tensor elements with the subscript 's' are appropriate to a plasma with the density $N = \tilde{N}_s$, while the tensor elements with the subscript 'a' are used, as earlier, for an ambient plasma with the density $N = N_a$; the other notations were defined in section 4.2.3 of § 4.2. We emphasize that in each of the expressions (8.12), we have omitted the term that accounts for the escaping fine-scale part of the mode field because even in the domain $a_s < \rho < \rho^{(r)}(z_s)$ this term is much smaller than the retained term that describes the "trapped" large-scale part of the field. In the domain $\rho > \rho^{(r)}(z_s)$, where each field component is to be approximated by some function vanishing rather rapidly with ρ, we may use without loss of generality the expressions (8.12) to simulate the requisite behavior of the field. Note that only the expressions for transverse components are given since it is these components that will be used for further calculations of the characteristics of radiation emitted from the aperture plane.

We stress that coefficients $B_{1,2}^{(\nu)}$ and $C^{(\nu)}$ must provide the continuity of the components $E_{\phi;\nu}^{(s)}$, $\mathcal{H}_{\phi;\nu}^{(s)}$, and $\mathcal{H}_{z;\nu}^{(s)}$ at $\rho = a_s$. This point is of essential importance, because discontinuities in certain components (in particular, in the azimuthal magnetic field $\mathcal{H}_{\phi;\nu}^{(s)}$) lead to unphysical results when calculating the power radiating from the aperture, as explained in § 8.3. On finding $\mathcal{H}_{z;\nu}^{(s)}$ via the evident relation

$$\mathcal{H}_z = \frac{i}{k_0 \rho} \frac{d}{d\rho}(\rho E_\phi)$$

and satisfying the boundary conditions for the components mentioned, it is deduced that the approximate dispersion equation and the expressions for coefficients $B_{1,2}^{(s)}$

and $C^{(1)}$ can be written as follows:

$$\left[J(\tilde{Q}_1^{(s)}) + \frac{\tilde{n}_2^{(s)} - n_1}{\tilde{n}_2^{(s)} - \tilde{n}_1^{(s)}} \, K(S_1^{(s)}) \right] J(\tilde{Q}_2^{(s)}) = \frac{\tilde{n}_1^{(s)} - n_1}{\tilde{n}_2^{(s)} - \tilde{n}_1^{(s)}} \, J(\tilde{Q}_1^{(s)}) \, K(S_1^{(s)}), \quad (8.14)$$

where

$$J(\zeta) = \frac{J_1(\zeta)}{\zeta \, J_0(\zeta)}, \quad K(\zeta) = \frac{K_1(\zeta)}{\zeta \, K_0(\zeta)},$$

and

$$B_1^{(s)} = E_\nu^{(s)} (\tilde{n}_2^{(s)} - n_1) \, J_1(\tilde{Q}_2^{(s)}) \, K_1(S_1^{(s)}) \Big|_{p=p_\nu^{(s)}},$$

$$B_2^{(s)} = E_\nu^{(s)} (n_1 - \tilde{n}_1^{(s)}) \, J_1(\tilde{Q}_1^{(s)}) \, K_1(S_1^{(s)}) \Big|_{p=p_\nu^{(s)}}, \qquad (8.15)$$

$$C^{(s)} = E_\nu^{(s)} (\tilde{n}_2^{(s)} - \tilde{n}_1^{(s)}) \, J_1(\tilde{Q}_1^{(s)}) \, J_1(Q_2^{(s)}) \Big|_{p=p_\nu^{(s)}}.$$

In the preceding, $p = p_\nu^{(s)}$ are roots of the approximate dispersion equation (8.14). Note that the remaining quantities on the right sides of (8.11) and (8.12) are also taken for $p = p_\nu^{(s)}$. It can easily be shown that the roots $p_\nu^{(s)}$ are very close to the exact roots of the rigorous dispersion equation appropriate to parameters of the duct at $z = z_s$. This becomes evident from the fact that Equation (8.14) readily reduces to the form which was earlier used in the approximate dispersion equation (4.101) for a uniform duct. The complex constants $E_\nu^{(s)}$ characterizing amplitudes and phases of leaky-mode contributions at the duct end (see (8.9)) are taken from the solution to the problem of the mode adiabatic propagation, which was discussed in §8.2.

We notice that the approximate approach suggested for specifying the modal fields on the aperture plane does not provide continuity of the longitudinal field $E_{z;\nu}^{(s)}$ at $\rho = a_s$. However, this longitudinal component is, in general, small compared to the other components and does not affect the surface currents (8.8); therefore, this impression in specifying $E_{z;\nu}^{(s)}$ is insignificant.

To obtain the resulting field $\mathbf{E}^{(s)}$, $\mathcal{H}^{(s)}$ at the radiating aperture, the contributions from different modes are added, as indicated in (8.9). This above for representing the field at the duct end can be generalized to a more realistic radially-nonuniform density profile if the field in the core $\rho < a_s$ is determined by using numerical computations though in the outer region $\rho > a_s$ the previous expressions (8.12) may be used. In what follows, however, we employ the step-like approximation (8.10) for the plasma density in order to introduce the analytical methods for the problem in question in a self-contained and explicit fashion.

8.5. The characteristics of radiation re-emitted from the duct end

Once the distribution of fictitious surface currents over the radiating aperture is known, the field radiated from the duct end can be calculated by an application of the theory presented in §§ 3.2 and 3.3. Since we are mainly interested in the radiated power and its distribution over the spatial spectrum of excited waves, we have not analyzed the behavior of radiation fields but proceed to computing the radiating power. In general, the total (integral) radiated power of a source immersed in an elongated density duct can be calculated using the theory in § 6.11. One has only to substitute the duct parameters which correspond to the immediate vicinity of the radiator into the appropriate formulas. However, seeking the differential radiation characteristics (for example, the distribution of re-emitted power over the spatial spectrum) requires special consideration. To this effect, for the case of a lossless medium, the time-averaged power radiated from the duct end can obviously be written

$$
\begin{aligned}
P &= -\frac{1}{2}\,\mathrm{Re}\,\int \left[\left(\mathbf{J}_s^e(\mathbf{r})\right)^* \cdot \mathbf{E}(\mathbf{r}) + \left(\mathbf{J}_s^m(\mathbf{r})\right)^* \cdot \mathbf{H}(\mathbf{r}) \right]\,\mathrm{d}\mathbf{r} \\
&= -\frac{1}{2}\,Z_0^{-1}\,\mathrm{Re}\,\int \left[\left(\boldsymbol{\mathcal{J}}_s^e(\mathbf{r})\right)^* \cdot \mathbf{E}(\mathbf{r}) + \left(\boldsymbol{\mathcal{J}}_s^m(\mathbf{r})\right)^* \cdot \boldsymbol{\mathcal{H}}(\mathbf{r}) \right]\,\mathrm{d}\mathbf{r}, \quad (8.16)
\end{aligned}
$$

where $\mathbf{E}$ and $\boldsymbol{\mathcal{H}}$ are the fields which would be excited in the homogeneous unbounded plasma with the density N_a if the currents (8.8) were immersed in it. The total power P_Σ radiated from both ends of the duct is given, evidently, by $P_\Sigma = 2P$.

In order to represent the power P_Σ in a form which will be convenient for further computations, we can follow the steps described in § 3.3 and express first this quantity by the integral in $\mathbf{n}$-space. Integrating in this space over $p = (\mathbf{n} \cdot \hat{\mathbf{z}}_0)$ with the use of Cauchy's residue theorem and then over the azimuthal angle yields

$$
\begin{aligned}
P_\Sigma = \; & Z_0^{-1}\,\frac{\pi\left(k_0 a_s^2\right)^2}{|\eta_a|}\int_0^\infty \frac{q}{p_{\mathrm{x}}(q)}\left[q^2 + p_{\mathrm{x}}^2(q) - \varepsilon_a\right] \\
& \times \frac{(q^2 + |\eta_a|)(|f^e|^2 + |f^m|^2)\,\mathrm{d}q}{\left[q^4\left(1 + \dfrac{\varepsilon_a}{|\eta_a|}\right)^2 + 4q^2\,\dfrac{g_a^2}{|\eta_a|} + 4g_a^2\right]^{1/2}},
\end{aligned}
\qquad (8.17)
$$

where

$$
f^e = \sum_\nu f_\nu^e = \sum_\nu \left[\left(\tilde{n}_1^{(s)} - \frac{p g_a}{q^2 + p_{\mathrm{x}}^2(q) - \varepsilon_a}\right) B_1^{(s)}\,\tilde{F}_1 \right.
\qquad (8.18)
$$

$$
\left. + \left(\tilde{n}_2^{(s)} - \frac{p g_a}{q^2 + p_{\mathrm{x}}^2(q) - \varepsilon_a}\right) B_2^{(s)}\,\tilde{F}_2 + \left(n_1 - \frac{p g_a}{q^2 + p_{\mathrm{x}}^2(q) - \varepsilon_a}\right) C^{(s)} F\right]\Bigg|_{p = p_\nu^{(s)}},
$$

$$
\begin{aligned}
f^m &= \sum_\nu f_\nu^m = \sum_\nu \left\{ \frac{p_{\mathrm{x}}(q)\,|\eta_a|}{q^2 + |\eta_a|} \right. \\
&\quad \times \left[\frac{\tilde{n}_1^{(s)}p + \tilde{g}_s}{\tilde{\varepsilon}_s}\, B_1^{(s)}\tilde{F}_1 + \frac{\tilde{n}_2^{(s)}p + \tilde{g}_s}{\tilde{\varepsilon}_s}\, B_2^{(s)}\tilde{F}_2 + \frac{n_1 p + g_a}{\varepsilon_a}\, C^{(s)}F \right] \\
&\quad \left. - \frac{p_{\mathrm{x}}(q)\,g_a}{q^2 + p_{\mathrm{x}}^2(q) - \varepsilon_a}\, \left(B_1^{(s)}\tilde{F}_1 + B_2^{(s)}\tilde{F}_2 + C^{(s)}F \right) \right\}\Bigg|_{p=p_\nu^{(s)}}, \qquad (8.19)
\end{aligned}
$$

$$
\begin{aligned}
\tilde{F}_{1,2} &= \left[\tilde{Q}_{1,2}^{(s)}\, J_0(\tilde{Q}_{1,2}^{(s)})\, J_1(k_0 a_s q) - k_0 a_s q\, J_0(k_0 a_s q)\, J_1(\tilde{Q}_{1,2}^{(s)}) \right] \\
&\quad \times \left[(k_0 a_s q)^2 - \tilde{Q}_{1,2}^{(s)^2} \right]^{-1}, \qquad (8.20)
\end{aligned}
$$

$$
\begin{aligned}
F &= \left[S_1^{(s)}\, K_0(S_1^{(s)})\, J_1(k_0 a_s q) + k_0 a_s q\, J_0(k_0 a_s q)\, K_1(S_1^{(s)}) \right] \\
&\quad \times \left[(k_0 a_s q)^2 + S_1^{(s)^2} \right]^{-1}. \qquad (8.21)
\end{aligned}
$$

In the preceding expressions, the summation is over leaky modes, and the function $p_{\mathrm{x}}(q)$ appropriate to an ambient plasma is defined by (6.2).

Taking into account Equation (8.14) and relations (8.15), the expressions (8.18) to (8.21) can be rewritten

$$
\begin{aligned}
f^e &= \sum_\nu f_\nu^e = \sum_\nu \frac{1}{k_0 a_s q} \left[\left(\tilde{n}_1^{(s)} - \frac{p g_a}{q^2 + p_{\mathrm{x}}^2(q) - \varepsilon_a} \right) B_1^{(s)}\tilde{G}_1 \right. \\
&\quad + \left(\tilde{n}_2^{(s)} - \frac{p g_a}{q^2 + p_{\mathrm{x}}^2(q) - \varepsilon_a} \right) B_2^{(s)}\tilde{G}_2 \\
&\quad \left. + \left(n_1 - \frac{p g_a}{q^2 + p_{\mathrm{x}}^2(q) - \varepsilon_a} \right) C^{(s)} G \right]\Bigg|_{p=p_\nu^{(s)}}, \qquad (8.22)
\end{aligned}
$$

$$
\begin{aligned}
f^m &= \sum_\nu f_\nu^m = \sum_\nu \frac{1}{k_0 a_s q} \left\{ \frac{p_{\mathrm{x}}(q)\,|\eta_a|}{q^2 + |\eta_a|} \left[\frac{\tilde{n}_1^{(s)}p + \tilde{g}_s}{\tilde{\varepsilon}_s}\, B_1^{(s)}\tilde{G}_1 \right. \right. \\
&\quad + \frac{\tilde{n}_2^{(s)}p + \tilde{g}_s}{\tilde{\varepsilon}_s}\, B_2^{(s)}\tilde{G}_2 + \frac{n_1 p + g_a}{\varepsilon_a}\, C^{(s)} G - J_0(k_0 a_s q) \\
&\quad \times \left(\frac{\tilde{n}_1^{(s)}p + \tilde{g}_s}{\tilde{\varepsilon}_s}\, B_1^{(s)} J_1(\tilde{Q}_1^{(s)}) + \frac{\tilde{n}_2^{(s)}p + \tilde{g}_s}{\tilde{\varepsilon}_s}\, B_2^{(s)} J_1(\tilde{Q}_2^{(s)}) \right. \\
&\quad \left. \left. - \frac{n_1 p + g_a}{\varepsilon_a}\, C^{(s)} K_1(S_1^{(s)}) \right) \right] \\
&\quad \left. - \frac{p_{\mathrm{x}}(q)\,g_a}{q^2 + p_{\mathrm{x}}^2(q) - \varepsilon_a}\, \left(B_1^{(s)}\tilde{G}_1 + B_2^{(s)}\tilde{G}_2 + C^{(s)}G \right) \right\}\Bigg|_{p=p_\nu^{(s)}}, \qquad (8.23)
\end{aligned}
$$

where

$$
\tilde{G}_{1,2} = \tilde{Q}_{1,2}^{(s)}\Big[k_0 a_s q\, J_1(k_0 a_s q)\, J_0(\tilde{Q}_{1,2}^{(s)}) - \tilde{Q}_{1,2}^{(s)}\, J_1(\tilde{Q}_{1,2}^{(s)})\, J_0(k_0 a_s q)\Big]
$$
$$
\times\ \Big[(k_0 a_s q)^2 - \tilde{Q}_{1,2}^{(s)^2}\Big]^{-1}, \tag{8.24}
$$

$$
G = S_1^{(s)}\Big[k_0 a_s q\, J_1(k_0 a_s q)\, K_0(S_1^{(s)}) - S_1^{(s)}\, K_1(S_1^{(s)})\, J_0(k_0 a_s q)\Big]
$$
$$
\times\ \Big[(k_0 a_s q)^2 + S_1^{(s)^2}\Big]^{-1}. \tag{8.25}
$$

Expressions (8.22) to (8.25) are the most convenient when considering the integrand of (8.17) in the limiting case $q \to \infty$. In particular, it follows from (8.22) to (8.25) that the integral (8.17) is convergent, so that the total power P_Σ takes the finite value.

The spectral representation (8.17) of the radiated power enables us to investigate the power distribution over the continuous spectrum of waves which contribute to the radiation field in the surrounding medium. To do this, we divide the entire domain of integration in (8.17) into three parts defined by the inequalities (3.44) in section 3.3.2 of § 3.3. By evaluating the separate integrals in the limits denoted in (3.44) as I, II, and III, we find the partial powers P_W, P_Q, and P_I, respectively, which go the long-wavelength part (I), short-wavelength (quasi-electrostatic) part (II), and intermediate part (III) of the spatial spectrum of whistler-mode waves. Along with the total power P_Σ and the partial powers P_W, P_I, and P_Q for the duct with the ring source (5.14) carrying the current I_0^e, we also introduce the total radiation resistance $R_\Sigma = 2P_\Sigma/|I_0^e|^2$ and the partial quantities $R_W = 2P_W/|I_0^e|^2$, $R_I = 2P_I/|I_0^e|^2$, and $R_Q = 2P_Q/|I_0^e|^2$. Recall that although the above partitioning was introduced conventionally in § 3.3, this procedure is physically relevant to the problem in question.

For the later work, it is convenient to use the normalized quantities $\xi_W = R_W/R_\Sigma$, $\xi_I = R_I/R_\Sigma$, and $\xi_Q = R_Q/R_\Sigma$. Careful tests show that the best means of accurately determining the quantities $R_W = \xi_W R_\Sigma$, $R_I = \xi_I R_\Sigma$, and $R_Q = \xi_Q R_\Sigma$ is to find the values of ξ_W, ξ_I, and ξ_Q by using the formulation based on the integral (8.17), and to take the value of R_Σ from the exact theory of Chapter 6 for an infinitely long, z-independent duct (the duct parameters must, of course, coincide with those in the radiator vicinity).

As is clear from the above expressions, the integrand of (8.17) is rather complicated, and the quantities ξ_W, ξ_I, and ξ_Q can therefore be evaluated only numerically. Before proceeding in this way, we shall make some remarks concerning the behavior of the integrand in the special case when the field at the duct radiating aperture is determined by only one leaky mode, say of number ν, in which case $f^{e,m} = f_\nu^{e,m}$.

From (8.18) to (8.21), and (8.22) to (8.25), we see that the integrand of (8.17) takes the maximum values at the points $q = \tilde{Q}_1^{(s)}/(k_0 a_s)$ and $q = \tilde{Q}_2^{(s)}/(k_0 a_s)$. Indeed, under these circumstances, the inequality $\tilde{Q}_2^{(s)} \gg 1$ is always satisfied for leaky modes, in which case the function $\tilde{F}_2$ simplifies to

$$
\begin{aligned}
\tilde{F}_2 &\approx \frac{1}{\pi(k_0 a_s q \tilde{Q}_2^{(s)})^{1/2}} \left[\frac{\sin(k_0 a_s q - \tilde{Q}_2^{(s)})}{k_0 a_s q - \tilde{Q}_2^{(s)}} + \frac{\cos(k_0 a_s q + \tilde{Q}_2^{(s)})}{k_0 a_s q + \tilde{Q}_2^{(s)}} \right] \\[2mm]
&\approx \frac{1}{\pi(k_0 a_s q \tilde{Q}_2^{(s)})^{1/2}} \frac{\sin(k_0 a_s q - \tilde{Q}_2^{(s)})}{k_0 a_s q - \tilde{Q}_2^{(s)}} .
\end{aligned}
\tag{8.26}
$$

From this it is immediately evident that the integrand reaches a maximum value at $q = \tilde{Q}_2^{(s)}/(k_0 a_s)$. By analogy with the above, we can prove that the other maximum of the function $\tilde{F}_1$ will occur at $q = \tilde{Q}_1^{(s)}/(k_0 a_s)$. The radiated power distribution over the spatial spectrum would thus be determined by the locations of the two maxima of the integrand.

It should be noted that for realizable duct parameters whose feasibility has been demonstrated in the ionospheric experiments mentioned in § 1.3, the relations $k_0 a_s p_\nu^{(s)} \sim 1$ and $\tilde{N}_s - N_a \sim N_a$ are normally valid. From these, one can deduce that the points $q = \tilde{Q}_2^{(s)}/(k_0 a_s)$ and $q = \tilde{Q}_1^{(s)}/(k_0 a_s)$ would lie in the ranges II and III, respectively. This means that the major portion of the radiated power goes to the quasi-electrostatic whistler-mode waves (range II) and the "intermediate" whistler-mode waves (range III). Note that by increasing the radiating aperture radius it is possible, in principle, to achieve a noticeable increase in the power going to the whistler-mode waves of range I. In particular, if the inequality $k_0 a_s p_\nu^{(s)} \gg 1$ is satisfied, the maximum $q = \tilde{Q}_1^{(s)}/(k_0 a_s)$ shifts to range I, thereby leading to some increase in the relative fraction of the power radiated in this part of the spatial spectrum.

More specific and refined conclusions can be drawn based on the results of numerical computations which will be discussed below.

8.6. Some numerical examples

Here we give selected results of the numerical calculations showing the influence of the density duct on the distribution of radiated power over the spatial spectrum. The calculations were performed for the ambient plasma parameters typical of the earth's ionosphere: $N_a = 10^6\,\mathrm{cm}^{-3}$ ($\omega_\mathrm{p} = 5.64 \times 10^7\,\mathrm{s}^{-1}$), $B_0 = 0.5\,\mathrm{Gs}$ ($\omega_\mathrm{H} = 8.78 \times 10^6\,\mathrm{s}^{-1}$); the chosen radius of the source and the frequency of the radiated signal were $b = 2.5\,\mathrm{m}$ and $\omega = 1.88 \times 10^5\,\mathrm{s}^{-1}$. To simplify our calculations, we used an approximation (8.10) for the plasma density. The density and the radius at the duct center were $N_m = 5 \times 10^7\,\mathrm{cm}^{-3}$ and $a(0) = 5\,\mathrm{m}$. To provide the conditions of the adiabatic propagation for leaky modes, we constructed a special approximation of the functions $\tilde{N}(z)$ and $a(z)$ for all points in the interval $0 \leq z \leq z_s$. A change in z_s was taken to be within the limits of $3.5\,\mathrm{km} < z_s < 6\,\mathrm{km}$. Within the approximation described, the local density $\tilde{N}(z)$ was decaying and the local radius $a(z)$ was increasing with distance z along the duct axis. They reached the values $\tilde{N}_s = 5 \times 10^6\,\mathrm{cm}^{-3}$ and $a_s = 25\,\mathrm{m}$ at the position $z = z_s$ of the radiating aperture.

For the chosen parameters, it appears within the framework of this model that the total radiation resistance of the ring electric current depends only slightly on the duct extent and practically coincides with that for the case of an infinitely long duct when the same parameters $\tilde{N} = N_m$, $a = a(0)$, N_a, B_0, b, and ω are used. Recall that our previous calculations (see §6.13) for these parameters show that the total radiation resistance of the ring source takes the value $R_\Sigma = 1.19 \times 10^{-1}\,\Omega$ in the presence of the duct, the radiation resistance into whistler leaky modes being equal to $1.09 \times 10^{-1}\,\Omega$. We regard only slightly leaky modes, whose real propagation constants p'_ν satisfy the relations (8.2). In those relations, for $z = 0$, $\tilde{\mathcal{P}}_c(0) \approx 2\tilde{\varepsilon}_m^{1/2}$ and $\tilde{\mathcal{P}}(0) = (\tilde{\varepsilon}_m + |\tilde{g}_m|)^{1/2}$ where the quantities $\tilde{\varepsilon}_m$ and $\tilde{g}_m$ refer to a magnetoplasma with the maximum density $N_m = \tilde{N}(0)$. At the aperture $z = z_s$, the modes transferring power from the source satisfy the analogous conditions

$$\max\left\{\mathcal{P}, \tilde{\mathcal{P}}_c^{(s)}\right\} < p_\nu^{(s)} < \tilde{\mathcal{P}}^{(s)}, \tag{8.27}$$

where $\tilde{\mathcal{P}}_c^{(s)} \equiv \tilde{\mathcal{P}}_c(z_s) \approx 2\tilde{\varepsilon}_s^{1/2}$ and $\tilde{\mathcal{P}}^{(s)} \equiv \tilde{\mathcal{P}}(z_s) = (\tilde{\varepsilon}_s + |\tilde{g}_s|)^{1/2}$. Generally, all the modes whose real propagation constants lie in the interval (8.27) contribute to the field radiated from the duct endface.

We begin with the simplest situation when the fields $\mathbf{E}^{(s)}$ and $\mathcal{H}^{(s)}$ are determined by the leaky mode of order ν only. Thus we set $f^{e,m} = f_\nu^{e,m}$. The results of numerical computations for this case are presented in Table 8.1 where we examine the dependence of relative portions $\xi_{W,\nu}$, $\xi_{I,\nu}$ and $\xi_{Q,\nu}$ on ν. Here, the subscript ν indicates that the corresponding quantities refer to the case where $f^{e,m} = f_\nu^{e,m}$. In Table 8.1 we also give the values of the real propagation constants $p = p_\nu^{(s)}$. Note that the whistler-mode propagation constant along $\mathbf{B}_0$ in the ambient plasma is $\mathcal{P}{=}44.30$.[*]

We recall that the leaky modes guided by a density enhancements in the whistler range can be distinguished from each other by both the number of large-scale variations over the radius and the number of fine-scale variations. In the particular case presented in Table 8.1, all of the modes have one large-scale variation, but different numbers of fine-scale oscillations. Among these modes, we distinguish especially the least attenuated mode. In the case considered, it has the number $\mu = 4$ and its attenuation constant p''_μ satisfies the condition $p''_\mu \ll p''_\nu\ (\nu \neq \mu)$.

---

[*] As calculations show, the major portion of power is spent for modes whose propagation constants $p_\nu^{(s)}$ are close to the lower limit of (8.27). Therefore, the model used, in which the backward reflection of modes is neglected, is here applicable though the ratio $(\tilde{N}_s - N_a)/N_a$ is not too small.

As seen from Table 8.1, we have

$$\xi_{W,\nu} \ll \xi_{I,\nu} \ll \xi_{Q,\nu}, \tag{8.28}$$

so that the major part of the radiated power goes to quasi-electrostatic whistler-mode waves. The values of $\xi_{W,\nu}$ and $\xi_{I,\nu}$ decrease with ν because as $p_\nu^{(s)}$ tends to the upper limit $p = \tilde{\mathcal{P}}^{(s)}$ of the interval (8.27), the maximum of the integrand of (8.17) at the point $q = \tilde{Q}_1^{(s)}/(k_0 a_s)$ gradually fades out.

Table 8.1. $p_\nu^{(s)}$ and $\xi_{W,\nu}$, $\xi_{I,\nu}$, $\xi_{Q,\nu}$ versus ν

ν	$p_\nu^{(s)}$	$\xi_{W,\nu}$	$\xi_{I,\nu}$	$\xi_{Q,\nu}$
1	48.19	7.59×10^{-4}	1.04×10^{-1}	8.95×10^{-1}
2	51.61	1.41×10^{-3}	1.19×10^{-1}	8.80×10^{-1}
3	55.05	1.80×10^{-3}	1.18×10^{-1}	8.80×10^{-1}
4	58.57	1.79×10^{-3}	1.01×10^{-1}	8.98×10^{-1}
5	62.20	1.49×10^{-3}	7.51×10^{-2}	9.23×10^{-1}
6	65.95	1.14×10^{-3}	5.28×10^{-2}	9.46×10^{-1}
7	69.81	8.49×10^{-4}	3.69×10^{-2}	9.62×10^{-1}
8	73.75	6.37×10^{-4}	2.62×10^{-2}	9.73×10^{-1}
9	77.75	4.85×10^{-4}	1.91×10^{-2}	9.80×10^{-1}
10	81.79	3.78×10^{-4}	1.43×10^{-2}	9.85×10^{-1}
11	85.87	3.00×10^{-4}	1.10×10^{-2}	9.89×10^{-1}
12	89.97	2.43×10^{-4}	8.62×10^{-3}	9.91×10^{-1}
13	94.10	2.00×10^{-4}	6.90×10^{-3}	9.93×10^{-1}
14	98.24	1.67×10^{-4}	5.63×10^{-3}	9.94×10^{-1}

The data presented in Table 8.1 enable us to estimate without difficulty the values of R_W, R_I and R_Q in a rather "exotic" case of a very extended duct when the inequality (8.3) is valid only for the least attenuated mode, while for other modes the opposite condition $k_0 p_\nu'' z_s \gg 1$ is fulfilled. In this case, the only mode that re-emits from the duct end is the least attenuated mode since other modes would radiate from the duct walls directly in quasi-electrostatic whistler-mode waves of the surrounding plasma. We thus have $R_W = \xi_{W,\mu} R_\mu$ and $R_I = \xi_{I,\mu} R_\mu$ where R_μ is the radiation resistance in the least attenuated mode. The radiation resistance in quasi-electrostatic whistler-mode waves R_Q consists now of $\xi_{Q,\mu} R_\mu$ and $\sum_{\nu \neq \mu} R_\nu$, where R_ν stands for the radiation resistance in the νth mode. It is apparent that

$$R_Q = \xi_{Q,\mu} R_\mu + \sum_{\nu \neq \mu} R_\nu = R_\Sigma - R_W - R_I.$$

Since, according to computations, $R_\mu = 1.88 \times 10^{-2}\ \Omega$, we arrive at $R_W = 3.37 \times 10^{-5}\ \Omega$, $R_I = 1.90 \times 10^{-3}\ \Omega$, and $R_Q = 1.17 \times 10^{-1}\ \Omega$. For comparison, we give the analogous values $R_{W,a}$, $R_{I,a}$, and $R_{Q,a}$, which correspond to the case when the same source is immersed in a uniform unbounded plasma with the ambient density. They are $R_{W,a} = 1.80 \times 10^{-5}\ \Omega$, $R_{I,a} = 7.91 \times 10^{-4}\ \Omega$, and $R_{Q,a} = 2.72 \times 10^{-3}\ \Omega$, whence $R_{\Sigma,a} = R_{W,a} + R_{I,a} + R_{Q,a} = 3.53 \times 10^{-3}\ \Omega$. It is evident that the increase in the total radiated power is connected mainly with increasing the partial power going to quasi-electrostatic whistler-mode waves: $R_Q/R_{Q,a} = 43$, $R_I/R_{I,a} = 2.4$, and $R_W/R_{W,a} = 1.9$.

For artificial ducts created in the vicinity of electromagnetic sources in the earth's ionosphere, the case when the inequality (8.3) is fulfilled for a few modes seems to be more realistic. Under these circumstances, the quantities ξ_W, ξ_I, and ξ_Q vary with longitudinal stretching of the density distribution. To show this effect more easily, we note that for the parameters given above, the major part of the total radiated power goes to the first six modes with numbers $\nu = 1, \ldots 6$. The radiation resistance in these modes $R_{1-6} = 0,10\,\Omega$ that is close to the value of $R_\Sigma = 1.19 \times 10^{-1}\,\Omega$. Hence, we confine ourselves to only these six modes and put further $R_\Sigma \approx R_{1-6}$. The dependences of ξ_W, ξ_I, and ξ_Q on the duct extent (in fact, on z_s) are distinctly oscillatory. At some values of z_s, the quantities ξ_W and ξ_I simultaneously take the maximum values $\xi_W^{(\max)} = 7.63 \times 10^{-3}$ and $\xi_I^{(\max)} = 5.14 \times 10^{-1}$; the quantity ξ_Q is then equal to 4.78×10^{-1}. Multiplying these values by the total radiation resistance, we find the partial radiation resistances $R_W^{(\max)} = 7.63 \times 10^{-4}\,\Omega$, $R_I^{(\max)} = 5.14 \times 10^{-2}\,\Omega$, and $R_Q = 4.78 \times 10^{-2}\,\Omega$. As an example, Figure 8.1 shows the dependence of R_W normalized to $R_{W,a}$ on the duct extent z_s. The corresponding dependence for the ratio $R_I/R_{I,a}$ is similar and is therefore omitted here. At points where R_W and R_I take the minimum value, virtually all the power re-emitted from the duct ends goes to quasi-electrostatic waves, so that $\xi_Q^{(\max)} \approx 1$ and $R_Q^{(\max)} \approx R_\Sigma$.

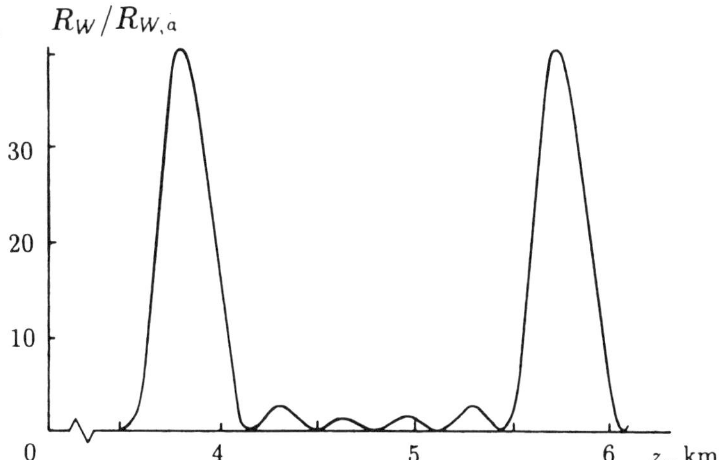

Figure 8.1. Dependence of the normalized radiation resistance R_W on duct extent; see text for discussion.

The above picture of the distribution of radiated power over the spatial spectrum can easily be explained from the physical viewpoint. It appears that the modes which give the main contribution to the total field at the duct end have close amplitudes ($|E_\nu^{(s)}| \simeq |E_{\nu\pm1}^{(s)}|$) and close transverse wavenumbers $\tilde{Q}_1^{(s)}(p_\nu^{(s)})$ for the large-scale part of their fields:

$\tilde{Q}_1^{(s)}(p_\nu^{(s)}) - \tilde{Q}_1^{(s)}(p_{\nu+1}^{(s)}) \ll \pi$ (recall that $|\tilde{Q}_1^{(s)}| \ll |\tilde{Q}_2^{(s)}|$). At the same time, the transverse wavenumbers $\tilde{Q}_2^{(s)}(p_\nu^{(s)})$, which correspond to the fine-scale part of mode fields, are considerably different from each other: $\tilde{Q}_2^{(s)}(p_{\nu+1}^{(s)}) - \tilde{Q}_2^{(s)}(p_\nu^{(s)}) \simeq \pi$ (see expressions (8.11)). Therefore, if the modes are in phase for some z_s, then the large-scale parts of their fields are merely summed up at the duct end. This is why the radiated power is redistributed to the long-wavelength part of the spatial spectrum. As a result, the maxima of R_W and R_I are observed. It is now clear that the form of the curve presented in Figure 8.1 would be approximately described by the function

$$F^{(M)} = \frac{1}{M^2} \left| \sum_{\nu=1}^{M} \exp\left(-\mathrm{i}k_0 \int_0^{z_s} p_\nu'(\zeta)\,\mathrm{d}\zeta \right) \right|^2, \qquad (8.29)$$

where M is the number of modes which are taken into account (in our case $M = 6$). As calculations show, curves $p_\nu'(z)$ remain nearly equidistant along the entire length from $z = 0$ to $z = z_s$, so that $p_\nu'(z) \simeq (1 + \nu\Delta)\,\bar{\mathcal{P}}(z)$, where Δ is a constant and $\bar{\mathcal{P}}(z) = \max\{\mathcal{P}, \tilde{\mathcal{P}}_c(z)\}$. Bearing this in mind, we find, from (8.29),

$$F^{(M)} \simeq \frac{1}{M^2} \frac{\sin^2(k_0\bar{\gamma}M\Delta/2)}{\sin^2(k_0\bar{\gamma}\Delta/2)}, \qquad (8.30)$$

where

$$\bar{\gamma} = \int_0^{z_s} \bar{\mathcal{P}}(\zeta)\,\mathrm{d}\zeta.$$

It is readily seen that expression (8.30) indeed describes the behavior of R_W versus z_s.

We have thus seen that when a near-antenna duct with enhanced density is used, the radiation resistance R_W for of the duct endface can, at times, considerably exceed the quantity $R_{W,a}$, which corresponds to a ring electric current in the ambient plasma $(R_W^{(\mathrm{max})}/R_{W,a} = 42)$, although the relative contribution of the long-wavelength part of the spatial spectrum to the total radiated power is, as previously, small $(\xi_W^{(\mathrm{max})} \ll 1)$. In principle, a rise in ξ_W can be achieved by increasing the duct width in the vicinity of the end $z \simeq z_s$. Specifically, if the duct radius at the end is large enough, so that $a_s > k_0^{-1}(\varepsilon_a + |g_a|)^{-1/2}$, the contribution from the long-wavelength part of the spatial spectrum will exceed that from the intermediate part.

Another way to increase ξ_W may be associated with the use of non-symmetric modes. It is clear that such modes will contribute much more efficiently to the long-wavelength part of the spatial spectrum, compared to the case of purely axisymmetric modes. Consideration of re-emission of non-symmetric modes from the duct end needs a very cumbersome mathematics though and is therefore omitted. Finally, we emphasize that the above discussion specialized to a simple axially symmetric case was given, not to establish a universal formalism but rather as a demonstration to the reader of how one can use the results of the accurate theory to formulate reasonable approximate approaches in the study of more complicated problems.

8.7. The use of artificial density ducts for increasing power radiated from VLF/ELF sources

The need to communicate over long distances requires the use of transmitting antennas operating in the very low (3–30 kHz) and extremely low (3–3000 Hz) frequency bands. A few types of transmitting systems used as antennas for VLF/ELF signals now exist (see, for example, Rycroft, 1985) but they are all relatively inefficient. A new unconventional means to improve the efficiency of VLF/ELF satellite transmitting systems in the near-Earth space is possible if an artificial duct with a strong enhancement of plasma density is used. As explained in Chapter 1, such ducts can be formed in a self-consistent way in the vicinity of the usual satellite antenna system as a result of the nonlinear interaction of the antenna field with the surrounding plasma. Releases of an easily ionized gas into the near-antenna zone facilitate creation of ducts with enhanced density.

We saw in Chapter 6 that when the plasma density in the duct is high enough, the wavelength of the VLF/ELF signal excited in the duct core becomes of the same order as the antenna size. This leads to a noticeable increase in the total radiated power due to the effective excitation of guided modes in the duct. We showed in Chapter 7 and the present chapter that these modes can be conveyed by the duct and can re-emit from its end, giving rise to an increase in power that goes to the long-wavelength part of the spectrum of waves excited in the surrounding medium. It is this part that is useful for communication through the ionosphere, enabling a VLF/ELF signal to be received anywhere on the surface of the Earth. In other words, the antenna plus the artificial duct can act as a *plasma-waveguide transmitting system* ('plasma antenna') which excites, as a whole, VLF/ELF signals in the ambient medium. Of course, the radiation characteristics of such plasma-waveguide systems depend on various plasma parameters in a very complicated manner and, for the future, there is much to be done in improving our understanding of both the mechanisms of their formation and the methods of efficient control of wave emissions from them.

Although the theory of this chapter relates to artificial ducts, the approach itself and the relevant general expressions can be applied, under definite conditions, to natural ducts in the near-Earth space. A detailed study taking into account the theories in this book has not yet been done for magnetospheric ducts but it is possible that future investigations based on this material will reveal many interesting properties of ducted wave propagation in the magnetosphere.

Problem

8.1. Discuss the peculiarities of re-emission of particular non-symmetric modes from the end of a duct with enhanced density in the frequency interval (2.71). Show that such modes can contribute a higher portion of the re-emitted power to the whistler-mode waves belonging to range I of the spatial spectrum, compared with the case of re-emission of purely axisymmetric modes. (A full solution of this problem would probably involve extensive numerical computations.)

Bibliography

Abramowitz, M. and Stegun, I. A. (Editors) (1964) *Handbook of Mathematical Functions*, New York: National Bureau of Standards

Adachi, S. (1965) Study on the guiding mechanism of whistler radio waves. *J. Res. NBS*, **69D**, 493–502

Adachi, S. (1966) Theory of duct propagation of whistler radio waves. *Radio Science*, **1**, 671–678

Agafonov, Yu. N., Babaev, A. P., Bazhanov, V. S., Isyakaev, V. Ya., Markov, G. A., Namazov, S. A., Pokhunkov, A. A. and Chugunov Yu. V. (1989) Plasma-wave discharge in the ionosphere. *Soviet Technical Physics Letters*, **15**, 661–663

Agafonov, Yu. N., Bazhanov, V. S., Isyakaev, V. Ya., Markov, G. A., Pokhunkov, A. A., Chugunov, Yu. V. and Kulistikov, S. A. (1990) Stimulated emissions of energetic particles by plasma-wave discharge in the polar ionosphere. *JETP Lett.*, **52**, 530–533

Akhiezer, A. I., Akhiezer, I. A., Polovin, R. V., Sitenko, A. G. and Stepanov, K. N. (1975) *Plasma Electrodynamics*. Oxford: Pergamon Press

Allis, W. P., Buchsbaum, S. J. and Bers, A. (1963) *Waves in Anisotropic Plasmas*. Cambridge, Ma: M. I. T. Press

Al'pert, Ya. L., Budden, K. G., Moiseyev, B. S. and Stott, G. F. (1983) Electromagnetic radiation from a dipole source in a homogeneous magnetoplasma. *Phil. Trans. Roy. Soc. London A*, **309**, 503–557

Altman, C. and Cory, H. (1969) The generalized thin-film optical method in electromagnetic wave propagation. *Radio Science*, **4**, 459–470

Andronov, A. A. and Chugunov, Yu. V. (1975) Quasi-stationary electric fields of sources in a rarefied plasma. *Uspekhi Fizicheskikh Nauk*, **116**, 79–113

Angerami, J. J. (1970) Whistler duct properties deduced from VLF observations made with the OGO 3 satellite near the magnetic equator. *J. Geophys. Res.*, **75**, 6115–6135

Angulo, C. M. and Chang, W. S. C. (1959) A variational expression for the terminal admittance of a semi-infinite dielectric rod. *IRE Trans. Antennas and Propagation*, **AP-7**, 207–212

Arbel, E. and Felsen, L. B. (1963) Theory of radiation from sources in anisotropic media. Part II: Point source in infinite inhomogeneous medium. In *Proceedings of the Symposium on Electromagnetic Theory and Antennas* (Kopenhagen, 1962), Part I, edited by E. C. Jordan, pp. 421–459. New York: Pergamon Press

Balmain, K. G. (1964) The impedance of a short dipole antenna in a magnetoplasma. *IEEE Trans. Antennas and Propagation*, **AP-12**, 606–617

Bateman (manuscript project) See under Erdélyi

Bell, T. F. and Wang, T. N. C. (1971) Radiation resistance of a small filamentary loop antenna in a cold multicomponent magnetoplasma. *IEEE Trans. Antennas and Propagation*, **AP-19**, 517–522

Bellyustin, N. S. (1978) Wave emission of the whistler range in a plasma. *Radiophysics and Quantum Electronics*, **21**, 13–22

Bernhardt, P. A. and Park, C. G. (1977) Protonospheric-ionospheric modelling of VLF ducts. *J. Geophys. Res.*, **82**, 5222–5230

Booker, H. G. (1935) The application of the magneto-ionic theory to the ionosphere. *Proc. Roy. Soc. London A*, **150**, 267–286

Booker, H. G. and Dyce, R. B. (1965) Dispersion of waves in a cold magnetoplasma from hydromagnetic to whistler frequencies. *J. Res. NBS*, **69D**, 463–492

Braginskii, S. I. (1965) Transport processes in a plasma. In *Reviews of Plasma Physics*, vol. 1, edited by M. A. Leontovich, pp. 205–210. New York: Consultants Bureau

Brekhovskikh, L. M. (1960) *Waves in Layered Media*. New York: Academic Press

Bresler, A. D. and Marcuvitz, N. (1956) Operator methods in electromagnetic theory. *Report MRI-R-495-56*, Microwave Research Institute, Polytechnic Institute of Brooklyn, New York, USA

Brice, N. M. and Smith, R. L. (1971) Whistlers: diagnostic tools in space plasma. In *Plasma Physics, vol. 9, Methods of Experimental Physics*, edited by R. N. Lorberg and H. R. Grien. New York: Academic Press

Brodskii, Yu. Ya., Kondrat'ev, I. G. and Miller, M. A. (1969) Electromagnetic beams in anisotropic media. *Alta Frequenza*, **38**, Special Issue: Selected Papers from the URSI Symposium on Electromagnetic Waves (Stresa, Italy, 1968), 89–96

Brown, J. and Spector J. O. (1957) The radiating properties of endfire aerials. *Proc. IEE*, pt. B, 27–34

Budden, K. G. (1961 a) *Radio Waves in the Ionosphere*. Cambridge: University Press

Budden, K. G. (1961 b) *The Wave-Guide Mode Theory of Wave Propagation*. London: Logos Press

Bunkin, F. V. (1957) On radiation in anisotropic media. *Sov. Phys. JETP*, **5**, 277–283

Cerisier, J. C. (1974) Ducted and partly ducted propagation of VLF waves through the magnetosphere. *J. Atmos. and Terr. Phys.*, **36**, 1443–1467

Cole, K. D. (1971) Formation of field-aligned irregularities in the magnetosphere. *J. Atmos. and Terr. Phys.*, **33**, 741–750

Copson, E. T. (1935) *Theory of Functions of a Complex Variable*. Oxford University Press

Dokuchaev, V. P., Tamoikin, V. V. and Chugunov, Yu. V. (1976) Emission of spiral waves in a magnetoactive plasma by distributed sources. *Radiophysics and Quantum Electronics*, **19**, 792–798

Duff, G. L. and Mittra, R. (1970) Loop impedance in magnetoplasma: theory and experiment. *Radio Science*, **5**, 81–94

Egorov, S. V., Kostrov, A. V. and Tronin, A. V. (1988) Thermal diffusion and eddy currents in a magnetized plasma. *JETP Lett.*, **47**, 102–106

Erdélyi, A., Magnus, W., Oberhettinger, F. and Tricomi, F. G. (1953) *Higher Transcendental Functions* (Bateman manuscript project), vol. 1. New York: McGraw-Hill

Felsen, L. B. and Marcuvitz, N. (1966) Alternative representations of source-excited vector and scalar fields. *Radio Science*, **1**, 619–640

Felsen, L. B. and Marcuvitz, N. (1973) *Radiation and Scattering of Waves*. Englewood Cliffs, NJ: Prentice-Hall

Fisher, R. K. and Gould, R. W. (1971) Resonance cones in the field pattern of a radio frequency probe in a warm anisotropic plasma. *Phys. Fluids*, **14**, 857–867

Fröman, N. and Fröman, P. O. (1965) *J. W. K. B. Approximation*. Amsterdam: North-Holland

Gershman, B. N., and Ugarov, V. A. (1961) Propagation and generation of low-frequency electromagnetic waves in the upper atmosphere. *Sov. Phys. Uspekhi*, **3** (5), 743–763

GiaRusso, D. P. and Bergeson, J. E. (1970) Studies of VLF radiation patterns of a dipole immersed in a slightly lossy magnetoplasma. *Radio Science*, **5**, 745–766

Ginzburg, V. L. (1970) *The Propagation of Electromagnetic Waves in Plasmas*, 2nd edn. Oxford: Pergamon Press

Golant, V. E. and Fedorov, V. I. (1989) *RF Plasma Heating in Toroidal Fusion Devices*. New York: Consultants Bureau

Golubyatnikov, G. Yu., Egorov, S. V., Eremin, B. G., Litvak, A. G., Strikovskii, A. V., Tolkacheva, O. N. and Chugunov Yu. V. (1995) Lower-hybrid breakdown of gas in the field of a current-carrying loop in a plasma-filled magnetic confinement system. *JETP*, **80**, 234–239

Gradshteyn, I. S. and Ryzhik, I. M. (1965) *Tables of Integrals, Series and Products.* New York: Academic Press

Gurevich, A. V. (1978) *Nonlinear Phenomena in the Ionosphere*, in *Physics and Chemistry in Space*, vol. 10, edited by J. G. Roederer and J. T. Wasson. New York: Springer-Verlag

Hansen, H. J., Scourfield, M. W. J. and Rash, J. P. S. (1983) Whistler duct lifetimes. *J. Atmos. and Terr. Phys.*, **45**, 789–794

Heading, J. (1962) *An introduction to phase-integral methods.* London: Methuen

Hegai, V. V., Kim, V. P. and Illich-Switych, P. V. (1990) The formation of a cavity in the night-time midlatitude ionospheric E-region above a thundercloud. *Planet. Space Science*, **38**, 703–707

Helliwell, R. A. (1965) *Whistlers and Related Ionospheric Phenomena.* Stanford: Stanford University Press

Huddlestone, R. H. and Leonard, S. L. (Editors) (1965) *Plasma Diagnostic Techniques.* New York – London: Academic Press

Inoue, Y. and Horowitz, S. (1966) Numerical solution of full-wave equations with mode coupling. *Radio Science*, **1**, 957–970

Jackson, J. D. (1962) *Classical Electrodynamics.* New York: Wiley

Johler, J. R. and Harper, J. (1962) Reflection and transmission of radio waves at a continuously stratified plasma with arbitrary magnetoionic induction. *J. Res. NBS*, **66D**, 81–101

Karpman, V. I. and Kaufman, R. N. (1982 a) The self-focusing of whistler waves. *Physica Scripta*, **T2:1**, 252–261

Karpman, V. I. and Kaufman, R. N. (1982 b) Whistler wave propagation in density ducts. *J. Plasma Phys.*, **27**, 225–238

Karpman, V. I. and Kaufman, R. N. (1984) Whistler wave propagation in magnetospheric ducts (in the equatorial plane). *Planet. Space Science*, **32**, 1505–1511

Karpman, V. I., Kaufman, R. N. and Shagalov, A. G. (1984) Axially symmetric self-focusing of whistler waves. *J. Plasma Phys.*, **31**, 209–223

Katzenelenbaum, B. Z. (1961) *Theory of Irregular Waveguides with Slowly Varying Parameters.* Moscow: USSR Academy of Sciences

Kelso, J. M. (1964) *Radio Ray Propagation in the Ionosphere.* New York: McGraw-Hill

Kogelnik, H. (1960) On electromagnetic radiation in magnetoionic media. *J. Res. NBS* , **64D**, 515–523

Kondratenko, A. N. (1976) *Plasma Waveguides.* Moscow: Atomizdat

Kondrat'ev, I. G., Kudrin, A. V. and Zaboronkova, T. M. (1992) Radiation of whistler waves in magnetoactive plasma. *Radio Science*, **27**, 315–324

Kondrat'ev, I. G., Kudrin, A. V. and Zaboronkova, T. M. (1994) Radiation of whistler waves in a magnetized plasma medium in the presence of density ducts. *Radiophysics and Quantum Electronics* **37**, 579–592

Kondrat'ev, I. G., Kudrin, A. V. and Zaboronkova, T. M. (1996) Excitation and propagation of electromagnetic waves in nonuniform density ducts. *Physica Scripta*, **54**, 96–112

Kondrat'ev, I. G. and Talanov, V. I. (1965) Application of the Lorentz lemma to calculation of radiation fields of given sources in unbounded media. *Zhurnal Tekhnicheskoi Fiziki*, **35**, 571–573

Korn, G. A. and Korn, T. M. (1968) *Mathematical Handbook for Scientists and Engineers*, 2nd edn. New York: McGraw-Hill

Kostrov, A. V. and Kim, A. V. (1988) Thermal nonlinearity and small-scale stratification in a magnetized plasma showing coulomb collisions in a high-frequency field. *Radiophysics and Quantum Electronics*, **31**, 404–409

Krall, N. A. and Trivelpiece, A. W. (1973) *Principles of Plasma Physics*. New York: McGraw-Hill

Kudrin, A. V., Markov, G. A., Trakhtengertz, V. Yu. and Chugunov, Yu. V. (1991) Secondary emission induced by the action of an intense electromagnetic beam on the ionosphere. *Geomagnetism and Aeronomy*, **31**, 253–258

Kudrin, A. V. and Markov, G. A. (1991) Dispersion and matching properties of inhomogeneous plasma waveguides. *Radiophysics and Quantum Electronics*, **34**, 141–147

Kuehl, H. H. (1962) Electromagnetic radiation from an electric dipole in a cold anisotropic plasma. *Phys. Fluids*, **5**, 1095–1103

Laird, M. J. and Nunn, D. (1975) Full-wave VLF modes in a cylindrically symmetric enhancement of plasma density. *Planet. Space Science*, **23**, 1649–1657

Laird, M. J. (1992) On the guidance of whistlers by density gradients. *J. Atmos. and Terr. Phys.*, **54**, 1593–1594

Landau, L. D. and Lifshitz, E. M. (1960) *Electrodynamics of Continuous Media*. London: Pergamon Press

Lee, S. W. (1969) Cylindrical antenna in uniaxial resonant plasmas. *Radio Science*, **4**, 179–189

Lester, M. and Smith, S. J. (1980) Whistler duct structure and formation. *Planet. Space Science*, **28**, 645–654

Lorentz, H. A. (1896) Het theorema van Poynting over de energie in het electromagnetisch veld en een paar algemeene stellingen over de voortplanting van het licht. In *Verslagen der Zittingen van de Wissenschaftlichen Naturkundige Afdeeling der K. Academie van Wettenschappen*, bd. 4, p. 176

Lukin, D. S., Presnyakov, V. B. and Savchenko, P. P. (1988) Calculation of wave fields in the near zone of a frame VLF-emitter in the uniform magnetoactive plasma. *Geomagnetism and Aeronomy*, **28**, 218–221

Madelung, E. (1957) *Die Mathematischen Hilfsmittel des Physikers*, 6th edn. Berlin – Göttingen – Heidelberg: Springer-Verlag

Marcuse, D. (1972) *Light Transmission Optics*. New York: Van Nostrand

Marcuse, D. (1974) *Theory of Dielectric Optical Waveguides*. New York: Academic Press

Marcuvitz, N. (1956) On field representations in terms of leaky modes or eigenmodes. *IRE Trans. Antennas and Propagation*, **AP-4**, 192–194

Mareev, E. A. and Chugunov, Yu. V. (1991) *Antennas in Plasmas*. Nizhny Novgorod: Applied Physics Institute

Markov, G. A., Mironov, V. A. and Sergeev, A. M. (1979) Self-channeling of plasma waves in a magnetic field. *JETP Lett.*, **29**, 617–621

Markov, G. A., Mironov, V. A., Sergeev, A. M. and Sokolova, I. A. (1981) Multibeam self-channeling of plasma waves. *Sov. Phys. JETP*, **53**, 1183–1186

Markov, G. A. (1988 a) Ionization ducting of electromagnetic fields. *ScD Thesis*, Radiophysical Research Institute (NIRFI), Nizhny Novgorod, Russia

Markov, G. A. (1988 b) Observation of a resonant autotuning of a magnetic antenna by rf-discharge plasma. *Sov. J. Plasma Phys.*, **14**, 641–644

Mathews, J. and Walker, R. L. (1965) *Mathematical Methods of Physics*. New York: Benjamin

McIlwain, C. E. (1961) Coordinates for mapping the distribution of magnetically trapped particles. *J. Geophys. Res.*, **66**, 3681–3692

Mittra, R. and Deschamps, G. A. (1963) Field solution for a dipole in an anisotropic medium. In *Proceedings of the Symposium on Electromagnetic Theory and Antennas* (Kopenhagen, 1962), Part I, edited by E. C. Jordan, pp. 495–512. New York: Pergamon Press

Moiseyev, B. S. (1985) The peculiarities of radiation from an electric dipole in a magnetoactive plasma and the problem of the wave guidance in the magnetosphere of the Earth. *PhD Thesis*, Institute of Terrestrial Magnetism, Ionosphere and Radio Wave Propagation (IZMIRAN), Moscow, Russia

Nagano, I., Mambo, M. and Nutatsuishi, G. (1975) Numerical calculation of electromagnetic waves in an anisotropic multilayered medium. *Radio Science*, **10**, 611–617

Ohnuki, S., Sawaya, K. and Adachi, S. (1986) Impedance of a large circular loop antenna in a magnetoplasma. *IEEE Trans. Antennas and Propagation*, **AP-34**, 1024–1029

Ondoh, T. (1976) Field aligned irregularities in whistler ducts as observed by the ISIS satellites. In *Space Research XVI*, edited by M. J. Rycroft, pp. 555–559. Berlin: Academie-Verlag

Park, C. G. and Carpenter, D. L. (1970) Whistler evidence of large-scale electron irregularities in the plasmasphere. *J. Geophys. Res.*, **75**, 3825–3836

Park, C. G. and Helliwell, R. A. (1971) The formation by electric fields of field-aligned irregularities in magnetosphere. *Radio Science*, **6**, 299–304

Park, C. G. and Dejnakarintra M. (1973) Penetration of thundercloud electric fields into the ionosphere and magnetosphere. 1. Middle and subauroral latitudes. *J. Geophys. Res.*, **78**, 6623–6633

Pitteway, M. L. V. (1965) The numerical calculation of wave-fields, reflection coefficients and polarizations for long radio waves in the lower ionosphere. I. *Phil. Trans. Roy. Soc. London A.*, **257**, 219–262

Pitteway, M. L. V. and Jespersen, J. L. (1966) A numerical study of the excitation, internal reflection and limiting polarization of whistler waves in the lower ionosphere. *J. Atmos. and Terr. Phys.*, **28**, 17–44

Raitt, W. J. (1996) Active plasma experiments in space: steps towards a space laboratory facility. In *Review of Radio Science 1993-1996*, edited by W. Ross Stone, pp. 651–675. New York: Oxford University Press

Ratcliffe, J. A. (1959) *The Magneto-Ionic Theory and Its Applications to the Ionosphere*. London: Cambridge University Press

Richards, P. G. and Cole, K. D. (1979) A numerical investigation of the formation and evolution of magnetospheric irregularities by the interchange of magnetospheric flux tubes. *Planet. Space Science*, **27**, 1351–1360

Rycroft, M. J. (1985) How to make a long antenna. *Nature*, **317**, 114–115

Sagredo, J. L. and Bullough, K. (1973) VLF goniometer observations at Halley Bay, Antarctica–II. Magnetospheric structure deduced from whistler observations. *Planet. Space Science*, **21**, 913–923

Sammut, R. and Snyder, A. W. (1975) Leaky modes on circular optical waveguide. *Appl. Optics*, **15**, 477–482

Sazhin, S., Hayakawa, M. and Bullough, K. (1992) Whistler diagnostics of magnetospheric parameters: a review. *Ann. Geophys.*, **10**, 293–308

Scarabucci, R. R. (1969) Analytical and numerical treatment of wave propagation in the lower ionosphere. *Techn. Report No. 3412-11*. Stanford University, Stanford, USA

Scarabucci, R. R. and Smith, R. L. (1971) Study of magnetospheric field oriented irregularities — the mode theory of bell-shaped ducts. *Radio Science*, **6**, 65–86

Scarf, F. L. and Chappel, C. R. (1973) An association of magnetospheric whistler dispersion characteristics with change in local plasma density. *J. Geophys. Res.*, **78**, 1597–1602

Schwartz, L. (1950) *Théorie des Distributions*, vol. 1. Strasbourg: Publications de l'Institut de Mathématique de L'Université de Strasbourg

Seshadri, S. R. and Yip, G. L. (1966) Radiation from an electric dipole in an axially magnetised plasma column. *Electronics Lett.*, **2**, 30–33

Shevchenko, V. V. (1966) Electromagnetic waves in isotropic stratified plasma waveguide. *Izv. VUZov. Radiofizika*, **9**, 110–125

Shevchenko, V. V. (1971 a) *Continuous Transitions in Open Waveguides*. Boulder, Colo: The Golem Press

Shevchenko, V. V. (1971 b) The expansion of the fields in open waveguides in proper and improper modes. *Radiophysics and Quantum Electronics*, **14**, 972–977

Shvartsburg, A. B. and Stenflo, L. (1990) Waveguide properties of the ionospheric F-layer. *J. Electromagnetic Waves and Applications*, **4**, 1215–1221

Smith, R. L., Helliwell, R. A. and Yabroff, I. W. (1960) A theory of trapping of whistlers in field-aligned columns of enhanced ionization. *J. Geophys. Res.*, **65**, 815–823

Smith, R. L. (1961) Propagation characteristics of whistlers trapped in field aligned columns of enhanced ionization. *J. Geophys. Res.*, **66**, 3699–3707

Smith, R. L. and Angerami, J. J. (1968) Magnetospheric properties deduced from Ogo 1 observations of ducted and nonducted whistlers. *J. Geophys. Res.*, **73**, 1–20

Smith, G. H. and Pitteway, M. L. V. (1974) Fortran program for obtaining wave fields of penetrating, non-penetrating and whistler modes of radio waves in the ionosphere. In *ELF-VLF Radio Wave Propagation*, pp. 69–86. Boston: Dordrecht

Sneddon, I. N. (1951) *Fourier Transforms*. New York: McGraw-Hill

Snyder, A. W. and Love, J. D. (1983) *Optical Waveguide Theory*. London: Chapman and Hall

Stenzel, R. L. (1976) Filamentation instability of large amplitude whistler wave. *Phys. Fluids*, **19**, 865–871

Stenzel, R. L. (1977) Experiments on whistler wave filamentation and VLF hiss in a laboratory plasma. *J. Physique*, **38**, C6-89–C6-102

Stix, T. H. (1962) *The Theory of Plasma Waves*. New York: McGraw-Hill

Storey, L. R. O. (1953) An investigation of whistling atmospherics. *Phil. Trans. Roy. Soc. London A*, **246**, 113–141

Strangeways, H. J. (1981) Trapping of whistler-mode waves in ducts with tapered ends. *J. Atmos. and Terr. Phys.*, **43**, 1071–1079

Strangeways, H. J. (1982) The effect of multi-duct structure on whistler-mode wave propagation. *J. Atmos. and Terr. Phys.*, **44**, 901–912

Strangeways, H. J. (1996) Lightnings, trimpis and sprites. In *Review of Radio Science 1993–1996*, edited by W. Ross Stone. New York: Oxford University Press

Sugai, H., Maruyama M., Sato, M. and Takeda, S. (1978) Whistler wave ducting caused by antenna actions. *Phys. Fluids*, **21**, 690–694

Tamir, T. and Oliner, A. A. (1963) The spectrum of electromagnetic waves guided by a plasma layer. *Proc. IEEE*, **51**, 317–332

Thomson, R. J. (1978) The formation and lifetime of whistler ducts. *Planet. Space Science*, **26**, 423–430

Tiberio, R., Maci, S. and Toccafondi, A. (1996) High-frequency methods in electromagnetics. In *Modern Radio Science 1996*, edited by J. Hamelin, pp. 111–131. New York: Oxford University Press

Tsytovich, V. N. (1970) *Nonlinear Effects in Plasma.* New York: Plenum Press

Vaynshteyn, L. A. (1969) *Theory of Diffraction and Factorization Method.* Boulder, Colo: Golem Press

Vaynshteyn, L. A. (1988) *Electromagnetic Waves*, 2nd edn. Moscow: Radio i Svyaz'

Vdovichenko, I. A., Markov, G. A., Mironov, V. A. and Sergeev, A. M. (1986) Ionizational self-ducting of whistlers in a plasma. *JETP Lett.*, **44**, 275–279

Velinov, P. I. and Tonev, P. T. (1994) Penetration of multipole thundercloud electric fields into the ionosphere. *J. Atmos. and Terr. Phys.*, **56**, 349–359

Velinov, P. I. and Tonev, P. T. (1995) Modelling the penetration of thundercloud electric fields into the ionosphere. *J. Atmos. and Terr. Phys.*, **57**, 687–694

Vorob'ev, N. F. and Rukhadze, A. A. (1994) Excitation of a helicon in a plasma cylinder by surface current sources. *Plasma Physics Reports*, **20**, 955–958

Voskoboinikov, S. P., Gurvich, I. Yu. and Rozhanskii, V. A. (1989) Multidimensional thermal diffusion in magnetic field. *Sov. J. Plasma Phys.*, **15**, 479–484

Wait, J. R. (1962) *Electromagnetic Waves in Stratified Media.* New York: Pergamon Press

Wait, J. R. (1966) Transverse propagation of waveguide modes in a cylindrically stratified magnetoplasma. *Radio Science*, **1**, 641–654

Walker, A. D. M. (1971) The propagation of very low-frequency radio waves in ducts in the magnetosphere. *Proc. Roy. Soc. London A*, **321**, 69–93

Walker, A. D. M. (1972) The propagation of very low-frequency waves in ducts in the magnetosphere. II. *Proc. Roy. Soc. London A*, **329**, 219–231

Walker, A. D. M. (1976) The theory of whistler propagation. *Rev. Geophys. Space Phys.*, **14**, 629–638

Walker, A. D. M. (1978) Formation of whistler ducts. *Planet. Space Science*, **26**, 375–379

Wang, T. N. C. and Bell, T. F. (1969 a) Radiation resistance of a short dipole immersed in a cold magnetoionic medium. *Radio Science*, **4**, 167–177

Wang, T. N. C. and Bell, T. F. (1969 b) On VLF radiation fields along the static magnetic field from sources immersed in a magnetoplasma. *IEEE Trans. Antennas and Propagation*, **AP-17**, 824–827

Wang, T. N. C. and Bell, T. F. (1970) On VLF radiation resistance of an electric dipole in a cold magnetoplasma. *Radio Science*, **5**, 605–610

Wang, T. N. C. and Bell, T. F. (1972 a) VLF/ELF radiation patterns of arbitrarily oriented electric and magnetic dipoles in a cold lossless multicomponent magnetoplasma. *J. Geophys. Res.*, **77**, 1174–1189

Wang, T. N. C. and Bell, T. F. (1972 b) VLF/ELF input impedance of an arbitrarily oriented loop antenna in a cold collisionless multicomponent magnetoplasma. *IEEE Trans. Antennas and Propagation*, **AP-20**, 394–398

Wang, S., Wang, J. F. and Comfort, R. H. (1984) A magnetohydrodynamic model of whistler duct structure in the magnetosphere. *Planet. Space Science*, **32**, 143–150

Washimi, H. (1976) Wave-trapping in an inhomogeneous magnetoplasma. *J. Phys. Soc. Japan*, **41**, 2098–2104

Whittaker, E. T. and Watson, G. N. (1969) *A Course of Modern Analysis*. London: Cambridge University Press

Yabroff, I. (1961) Computation of whistler ray paths. *J. Res. NBS*, **65D**, 485–505

Yip, G. L. and Seshadri, S. R. (1967) Radiation from an electric dipole in an axially magnetized plasma column — dipolar modes. *Can. J. Phys.*, **45**, 3627–3648

Zaboronkova, T. M., Kondrat'ev, I. G. and Kudrin, A. V. (1991) Whistlers in magnetoactive plasma. I. *Radiophysics and Quantum Electronics*, **34**, 784–790

Zaboronkova, T. M., Kostrov, A. V., Kudrin, A. V., Tikhonov, S. V., Tronin, A. V. and Shaikin, A. A. (1992 a) Channeling of waves in the whistler frequency range within nonuniform plasma structures. *Sov. Phys. JETP*, **75**, 625–632

Zaboronkova, T. M., Kondrat'ev, I. G. and Kudrin, A. V. (1992 b) Radiation of whistler-band waves in magnetoactive plasma. II. *Radiophysics and Quantum Electronics*, **35**, 407–412

Zaboronkova, T. M., Kondrat'ev, I. G. and Kudrin, A. V. (1993 a) On the radiation directivity pattern of electric ring currents in a magnetoactive plasma in the whistler frequency range. *Journal of Communications Technology and Electronics*, **38** (15), 110–118

Zaboronkova, T. M., Kudrin, A. V. and Markov, G. A. (1993 b) Waves in the whistler range directed by channels containing high-density plasma. *Plasma Physics Reports*, **19**, 397–403

Zaboronkova, T. M., Kondrat'ev, I. G. and Kudrin, A. V. (1993 c) Radiation of given currents in the presence of a cylindrical plasma column immersed in a magnetoplasma. *Preprint NIRFI No. 375*, pp. 1–68, Radiophysical Research Institute (NIRFI), Nizhny Novgorod, Russia

Zaboronkova, T. M., Kostrov, A.V., Kudrin, A.V., Smirnov, A. I. and Shaikin, A.A. (1996 a) Structure of the electromagnetic fields of loop radiators in magnetized plasmas over the whistler frequency range. *Radiophysics and Quantum Electronics*, **39**, 132–139

Zaboronkova, T. M., Kondrat'ev, I. G. and Kudrin, A. V. (1996 b) Wave emission from a VLF plasma-waveguide antenna system under ionospheric conditions. *Radiophysics and Quantum Electronics*, **39**, 145–154

Zaboronkova, T. M., Kondrat'ev, I. G. and Kudrin, A. V. (1997) Radiation of ring sources of the very low frequency band in a magnetized plasma medium in the presence of a cylindrical plasma channel. *Journal of Communications Technology and Electronics*, **42**, 37–44

Zhilinsky, A. P. and Tsendin, L. D. (1980) Collisional diffusion of a partially ionized plasma in a magnetic field. *Uspekhi Fizicheskikh Nauk*, **131**, 343–385

Index of Definitions of the More Important Symbols

$\mathbf{H}$	magnetic field intensity·	17
	(used with subscripts to denote magnetic field	
	in specified waves)	
H_x, H_y, H_z	components of $\mathbf{H}$ in Cartesian coordinates	19
H_ρ, H_ϕ, H_z	components of $\mathbf{H}$ in cylindrical coordinates	19
$\mathcal{H}$	$Z_0\mathbf{H}$, scaled magnetic field	22
	(used with subscripts to denote scaled	
	magnetic field in specified waves)	
$\mathcal{H}_x$, $\mathcal{H}_y$, $\mathcal{H}_z$	components of $\mathcal{H}$ in Cartesian coordinates	22
$\mathcal{H}_\rho$, $\mathcal{H}_\phi$, $\mathcal{H}_z$	components of $\mathcal{H}$ in cylindrical coordinates	23
$H_m^{(1)}$, $H_m^{(2)}$	Hankel functions of the first and the second	51
	kind, respectively, of order m	
Im	imaginary part of	
I_m	modified Bessel function of the first kind of order m	102
i	$\sqrt{(-1)}$	
$\mathbf{J}^e$	electric current density in source	17
J_x^e, J_y^e, J_z^e	components of $\mathbf{J}^e$ in Cartesian coordinates	19
J_ρ^e, J_ϕ^e, J_z^e	components of $\mathbf{J}^e$ in cylindrical coordinates	19
$\mathcal{J}^e$	$Z_0\mathbf{J}^e$ scaled electric-current density in source	22
$\mathcal{J}_x^e$, $\mathcal{J}_y^e$, $\mathcal{J}_z^e$	components of $\mathcal{J}^e$ in Cartesian coordinates	22
$\mathcal{J}_\rho^e$, $\mathcal{J}_\phi^e$, $\mathcal{J}_z^e$	components of $\mathcal{J}^e$ in cylindrical coordinates	23
$\mathbf{J}^m$	magnetic current density in source	17
J_x^m, J_y^m, J_z^m	components of $\mathbf{J}^m$ in Cartesian coordinates	19
J_ρ^m, J_ϕ^m, J_z^m	components of $\mathbf{J}^m$ in cylindrical coordinates	19
J_m	Bessel function of the first kind of order m	44
j	used to denote an integer, and with other meanings	
K_m	modified Bessel function of the second kind	88
	of order m	
$\mathbf{k}$	propagation vector	23
k	used to denote an integer	
k_0	ω/c, wave number in free space	22
L	McIlwain parameter, used to specify a particular	3
	field line in the geomagnetic field	
	half length of a dipole source	50, 57
	(also used with other meanings)	
$L_\perp$, $L_\parallel$	transverse and longitudinal scales	7
	of the plasma perturbation	
ℓ	used to denote an integer	

Author Index

Subject Index*